Environment and Society in the Japanese Islands

Environment and Society
in the Japanese Islands

From Prehistory to the Present

EDITED BY BRUCE L. BATTEN AND PHILIP C. BROWN

Oregon State University Press *Corvallis*

Library of Congress Cataloging-in-Publication Data

Environment and society in the Japanese islands : from prehistory to the present / edited by Bruce L. Batten and Philip C. Brown.

    pages cm

    Papers from a conference that took place at the Tokai University Pacific Center in Honolulu on March 28-29, 2011.

    Includes bibliographical references and index.

    ISBN 978-0-87071-801-4 (original paperback) — ISBN 978-0-87071-802-1 (e-book)

1. Nature—Effect of human beings on—Japan—History—Congresses.
2. Human ecology—Japan—History—Congresses. 3. Climatic changes—Japan—History—Congresses. 4. Resilience (Ecology)—Japan—History—Congresses.
5. Environmental degradation—Risk assessment—Japan—History—Congresses.
6. Japan—Environmental conditions—Congresses. 7. Japan—History—Congresses. 8. Japan—Geography—Congresses. I. Batten, Bruce Loyd. II. Brown, Philip C., 1947-

  GF666.E68   2015

   304.20952—dc23

2014044730

Oregon State University Press
121 The Valley Library
Corvallis OR 97331-4501
541-737-3166 • fax 541-737-3170
www.osupress.oregonstate.edu

Contents

List of Plates, Figures, and Tables

LIST OF TABLES

Preface and Acknowledgments

The idea for this book was conceived at "Japan's Natural Legacies," a conference on Japanese environmental history held in Bozeman, Montana, in October 2008. The conference, organized by Brett L. Walker of Montana State University, Julia Adeney Thomas of the University of Notre Dame, and Ian Jared Miller of Harvard University, focused largely on the nineteenth and twentieth centuries and included contributions from scholars in anthropology and literature as well as history. Both of the present volume's editors took part and were stimulated to pursue further work in the area of Japanese environmental history. (The results of the Montana conference were later published under the title *Japan at Nature's Edge*.[1])

During and after "Japan's Natural Legacies," we began to consider the idea of a separate project on the same general topic but emphasizing developments over Japan's long history and calling on the skills and perspectives of technically disposed disciplinary specialists in addition to those of more traditional historians. Initial explorations with friends and acquaintances were met with enthusiastic expressions of interest among both younger and senior scholars. As "dry runs" for a book project, we organized panels at the annual meetings of the American Society for Environmental History (ASEH) in Tallahassee, Florida, in February 2009, and in Portland, Oregon, in March 2010. Various contributors to the present work presented papers and benefited from comments by John Brooke and Nicholas Breyfogle of The Ohio State University (in 2009) and Margaret McKean of Duke University (in 2010). At ASEH we also made the acquaintance of Mary Braun, acquisitions editor for Oregon State University Press, who expressed interest in publishing an edited volume on Japanese environmental history.

The next step was to hold our own dedicated conference. Following nearly a year of planning, "Environment and Society in the Japanese Islands: From Prehistory to the Present" took place at the Tokai University Pacific Center in Honolulu on March 28–29, 2011—just two and a half weeks after the earthquake and tsunami that struck Japan's eastern seaboard on March 11, even as the resulting nuclear crisis was still unfolding. The conference brought together all but three of the scholars represented in this book (Junpei Hirano, Osamu Saito, and Shizuyo Sano were unable to attend). John Brooke and two other invited guests, William Wayne Farris and Kieko Matteson of the University of Hawai'i, Mānoa, offered valuable comments. Informal activities also provided important forums for discussion and reflection. Following the conference, participants were asked to revise their work based on observations from the invited guests and direction from the two editors. The result is the volume that you now hold in your hands.

In organizing the conference and assembling this book we were guided by two seemingly antithetical principles: diversity and coherence. To create a showcase for the full range of possibilities for Japanese environmental history, we made a conscious attempt to include not only a range of topics, eras, methodologies, and perspectives but also a broad array of professional experience. Our authors range from graduate students to emeritus professors; they live and work in the United States, England, and Japan. We are particularly happy to have Japanese voices represented in this collection, partly for the obvious reason that the book is about Japan, and partly because Japanese scholarship on that country's past, despite its generally high quality, is poorly represented in English. Perhaps not surprisingly for a book on environmental history, the academic specialization of most of our contributors is history; however, other disciplines, including archaeology, geography, and climatology, are also represented. We made inclusion of these fields a priority as a way of bridging C. P. Snow's "two cultures" of science and the humanities.[2] If ever there was a field in need of interdisciplinary research, it is environmental history.

Our second major guiding principle was coherence, a rather stiff challenge given the diversity of authors and topics. We tried to achieve coherence from the very earliest stages of planning by compiling a reading list for participants and asking them to incorporate and build on the ideas represented. Special emphasis was placed on recent theoretical literature, particularly on the ecological concept of "resilience"—the ability of natural, or by extension social or socionatural, systems to continue functioning when disturbed. (We are deeply indebted to Mark J. Hudson of Nishikyushu University, one of the participants in the Honolulu conference, for introducing us to this literature.) In the end, some contributors addressed these issues more explicitly than others. We believe that we have achieved an acceptable level of coherence, but in the end our success (or lack thereof) is for the reader to judge.

In completing this project we have benefited from the help of many individuals and organizations in addition to those named above. For our 2011 Hawaii conference we would like to thank Chancellor Naoto Yoshikawa and Vice Chancellor

Douglas Fuqua of Hawaii Tokai International College for hosting us; Wanda Sako for logistic support; and the Northeast Asia Council of the Association for Asian Studies and the History Department, the East Asian Studies Center, and the College of Arts and Sciences at The Ohio State University for funding. (Thanks also to Randall Sasaki of Texas A&M University for his participation.) At Oregon State University Press, in addition to Mary Braun, whose enthusiasm sustained the project to completion, we are indebted to Micki Reaman, Sally Atwater, and the two anonymous readers whose constructive comments helped us clarify our thinking, resulting in (we hope) significant improvements, particularly to the introduction. We would also like to thank all of our contributors for putting up with our requests for rewrites and for waiting patiently over the course of several years to see their research results published. And finally, we are grateful to our respective families and employers for giving us the latitude to pursue this time-consuming bit of research and writing.

NOTES

1   Ian Jared Miller, Julia Adeney Thomas, and Brett L. Walker, eds., *Japan at Nature's Edge: The Environmental Context of a Global Power* (Honolulu: University of Hawai'i Press, 2013).

2   C. P. Snow, *The Two Cultures* (London: Cambridge University Press, 1993 [1959]).

A Note to the Reader

All Japanese words and names in the text are romanized according to the standard, modified Hepburn system. Consonants in modern Japanese are generally pronounced as a native English speaker would guess, except for *g,* which is always hard, and *r,* which sounds like a cross between an English *r* and *l*. Vowels are pronounced as in Italian. Long or double vowels are indicated with a macron (e.g., Jōmon), except in the case of place-names with which most readers will already be familiar (e.g., Tokyo, Kyoto).

The few Chinese terms and names are given in *pinyin* transcription, except for Taiwanese personal names, which are transliterated, as customary, using the older Wade-Giles system. (In a very few cases, exceptions have been made to accommodate the preferences of the individuals mentioned.)

Also regarding personal names, in both Japanese and Chinese the surname customarily precedes the given name. We have honored this convention in the text and notes, with two exceptions: cases where an individual's name is written in the English order (given name first, surname last) in a cited publication, and the names of our own Japanese and Chinese contributors, which are likewise given in the English order.

Premodern dates, where they occur, are given in the hybrid form standard among historians of Japan. The Japanese calendar is keyed to "era names" (*gengō* or *nengō*); thus 2015, for example, is the 27th year of Heisei. Also, in premodern times the Japanese used a lunar calendar (ultimately Chinese in origin), whose months and days do not correspond to the solar Gregorian calendar in use today. In the text we have converted Japanese years to the nearest Gregorian equivalent but have generally left months and days as they appear in the lunar calendar. To make it clear that these are early Japanese dates, not Western ones, we refer to the number, not the name, of the month: hence, the "fifth month," not "May." The one exception to this rule is Chapter 10, on climatology, where Takehiko Mikami and his colleagues have converted all Japanese dates to their Gregorian equivalents.

In the documentation, both chapter endnotes and the bibliography, the place of publication for all Japanese-language works is Tokyo unless otherwise noted.

Introduction

Green Perspectives on the Japanese Past

BRUCE L. BATTEN   PHILIP C. BROWN

Research trends come and go, in history as in other disciplines. As scholars we read the work of our peers and strive to engage them with our own contributions. Sometimes such scholarly discourse seems to have a life of its own, one hypothesis or analysis guiding the direction of future research sans explicit reference to contemporary events. However, scholars also live in the real world and are frequently influenced, whether directly or indirectly, by the events and problems of their own time and place.

Even historians, whose work by definition focuses on the past, feel the pull of current events; the subfield of environmental history provides an excellent case in point. Early works on the subject were an outgrowth of the environmentalist movement of the 1960s and 1970s. Since then, the number of books and articles has increased exponentially, reflecting scholars' continuing concern with contemporary pollution, resource depletion, biological extinctions, climate change, and other anthropogenic environmental issues.

As the term is generally used, environmental history is not the history of the natural environment per se. That is the province of climatology, geology, biology, and other such fields. Rather, it is the history of the relationship, broadly defined, between human society and the natural environment.[1] Historians explore the roots of observed problems or conditions (whether contemporary or past), and that is as true of their efforts to address issues related to the environment as it is for political, social, or economic dilemmas. What, for example, are the causes of climate change and what role has been exercised by human activity? More generally, how has the natural environment influenced the development of human society, and how have human activities transformed nature? What does the past tell us about the ability of our species to survive, or even thrive, in the face of changing or adverse environmental conditions? Conversely, what does it tell us about the resilience (or lack thereof) of the natural environment in the face of human disruption? What, if any, lessons does the past contain to guide our actions today and perhaps improve our future and that of the planet?[2]

The environmental issues we face today are in many cases international in scope, and so it is not surprising that many works on environmental history encompass a broad, even global perspective. A few authors have taken this approach to an extreme, attempting to synthesize the entire span of human prehistory and history from an environmental point of view. Clive Ponting's *A Green History of the World*, first published in 1991, was the pioneering work in this genre, but more recently Neil Roberts, J. Donald Hughes, I. G. Simmons, and Anthony N. Penna have also contributed ambitious surveys.[3] (Though not environmental history per se, works in the related genre of "big history" take this approach even further by extending coverage backward to the beginning of the universe.[4])

Of course, grandiose syntheses are the exception, not the rule: most studies of environmental history focus on a particular historical period, a particular theme or topic, a particular place, or some combination of the three. Among works with a chronological focus, two that stand out in our minds are John F. Richards's *The Unending Frontier*, on the environmental consequences of state expansion in the early modern period, and J. R. McNeill's *Something New Under the Sun*, an environmental history of the twentieth century.[5] Works with a thematic focus are legion, so any list will be arbitrary, but as scholars we have been particularly influenced by Alfred Crosby's books on the ecological consequences of European expansion into the New World; Jared Diamond's insights into the role of geography in world history; research by Diamond and others on possible environmental factors in social collapse; Brian Fagan's books on climate in world history; studies by Richard P. Tucker, Edmund Russell, and others on war and the environment; and the work of Donald Worster, Richard White, and others on rivers and human society.[6] Worster and White also exemplify the third category of research on environmental history, that with a specific geographic focus—in their case, North America. Published research on the environmental history of North America and of Europe and the Mediterranean seems to dwarf that on other regions, probably because of their outsized impact on modern world history (as well as the current prevalence of English as an international language).[7] However, other regions that have received detailed treatment are Meso-America, largely in connection with the Classic Maya collapse of ca. 900 CE, and China, in reflection of that country's long continuous history and growing influence on the world stage.[8]

It is in the spirit of such geographically focused works that we offer the present volume on Japanese environmental history. We believe that regional studies offer several important benefits. A relatively tight focus, in this case on a single archipelago, makes it possible to examine important issues in more detail than would otherwise be possible. We can also ask whether the historical experience of our target location follows the trajectory suggested by more global studies. If it does, the conclusions of previous studies are strengthened; if it does not, we may open up new avenues of exploration by asking, Why not? We believe Japan to be a particularly attractive case study for at least three reasons.

First, Japan presents an almost textbook example of many of the environmental issues that vex the world today. Japan is a densely populated industrial nation with the world's largest megacity and third-largest economy. It has a grim legacy of domestic industrial pollution and industrial accidents; one has only to think of Minamata disease, caused by mercury poisoning, or more recently, the radioactive contamination from the 2011 meltdowns at the Fukushima Daiichi nuclear plant. The Fukushima accident, of course, illustrates another type of environmental issue, cataclysmic natural events and the human response to them. This time it was an earthquake and resulting tsunami, but Japan is also subject to volcanic eruptions and typhoons; its experiences with all of these offer parallels to, and potential comparisons with, those of other regions prone to similar events. Not quite as dramatically but perhaps equally important, Japan's low-lying major cities, like their equivalents elsewhere, are at risk from sea level rise over the coming decades and centuries. Japan's scarcity of natural resources relative to the size of its population and industrial demand has resulted in aggressive exploitation of resources outside its borders, contributing to problems ranging from fisheries depletion in the open oceans to deforestation in Southeast Asia. Japan's exports of used electronic products and industrial waste also contribute to pollution overseas. Yet not all is bad. In the early 1960s, Tokyo had some of the worst air pollution in the world; today Tokyo's air is refreshingly clean. A similar "success story" could be told about the waters of Tokyo Bay. (Of course, to the degree that Japan is externalizing its environmental problems, these and other successes have come at the expense of other countries or regions.) More recently, the Fukushima nuclear meltdowns have led to serious rethinking inside Japan about how the country should and can meet its energy needs. "Sustainability," a concept once unknown to the general public, has become an important topic of discussion, from the street corner to the highest levels of government. Thus, beyond warnings and cautions based on its past, Japan offers glimmers of hope for the future.

Second, study of Japan permits us to take a long view. The Japanese islands have been inhabited for many millennia and boast a rich archaeological and written record that has been intensely studied by generations of Japanese (and some foreign) scholars. This evidence enables us to examine events and processes at a variety of temporal scales, ranging from the very short term to the ultralong, and to do so with a good degree of confidence. Understandably, many works on environmental history tend to focus on relatively recent events and trends, particularly the negative environmental consequences of commercialization and industrialization. Although we believe that an important purpose of writing history is to gain a perspective on present-day events and trends, our professional experience as historians of premodern Japan tells us that long Braudelian perspectives are frequently as good as or better than shorter ones, and that they provide alternative insights of great value. The case of Japan offers an excellent opportunity to introduce that valuable long-term perspective.

Third, Japan is interesting because it is an island country, and consequently, there has been a tendency to see preindustrial Japan as a closed system. Based on this view, world environmental historians such as Jared Diamond and John F. Richards have portrayed early modern Japan as a rare example of a sustainable society, one that somehow managed to survive, and even prosper, within fixed environmental limits. This is an attractive hypothesis that has obvious significance for thinking about the world we live in today. But is it really true? As specialists, we feel a need for experts on Japan to weigh in on this important question.

Japan's History: An Overview

One or more of these three issues is addressed by each essay in this volume, but before diving in, we want to provide some important background information. Quite a bit is already known about the historical relationship between humans and their natural environment in the Japanese archipelago, but much of that knowledge is scattered and inaccessible to nonspecialists, not least for reasons of language—most has appeared only in Japanese publications. On the assumption that not all of our readers will have intimate knowledge of Japan, here we provide a brief overview of its geography and historical development. Then, in the following section, we will introduce some of the themes explored in previous, especially Japanese-language, environmentally focused studies.[9]

Geographically, the Japanese archipelago is an elongate volcanic arc located on the eastern fringe of Eurasia (plate 1). Japan as a country currently occupies most of the chain, although some of the peripheral islands are the focus of territorial disputes with neighboring states. Japan consists of four main islands—from north to south, Hokkaido, Honshu (the largest island), Shikoku, and Kyushu—and thousands of smaller ones, as well as adjacent territorial seas. The country's total land area is 364,485 km², slightly smaller than California. Its climate is varied, ranging from cool temperate in the north to subtropical in the south, with four distinct seasons in most areas. Approximately 80 percent of the country is mountainous and unsuitable for human habitation. The remainder consists of inhabitable uplands (some quite elevated and challenging nonetheless) and low-lying plains near river mouths, mostly along the eastern coast of Honshu. Although contemporary Japan is far from self-sufficient in many important resources, it formerly had abundant mineral resources and even today is blessed with bountiful oceans and streams, abundant rainfall, and lush forests. However, as noted previously, it is also subject to significant natural dangers, including volcanic eruptions, earthquakes, tsunami, typhoons, and floods.

Human engagement with the natural environment of the Japanese islands began perhaps 35,000 years before present (BP), during the last ice age. Early settlers arrived from the north, through Hokkaido, from the west via the Korean Peninsula, and from the south via Taiwan and the Ryukyu Islands. They were hunters, fishers, and gatherers who lived in small, mobile bands. They made use of stone tools but had no pottery, and their culture is commonly referred to as Paleolithic or Preceramic.

Around 14,500 BCE, roughly the end of the glacial period, a new culture began to emerge. This was the Jōmon culture, named for its typical "cord-marked" pottery, in itself a material manifestation of increased environmental exploitation by these groups. The Jōmon people, like the Northwest Coast tribes of what is now the United States and Canada, were "affluent foragers" living in an unusually rich natural environment. They had a sedentary lifestyle and manipulated nature on a limited scale to practice horticulture in addition to their primary subsistence activities of hunting, fishing, and gathering.

A new culture, the Yayoi, was introduced to the islands in the first millennium BCE—not incidentally, a time of significant climatic cooling and human migrations throughout Eurasia. The Yayoi culture was based on wet-rice agriculture and encompassed the use of bronze and iron as well as new styles of pottery, both of which required increased use of fuel. It first appeared in northern Kyushu among immigrants (perhaps climatic refugees) from the Korean Peninsula. The new agricultural regime facilitated rapid population growth and increased social stratification. Within short order the immigrants had displaced or assimilated the aboriginal Jōmon population in most of the archipelago and created Japan's first complex chiefdoms in Kyushu and western Honshu. Lasting until around 250 CE, the Yayoi culture evolved into the succeeding Kofun culture.

Many scholars argue that the Kofun, or "Tumulus," period (roughly 250–600s) marked the emergence of Japan's first state (or according to some, states), whose leaders were capable of coordinating large-scale activities, including those related to resource utilization and environmental threats. The paramount Yamato kingdom was based not in Kyushu but in west-central Honshu, near the eastern terminus of the Inland Sea. Evidence for political consolidation over much of Japan's three main islands is found in the form of the giant grave mounds that give the period its name. Their distribution extends from the Tohoku region of northeastern Honshu to southern Kyushu—roughly the area suitable for wet-rice cultivation. In addition to state formation and monumental construction, the Kofun period was notable for continued immigration and growing cultural influence from China and the Korean Peninsula, including the diffusion of Chinese written characters and Buddhism.

The Kofun period was followed by Japan's ancient or classical era, *kodai,* lasting from roughly the seventh through the twelfth centuries. During this period a geographically expansive state emerged, one capable of directly affecting Japan's natural environment in many ways, including construction of large capital cities and systematic reclamation of arable land. The seventh-century luminescence of the Tang dynasty stimulated the Yamato court to emulate the Chinese aristocracy both culturally and politically. The imperial polity of the Nara period (710–794) was in some respects strikingly centralized. "Divine" monarchs (women as well as men) and their aristocratic entourage inhabited the capital of Heijō-kyō (modern Nara), from which they ruled the overwhelmingly rural, agricultural population of Honshu, Shikoku, and Kyushu. Although the capital moved in 794, the basic structure of state and powerful lineages that controlled it continued, evolving gradually during the Heian period (794–1185, named for the capital city that is now known as Kyoto). During Heian times, Japan's relations with the continent evolved from prickly diplomacy to peaceful trade. Partly because of decreased geopolitical tensions and partly because of domestic developments, such as tax exemptions dispensed as rewards for land reclamation and frontier expansion, the state gradually assumed a more relaxed, decentralized form. Power and wealth were increasingly monopolized by individual courtier families (including the imperial line) and religious establishments. Away from the capital, particularly in the rough-and-tumble frontier environment of eastern Honshu, warriors (samurai) came to dominate rural society.

The medieval era, or *chūsei,* lasting roughly from the end of the twelfth century to the end of the sixteenth, was characterized politically by the increasing dominance of a military aristocracy but also, paradoxically, by progressive decentralization. Japan's first military regime, the Kamakura shogunate (1185–1133, headquartered southwest of modern Tokyo at Kamakura), shared power with the old civilian aristocracy. The military assumed a clearly preeminent position vis-à-vis the court as it managed Japan's defense against the late-thirteenth-century Mongol invasions and their aftermath. The shogunate, an alliance of local military powers led by Minamoto Yoritomo, had contended with centrifugal forces following its founder's death in 1199; however, its cohesion was finally shattered in 1333 when Emperor Go-Daigo sought unsuccessfully to restore imperial preeminence. The succeeding Muromachi shogunate (1336–1573, based in Kyoto and also known as the Ashikaga shogunate, after its founding family) never achieved the level of authority held by its predecessor, much less that of the Nara-Heian state at its peak. Although in name the Muromachi shogunate was longer lived than the Kamakura regime, by the late fifteenth century its effective control was restricted to Kyoto and its immediate environs. The rest of Japan sank into a state of pervasive disorder punctuated by intermittent warfare among a large number of virtually autonomous feudal domains. This was Japan's "wild West" era, the so-called Sengoku or Warring States period (late fifteenth–late sixteenth centuries). The increasing fragmentation of authority throughout Japan's medieval centuries meant that central governments, such as they were, had limited ability to implement policies with significant consequences, environmental or otherwise.

Despite increasing political decentralization, the Kamakura, Muromachi, and Warring States periods saw tremendous economic growth and social diversification. Spurred by local efforts at land reclamation, expansion of trade, and monetization, population increased along with social mobility and cultural vitality. Although central authorities were increasingly powerless, regional lords were not, and their activities had important environmental consequences; by the sixteenth century many took aggressive measures to reclaim land, limit flood damage, and extract precious metals. Incessant fighting and scorched-earth tactics also took their toll on the environment. Finally, the sixteenth century was marked by the first arrival of European traders and missionaries. The firearms they brought encouraged the emergence of new styles of battle and of fortified settlements surrounded by growing urban populations.

During the subsequent early modern era, *kinsei* (latest sixteenth–early nineteenth centuries), distinctive patterns of population and economic growth and of political control led to more dramatic environmental effects and challenges than in earlier periods. The three islands of Honshu, Shikoku, and Kyushu were rapidly pacified under the leadership of Oda Nobunaga, Toyotomi Hideyoshi, and Tokugawa Ieyasu. The last of these founded the third of Japan's shogunates, the Tokugawa, which remained in power for two and a half centuries (1603–1867, also known to historians as the Edo period). Ieyasu established his capital at Edo (modern Tokyo), transforming a small, marshy, lowland town into a city that in just over a century ballooned to a million inhabitants. The shogunate restricted international contacts to those parts of the world with which it could establish peaceful economic or diplomatic relations (China, Korea, the Netherlands, the Ryukyu kingdom, and the Ainu people of Hokkaido); it left most matters of domestic governance to some 250-plus baronial figures, the daimyo. Socially, the period was legally characterized by a rigid status system that (in theory, at least) brooked no social mobility. Despite legal ideals, restoration of peace and stability fostered broad economic, population, and urban growth greased by migration and information flows over increasing distances. Rapid economic and demographic growth in the seventeenth and early eighteenth centuries tapered off thereafter, in part the consequence of several widespread famines but also as a result of deliberate family population controls correlated with signs of improved standards of living for significant segments of the population. Especially in the late eighteenth and early nineteenth centuries, Japan experienced mounting social unrest, some of which involved privatization of communally managed forest and common lands. More problematic from an environmental perspective were the demands for food, fuel, and building materials posed by larger and more urban populations and the movement of people into more and more marginal lands subject to greater risk from vagaries of temperature and rainfall.

In the mid-nineteenth century, American gunboat diplomacy sparked dramatic political change and drew Japan into the modern world system in ways that transformed natural resource use, introduced polluting production methods, and encouraged externalization of environmental impacts. The Meiji Restoration of 1868, ostensibly intended to restore power to the emperor, in fact placed it in the hands of an oligarchy drawn from the old samurai class. Despite a widespread push to revert to an earlier order, Japan's new rulers did away with the governments of the Edo period and systematically created a centralized Western-style nation-state, the first in East Asia. The Meiji state grew in power both economically and politically and before long began to harbor, and then realize, imperialist ambitions. Hokkaido had been annexed at the end of the Edo period; the former Ryukyu kingdom now followed, becoming Okinawa prefecture. Japan fought with its Chinese, Russian, and Korean neighbors, ultimately acquiring colonial territories in Taiwan, Korea, and Manchuria. As the Meiji oligarchs died off, power fitfully drifted into the hands of the imperial army and navy, leading to further depredations in Asia and eventually Japan's participation in World War II on the side of the Axis powers. The country was soundly defeated in 1945, its hard-pressed population having been pushed to extreme overexploitation of forests and other resources by the time the United States dropped two atomic bombs on Japanese soil.

To the surprise of many, postwar Japan rebounded from total defeat to transform itself into a world-class economic power, complete with all the environmental accoutrements of that status. An American-led occupation lasting until 1952 resulted in a new constitution: the emperor was transformed from a divine figure into a mere symbol of the Japanese nation, and Japan—under the leadership of the Liberal Democratic Party, or LDP—became a democratic society focused on "catching" up with the United States and other advanced industrial economies. This goal was achieved within a few decades by dint of hard work of the war's survivors and later efforts of the baby boom generation—all at an extraordinary environmental cost, measured among other things in industrial disease and pollution that by the 1960s could no longer be ignored. Japanese society became increasingly Westernized in appearance while displaying a profound originality, first in industrial processes and manufactured products, ranging from cars to electronics, and then in popular culture, such as *anime* and *manga*—"cool Japan." The economy, accelerated to "bubble" levels during the 1980s, later lapsed into the doldrums. Birthrates declined precipitously and population growth—constant since the Meiji era, with a brief interruption at World War II—ceased. In the realm of politics, the leading parties (of which the LDP was now only one) exhibited a chronic and increasingly problematic deficit of leadership, most evident in the poor response to the triple disaster of earthquake, tsunami, and nuclear meltdowns that began on March 11, 2011. Despite the dissipation of political finesse, Japan's citizens and government have increasingly taken note of environmental issues.

Previous Research on Japanese Environmental History

The above account represents, we think, general consensus on the basic facts of Japan's geography, prehistory, and history. Next let us focus more closely on what previous scholarship has had to say about Japan's environmental history.

The first point to be made is that there is a huge volume of publications that are not specifically about environmental history but are nonetheless highly relevant. For example, a substantial literature in Japanese addresses natural disasters, and for obvious reasons this is likely to increase exponentially over the next few years. Japan also boasts strong, internationally recognized scholarship on historical demography, including epidemic disease. Less well known, but equally solid, are Japanese scholarly accomplishments in historical geography, including land tenure; economic history, encompassing studies on agriculture, fishing, forestry, mining, and industry; intellectual and religious history, including studies of attitudes toward nature; and many other topics, some of them in allied disciplines such as environmental archaeology and historical climatology. Although we are thinking primarily of Japanese-language scholarship, that on Japan in English and other languages is similarly diverse and contains much of relevance to the study of human-environmental interactions. Regardless of language, however, it seems fair to say that not until the 1980s and 1990s did these strands of scholarship come to be woven into what today would be called environmental history.

Incongruously, some of the pioneering work in this emerging field came from America, not Japan. Conrad Totman, of Yale University, could well be described as the father of Japanese environmental history. Totman's research on the lumber industry in the Edo period convinced him that sustainable forestry practices saved Japan from becoming a barren wasteland.[10] He later authored a comprehensive account of Japanese and Korean environmental history as well as a textbook of Japanese history that refreshingly applied a new periodization based on humans' relationship to their natural environment.[11] Jared Diamond and John F. Richards derived many of their ideas about Japan from Totman's books. In addition to these world historians, Totman also stimulated other specialists on Japan in the English-speaking world to work on environmental issues. Perhaps the most prolific is Brett L. Walker, whose various books have dealt with Japanese ecological imperialism (to borrow Alfred Crosby's term) in early-modern Hokkaido, the extermination of the Japanese wolf, and the "toxic archipelago" that resulted from industrial pollution in the nineteenth and twentieth centuries.[12] Other individuals could be mentioned, but here we choose to focus on Japanese-language scholarship, which will be less familiar, and less accessible, to most readers of this book.

Needless to say, the vast majority of research on Japanese environmental history has been by Japanese scholars. Although the field emerged only in the 1990s, well after Totman's groundbreaking efforts, the pace has picked up and now there is a large corpus of research. Rather than attempting to summarize all of it, here we simply describe three benchmark publications, each a multiauthor, multivolume series on environmental history, that convey some of the characteristics and foci of recent Japanese scholarship.

The first of these, which appeared in 1995–96, is the 15-volume *Essays on Civilization and the Environment*, edited by three scholars associated with the International Research Center for Japanese Studies (Kokusai Nihon Bunka Kenkyū Sentā) in Kyoto: philosopher Umehara Takeshi, historian of science and comparative civilizations Itō Shuntarō, and environmental archaeologist Yasuda Yoshinori.[13] Their series covered not just Japan but the entire world. In terms of overall vision, it tends toward environmental determinism of the type sometimes associated with the early-twentieth-century Japanese philosopher Watsuji Tetsurō.[14] Volume titles include *The Earth and Civilizational Cycles, Civilizational Crisis, Environmental Crisis and Contemporary Civilization,* and others that hint at the editors' point of view: the series tends to ascribe differences in culture to differences in natural conditions (e.g, forest versus desert environments) and connects turning points in the development of civilization to changes in climate. Another feature of the series is the idea that people formerly lived in harmony with nature, but that this balance was upset, first by the emergence of agricultural civilizations, and then more profoundly by industrialization. Western civilization is painted in generally negative terms as a destroyer of nature. The series takes a somewhat utopian view of the ancient Japanese past, particularly the preagricultural Jōmon period. That said, the editors and authors, who range from historians to archaeologists, biologists, and other disciplinary specialists with finely honed technical expertise, did their homework well: the offerings were based on the latest science of the time and a thorough reading of the relevant literature in both Japanese and Western languages. The result is at times prescient, prefiguring later (independent) work by Jared Diamond and others on civilizational collapse.

Next, after a hiatus of fifteen years, came the six-volume *Thirty-five Thousand Years of the Japanese Archipelago: An Environmental History of Humans and Nature*, published in 2011 and edited by ecologist Yumoto Takakazu of the Research Institute for Humanity and Nature (Sōgō Chikyū Kankyōgaku Kenkyūsho), also in Kyoto.[15] As the title implies, this series covers only the Japanese islands and, not surprisingly, features a strong ecological perspective. Indeed, two volumes, *What Is Environmental History?* and *Techniques for Assessing Environmental History*, are devoted to fundamental conceptual and methodological concerns of the field. The latter volume, in particular, presents a selection of striking studies that employ DNA analysis to explore transformations of biota, isotope analysis to uncover changes in eating habits, and much more. In *Thirty-five Thousand Years of the Japanese Archipelago*, as in *Civilization and the Environment*, authors show considerable creativity in unearthing ways to understand environmental issues, even where traditional historical documentation does not directly address them. Overall, the focus of this series is on sustainability and "wise use" of natural resources, particularly as seen in *satoyama*—semiartificial, seminatural environments and agricultural landscapes engineered by rural villagers (see especially volume 3, *The Environmental History of Rural Villages and Forests*). Somewhat oddly, the books contain almost nothing on the modern industrial period, and little attempt is made to compare conditions in Japan with those in other countries or regions.

Finally, there is the five-volume *Environmental Japanese History*, prepared under the supervision of Hirakawa Minami, director-general of the National Museum of History (Kokuritsu Rekishi Minzoku Hakubutsukan) in Chiba, and published in 2012–13.[16] This series is organized chronologically and attempts, generally successfully, to present a balanced and nuanced account of how humans interacted with the natural environment throughout the Japanese past, up to and including the 2011 disaster. The overall sense of the volumes is that people were at the mercy of, and in awe of, nature in ancient times (volume 2, *Ancient Life and Prayer*), that they made progress in medieval times amid the "ferocity" of nature (volume 3, *Reclamation, Livelihoods, and the Medieval Environment*), that they achieved a sustainable society in "coexistence" with nature in early modern times (volume 4, *Human Activities and Nature in the Early Modern Period*), and that there was widespread environmental destruction in modern times—but that in some respects people still remain at nature's mercy, as seen in the events of 2011 (volume 5, *Use and Destruction of Nature: The Modern and Contemporary Periods and Folklore*). In contrast with the first two collections, the articles in this series rely less on the scientific and technical skills of biologists, chemists, climatologists, and the like, and more on traditional historical scholarship, with an unusually strong emphasis on religion, folklore, and views of nature. The only real criticism we have of this series is that the authors make almost no use of scholarship originating outside Japan or published in languages other than Japanese, so their work is uninformed by, and isolated from, the larger international dialogue on environmental history.

Some observers have referred to *Garapogasu-ka*, or "Galapogos-ation" in contemporary Japanese society, most notably in the spectacular but entirely self-contained "evolution" of cellular phone technology. Much-touted cultural and intellectual globalization of Japan to the contrary, something of the same sort may be happening in Japanese historiography. However, important counterexamples can also be found, including the prominent participation of some Japanese scholars in the recently founded Association for East Asian Environmental History.[17] Over the long run, we expect such voluntary outreach, together with support from government and private funding agencies, to significantly expand intellectual exchanges between our Japanese colleagues and the broader international academic community.

What themes emerge from the existing literature on Japanese environmental history, whether by Japanese authors or foreigners? Perhaps not surprisingly, we find multiple conceptual binaries or dyads. Some authors emphasize the influence of nature over society, either for better (bountiful forests) or worse (frequent calamities); others emphasize human impacts on the natural environment, usually for the worse (pollution, resource depletion). Some authors stress sustainability or resilience; others focus on unsustainable practices or ways or life. Autarky or self-sufficiency is emphasized by some; others explore external connections and dependencies. Continuity is the theme of some studies; change occupies center stage in others.

The continuity-discontinuity binary is particularly important because in some ways it encompasses all of the others. Most of the literature surveyed above, to the extent that it addresses such large questions, can be read in support of the view that Japan, as elsewhere, has exhibited a general trend over time toward increasing human impacts on nature, expansion of unsustainable practices, and proliferation of external connections with environmental consequences. Authors further broadly sustain the view that a major point of inflection in

Japan's historical trajectories was the transition from preindustrial to modern times, a shift from a world in which the pace of change was often glacially slow to one in which it is increasingly, often blindingly fast—the world we live in today.

What This Book Is About

We, as editors, consider the above perspectives essential starting points for the present volume. However, we do wish to proffer an original "take" on Japanese environmental history—indeed, on environmental history in general. To that end, we choose to focus on the processes of historical, socionatural change.

Although history writing, almost by definition, is about how one thing leads to another, few practicing historians step back from their work to consider the nature of change in general. What exactly is the "system" we are examining? What are its components, its structure and scale? What sorts of interaction or feedback drive the evolution of the system? How does change occur? Recent developments in the science of ecology and in dynamical systems theory address precisely these questions.

Two important concepts in ecology are resilience and the adaptive cycle.[18] In an ecological context, resilience theory was developed to explain how in some cases, ecosystems appear to remain more or less stable in the face of drastic external perturbations, while in other cases, even a minor disturbance is sufficient to cause abrupt, seemingly irreversible charge—a "regime shift." Resilience is the capacity of a system to withstand or adapt to shocks without experiencing a fundamental regime shift. A related concept is that of the adaptive cycle, the idea that ecosystems tend to change in a four-step process, from growth or exploitation (r) to conservation (K) to collapse or release (Ω) to reorganization (α). The first two stages, sometimes referred to as the foreloop, are characterized by growth in organization and stability. Rapid exploitation or colonization of disturbed areas is followed by the gradual accumulation and storage of resources and energy. The third and fourth stages, the backloop, are characterized by disintegration and reconstitution. Major system thresholds are crossed, leading to collapse, and the old system gives way to a completely new configuration.[19]

Recently, some scholars—regrettably, none of them historians—have extended these ideas to social or socioenvironmental systems. C. S. Holling, Lance H. Gunderson, and Garry D. Peterson have developed the concept of panarchy, referring to "a nested set of adaptive cycles arranged as a dynamic hierarchy in space and time."[20] According to them, panarchies are characteristic of a wide range of human and natural systems and can be used to analyze pivotal shifts in them. Another example is Marten Scheffer's work on critical transitions, a mathematically sophisticated exploration of regime shifts.[21] Scheffer presents many fascinating case studies of critical transitions in lakes, climate, biological evolution, oceans, terrestrial ecosystems, and human society.

The details of these theories are complex, but as environmental historians we can take away the following core ideas. Society is not separate from nature; both are part of a larger socioenvironmental or socionatural system. That system has multiple components. All of them exhibit complex spatial organization at varied scales and operate on various time scales. Each component or subcomponent of the system affects and is affected by others; in other words, specific elements are linked by feedback cycles, some promoting change and others promoting stability. Historical change is the sum total of the interactive operation of these various cycles. Transformation may be rapid or slow; catastrophic or manageable; continuous or discontinuous; cyclic or one way; chaotic, contingent, or deterministic, depending on the components and pressures involved. Frequently, however, the process of change is characterized by punctuated equilibria—that is, rapid phase transitions between alternating stable and semistable states.

We have tried to incorporate this view of history in the present work, which consists of twelve essays by historians, climatologists, archaeologists, and geographers. To emphasize that socionatural systems contain various subsystems or components, we have organized the book not chronologically but thematically. Specifically, we examine how historical societies in Japan have related to the lithosphere, Japan's geology and physical landscape (Part I); the hydrosphere, the water in its soil, rivers, lakes, and oceans (Part II); the biosphere of plants, animals, and organic matter (Part III); and finally to the atmosphere that is host to Japan's weather and climate (Part IV).

The individual essays vary widely in terms of geographic scale. Those by Tatsunori Kawasumi (Chapter 2), Shizuyo Sano (Chapter 4), Philip C. Brown (Chapter 5), Eric G. Dinmore (Chapter 6), and Scott O'Bryan (Chapter 12) focus on individual cities or regions. By contrast, the essays by Gina L. Barnes (Chapter 1), Gregory Smits (Chapter 3), Takehiko Mikami, Masumi Zaiki, and Junpei Hirano (Chapter 10), and Osamu Saito (Chapter 11) are national in scope. Three essays, those by Toshihiro Higuchi (Chapter 7), Colin Tyner (Chapter 8), and Kuang-chi Hung (Chapter 9), offer an international perspective.

The essays also vary in their chronological coverage and the types of historical change they describe. At the long end, Barnes's study covers literally millions of years. At the short end, the chapters by Dinmore, Higuchi, Hung, and O'Bryan span "mere" decades. The essays by Brown, Kawasumi, Mikami et al., Saito, Sano, Smits, and Tyner all present stories spanning a century or more. Within this diversity we find examples of abrupt, convulsive changes or events (e.g., the earthquakes described by Barnes and Smits), long periods of stasis or stability (e.g., the littoral communities described by Sano), and everything in between. Our authors describe socionatural feedback loops promoting stability (see the chapters by Sano, Brown, and Saito) as well as those promoting change (see the chapters by Kawasumi, Dinmore, and O'Bryan). Some of the essays provide dramatic evidence of regime shifts, either human caused (see the Sano and Higuchi chapters) or climatically driven (see the chapter by Mikami et al.).

Taken as a whole, this book provides support for the broad arc of Japanese environmental history presented above—slow, geographically restricted change for the premodern eras, with increasingly rapid and wide-ranging transformation since the nineteenth century—while complicating it in important details. As one would expect, evidence abounds of increasing human domination over nature, to the point of actually changing climate (e.g., the chapter by O'Bryan). We find many examples of sustainable lifestyles and institutions in medieval and early modern times (e.g., the chapters by Sano and Brown) but few from the modern era. The essays on premodern times are relatively local in focus. Conversely, most of those on the nineteenth and twentieth centuries place Japan in a regional or international context, highlighting the historical shift from local autarky to global connectivity.

And yet there are important exceptions to this simple story. Kawasumi's essay shows that even ancient cities such as Nara could have large environmental footprints and were essentially unsustainable. The chapters by Barnes, on geology, and Mikami et al., on climate, make it clear that when the natural environment is taken into account, Japan has never been the self-contained, closed system envisioned by Diamond and others. Indeed, that same point is woven into the very framework of the book: how could the lithosphere, hydrosphere, biosphere, and atmosphere ever be confined to the Japanese islands?

In sum, the essays in this volume present a complex picture of Japan's environmental history. Parts of the story—the downsides of Japan's modern economy—are predictable. Others—for example, the discussions of sustainable resource management—help flesh out the rough outlines sketched by previous literature. However, we hope readers will also find some of the treatments completely new in both methodology and interpretation. We are the first to admit that more can be done to develop the subjects presented here, and that other important issues are not even addressed—examples include Japanese beliefs regarding nature, the role of animals in Japanese history, and the reasons behind Japan's relative successes in ameliorating modern environmental damage.[22] We will be pleased if this collection serves as a stimulus to others to undertake further work on the fascinating environmental history of one of the world's leading nations.

1   For a thoughtful book-length definition of environmental history, see J. Donald Hughes, *What Is Environmental History?* (Cambridge: Polity Press, 2006).

2   Note that many works on environmental history also devote attention to how people think about nature. While we acknowledge the importance of attitudes and values in shaping how people relate to nature, in the present work we are largely concerned with actions rather than words.

3   Clive Ponting, *A Green History of the World* (London: Penguin Books, 1991); Ponting, *A New Green History of the World: The Environment and the Collapse of Great Civilizations* (New York: Penguin Books, 2007); Neil Roberts, *The Holocene: An Environmental History*, 2nd ed. (Oxford: Blackwell Publishers, 1998); J. Donald Hughes, *An Environmental History of the World: Humankind's Changing Role in the Community of Life*, Routledge Studies in Physical Geography and Environment 2 (London and New York: Routledge, 2001); I. G. Simmons, *Global Environmental History: 10,000 BC to AD 2000* (Edinburgh: Edinburgh University Press, 2008); Anthony N. Penna, *The Human Footprint: A Global Environmental History* (Malden, Mass.: Wiley-Blackwell, 2010).

4   David Christian, *Maps of Time: An Introduction to Big History*, The California World History Library 2 (Berkeley: University of California Press, 2004); Cynthia Stokes Brown, *Big History: From the Big Bang to the Present* (New York: New Press, 2007).

5   John F. Richards, *The Unending Frontier: An Environmental History of the Early Modern World*, The California World History Library (Berkeley: University of California Press, 2003); J. R. McNeill, *Something New Under the Sun: An Environmental History of the Twentieth-Century World* (New York: W.W. Norton & Company, 2000).

6   Alfred W. Crosby, *The Columbian Exchange: Biological and Cultural Consequences of 1492* (Westport, Conn.: Greenwood Press, 1972); Crosby, *Ecological Imperialism: The Biological Expansion of Europe, 900–1900* (Cambridge: Cambridge University Press, 1986); Jared Diamond, *Guns, Germs, and Steel: The Fates of Human Societies* (New York and London: W.W. Norton & Company, 1997); Joseph A. Tainter, *The Collapse of Complex Societies*, New Studies in Archaeology (Cambridge: Cambridge University Press, 1988); Norman Yoffee and George L. Cowgill, eds., *The Collapse of Ancient States and Civilizations* (Tucson: University of Arizona Press, 1988); Jared Diamond, *Collapse: How Societies Choose to Fail or Succeed* (New York: Viking, 2005); Glen M. Schwartz and John J. Nichols, eds., *After Collapse: The Regeneration of Complex Societies* (Tucson: University of Arizona Press, 2006); Patricia A. McAnany and Norman Yoffee, eds., *Questioning Collapse: Human Resilience, Ecological Vulnerability, and the Aftermath of Empire* (Cambridge: Cambridge University Press, 2010); Brian Fagan, *Floods, Famines, and Emperors: El Niño and the Fate of Civilizations* (New York: Basic Books, 1999); Fagan, *The Little Ice Age: How Climate Made History 1300–1850* (New York: Basic Books, 2000); Fagan, *The Long Summer: How Climate Changed Civilization* (New York: Basic Books, 2004); Fagan, *The Great Warming: Climate Change and the Rise and Fall of Civilizations* (New York: Bloomsbury Press, 2008); Fagan, *The Attacking Ocean: The Past, Present, and Future of Rising Sea Levels* (New York: Bloomsbury Press, 2013); David Biggs, *Quagmire: Nation-Building and Nature in the Mekong Delta* (Seattle: University of Washington Press, 2011); Richard P. Tucker and Edmund Russell, eds., *Natural Enemy, Natural Ally: Toward an Environmental History of War* (Corvallis: Oregon State University Press, 2004); Donald Worster, *Rivers of Empire: Water, Aridity, and the Growth of the American West* (New York: Pantheon Books, 1985); Richard White, *The Organic Machine: The Remaking of the Columbia River* (New York: Hill & Wang, 1991); Christof Mauch and Thomas Zeller, *Rivers in History: Perspectives on Waterways in Europe and North America* (Pittsburgh, Pa.: University of Pittsburgh Press, 2008).

7   North America: classic studies include—in addition to the books by Worster and White cited in the previous note—Worster, *Dust Bowl: The Southern Plains in the 1930s* (New York: Oxford University Press, 1979); William Cronon, *Changes in the Land: Indians, Colonists, and the Ecology of New England* (New York: Hill & Wang, 1983); and Patricia Nelson Limerick, *Something in the Soil: Legacies and Reckonings in the New West* (New York: W.W. Norton & Company, 2000).

8   Meso-America: Richardson B. Gill, *The Great Maya Droughts: Water, Life, and Death* (Albuquerque: University of New Mexico Press, 2000); and Arthur A. Demarest, Prudence M. Rice, and Don S. Rice, eds., *The Terminal Classic in the Maya Lowlands: Collapse, Transition, and Transformation* (Boulder: University Press of Colorado, 2004). China: Mark Elvin and Ts'ui-jung Liu, eds., *Sediments of Time: Environment and Society in Chinese History*, Studies in Environment and History (New York, Cambridge University Press, 1998); Elvin, *The Retreat of the Elephants: An Environmental History of China* (New Haven: Yale University Press, 2004); Robert B. Marks, *China: Its Environment and History* (Plymouth, U.K.: Rowman & Littlefield, 2012); and Judith Shapiro, *China's Environmental Challenges* (Malden, Mass.: Polity Press, 2012).

9   Our summary of Japan's geography, prehistory, and history is presented sans footnotes because all of the information is well known and relatively uncontroversial. Readers seeking a more detailed account are referred to standard texts such as Conrad Totman, *A History of Japan*, 2nd ed., The Blackwell History of the World (Oxford: Blackwell Publishers, 2005); Karl F. Friday, ed., *Japan Emerging: Premodern History to 1850* (Boulder, Colo.: Westview Press, 2012); and Andrew Gordon, *A Modern History of Japan: From Tokugawa Times to the Present* (New York: Oxford University Press, 2003).

10 Conrad Totman, *The Green Archipelago: Forestry in Preindustrial Japan* (Berkeley: University of California Press, 1989).

11 Conrad Totman, *Pre-Industrial Korea and Japan in Environmental Perspective*, Handbook of Oriental Studies/Handbuch Der Orientalistik, Section Five: Japan, vol. 11 (Leiden: Brill, 2004); and Totman, *A History of Japan*.

12 Brett L. Walker, *The Conquest of Ainu Lands: Ecology and Culture in Japanese Expansion, 1590–1800* (Berkeley: University of California Press, 2001); Walker, *The Lost Wolves of Japan*, Weyerhaeuser Environmental Books (Seattle: University of Washington Press, 2005); Walker, *Toxic Archipelago: A History of Industrial Disease in Japan*, Weyerhaeuser Environmental Books (Seattle: University of Washington Press, 2010).

13 Umehara Takeshi, Itō Shuntarō, and Yasuda Yoshinori, eds., *Kōza: Bunmei to kankyō*, 15 vols. (Asakura Shoten, 1995–96). Note that our English rendering of the title is meant to convey the meaning; it is not a strictly literal translation.

14 His most famous work is Watsuji Tetsurō, *Fūdo: Ningengaku teki kōsatsu* (Iwanami Shoten, 1935).

15 Yumoto Takakazu, ed., *Shirīzu Nihon rettō no san-man go-sen nen: Hito to shizen no kankyōshi*, 6 vols. (Bun'ichi Sōgō Shuppan, 2011).

16 Hirakawa Minami, Miyake Kazuo, Ihara Kesao, Mizumoto Kunihiko, and Torigoe Hiroyuki, eds., *Kankyō no Nihonshi*, 5 vols. (Yoshikawa Kōbunkan, 2012–13).

17 See http://www.aeaeh.org/. The president of the association for 2013–2015 is Murayama Satoshi of Kagawa University, a specialist on the demographic and environmental history of early modern Germany whose work has now expanded to include research on, and comparisons with, early modern Japan.

18 These concepts have found homes in a very diverse array of disciplines in engineering, physical, and biological sciences, as well as the social sciences. Sources cited for the discussion below represent the tip of a very large body of literature.

19   Brian Walker and David Salt, *Resilience Thinking: Sustaining Ecosystems and People in a Changing World* (Washington, D.C.: Island Press, 2006), presents a highly readable and fully elaborated discussion of resilience theory with reference to key works in its development.

20   C. S. Holling, Lance H. Gunderson, and Garry D. Peterson, "Sustainability and Panarchies," in *Panarchy: Understanding Transformations in Human and Natural Systems*, ed. Lance H. Gunderson and C. S. Holling (Washington, D.C.: Island Press, 2002), 101.

21   Marten Scheffer, *Critical Transitions in Nature and Society*, Princeton Studies in Complexity (Princeton, N.J.: Princeton University Press, 2009). Similar ideas can be found in such popular books as Malcolm Gladwell, *The Tipping Point: How Little Things Can Make a Big Difference* (Boston: Little, Brown, 2000); and Philip Ball, *Critical Mass: How One Thing Leads to Another* (New York: Farrar, Straus and Giroux, 2004).

22   On Japanese views of nature, see (among many other works) Tessa Morris-Suzuki, *Re-Inventing Japan: Time, Space, Nation* (New York: M. E. Sharpe, 1998), 35–59; and Julia Adeney Thomas, *Reconfiguring Modernity: Concepts of Nature in Japanese Political Ideology* (Berkeley: University of California Press, 2001). On animals, see (again, among many others) Gregory M. Pflugfelder and Brett L. Walker, eds., *JAPANimals: History and Culture in Japan's Animal Life* (Ann Arbor: Center for Japanese Studies, University of Michigan, 2005). Possibly out of ignorance, we are unable to suggest readings on modern Japan's environmental successes.

PART I

Lay of the Land: Geology and Topography

Vulnerable Japan

*The Tectonic Setting of Life in the Archipelago**

GINA L. BARNES

The Great East Japan Earthquake and tsunami of March 2011 vividly illustrated just how vulnerable Japan is to repeated natural disasters. Most affected were flat, alluvial coastal plains, such as those now home to most of Japan's current population, where people lived despite stone markers set on high ground to warn about previous tsunami. One reads, "Remember the massive tsunami that caused disaster. . . . Do not build homes on sites lower than where you stand. . . . Bear this in mind even after many years have passed."[1] Though such disasters happen perhaps only once in a lifetime, any loss of life is tragic, and the events of 2011 were perhaps the first time that a localized occurrence had such wide-ranging ramifications in an advanced industrial economy.

Predicting earthquake and volcanic activity has become a demanding goal as the modern world tries to reduce risk and mitigate damage from natural processes. An important factor in risk is proximity to the hazard; however, throughout history, people have tended to recolonize areas affected by natural disasters, ignoring the personal risks of doing so. The alternative to removing populations from hazardous areas is developing an adequate prediction and warning system. This involves not only constantly monitoring earth movements today but also understanding the cyclicity of activity over time. New subdisciplines—palaeoseismology, archaeoseismology, and historic volcanology—are developing to take a long-range view of earthquake and volcanic occurrences.

This chapter concentrates on the volcanic setting of Japanese life, introducing the subdiscipline of *kazanbai kōkogaku* ("volcanic ash archaeology," or tephroarchaeology[2]) as a corollary to the subdiscipline of *jishin kōkogaku* ("earthquake archaeology,"[3] or archaeoseismology). These relatively new developments in Japanese archaeology are based in scientific disciplines rather than the humanities. For this reason, the first section of this chapter gives an overview from plate tectonics theory of the causes of earthquakes and volcanoes, both of which result from the archipelago's location within the Pacific Ring of Fire. The chapter closes with a consideration of modern efforts to monitor and mitigate volcanic activity and damage—a difficult and seemingly thankless task because public concern about volcanoes is very low compared with awareness of earthquake and tsunami hazards.

Land of Earthquakes and Volcanoes

Environmentally, the Japanese islands are a dangerous place to live. The archipelago is situated on four or maybe five different tectonic plates. Two of these are oceanic: the Pacific plate in the north and the Philippine plate in the south are being dragged down, or subducted, under the edge of the islands, which are sitting on the edge of the conjoined Eurasian and North American continental plates. This subduction of the oceanic plates gives rise to both earthquakes, as the plates rumble downward, and volcanoes, as the plate slabs dehydrate at depth. But earthquakes also have a second cause—compression brought about by distant continental collision resolved on megathrust faults.

Earthquake Hazards

The largest earthquakes are usually caused by compression rather than slab subduction. The collision of the Indian continent into the Eurasian continent is squeezing continental material north of the Himalayas out to the east in an example of escape tectonics. Japan is caught between the eastward movement of the Eurasian continent and the westward movement of the oceanic plates, giving rise to considerable compression faulting of the archipelago's landmass. Active faults, including megathrusts, in the shallow crust of the overlying plate are where stress builds up over time and is released over cycles that vary from hundreds to thousands of years on any one fault. The Kobe Earthquake of 1995 involved one such active fault, the Nojima fault; another was the Great Fushimi Earthquake of 1596, which caused much damage in the Kinai region. Shallow crustal earthquakes are seldom less than 6.5 in magnitude (M).

The Japan Meteorological Agency earthquake monitor does not distinguish between active fault and subduction earthquakes, but it shows that almost every day there are several quakes; these range from M2 to M7. Subduction earthquakes can rumble on continuously at hardly discernible levels, as under Shikoku. It is only when there is a blockage that does not allow the slab to subduct slowly and smoothly that a large subduction earthquake is felt, which happens in Japan about every one hundred years. Blockage is what caused the 2004 Sumatra Earthquake, when the down-going plate broke free after being stuck for some time, resulting in a M9.2 quake.[4] A similar scenario is proposed for the Kanto earthquake series. The Great Kanto Earthquake of 1923 was M7.9—the most recent of a series of large earthquakes affecting Tokyo, with earlier ones in 1855 (the Ansei Edo Earthquake, M7.2) and 1703 (the Genroku Earthquake, M8.2). A repeat performance—already overdue—is expected within this century.

The Great East Japan Earthquake of March 2011 was a subduction earthquake whose force and extent were unexpected and at M9.0 exceeded all those in the Kanto. Buildings constructed to modern earthquake standards generally survived the earthquake, but unfortunately many wooden structures were washed away in the following tsunami. The average height of the tsunami was 10 meters but reached 33 meters in places; a comparable earthquake and tsunami occurred in the same region in 1896, when waves reached 38 meters in height and caused 27,000 deaths.[5] The March 2011 death toll reached 15,781 with 4,086 missing, totaling 19,867, and there were 97,183 refugees from the Fukushima nuclear plant emissions.[6] In addition to actual damage to properties, running to trillions of yen, the world's increased vulnerability to "systemic disruption" was well demonstrated as the economic and nuclear effects rippled outward.[7]

Volcanic Hazards

Unlike earthquakes, whose cyclicity can be monitored (though not predicted) through palaeoseismology, volcanic eruption cycles are not regular.[8] Thus, research concentrates not on long-term prediction but on recognition of imminent activity. Earth tremors and expansion are detectable by seismology, global positioning system (GPS) readings, and changes in gas emissions. Nevertheless, it is impossible to predict what form the eruption will take, though most volcanic events in Japan are of the explosive kind—which again relates to the archipelago's position in a subduction zone. The causes of damage can be numerous, both direct and indirect: lava flows, pyroclastic surges (including ash cloud surges and base surges), mudflows (lahars), debris avalanches, landslides, lightning, forest fires, floods, earthquakes, tsunami, ash fall, and volcanic gas emissions.[9]

Subduction volcanoes can be extremely violent, as seen around the Pacific Ring of Fire and in the Mediterranean. The water dehydrated from the down-going slab lowers the melting point of the continental crust above it, and then dissolves into the magma as a gas. As the magma rises toward the earth's surface, the gas would normally escape. However, when increasing magma viscosity impedes the release of gas, pressure builds up inside the magma and causes it to decompress so quickly upon reaching normal air pressure at the earth's surface that it explodes in fragments (ash) that may be extruded into the atmosphere or roll down the volcanic flanks as pyroclastic flows (figure 1.1).

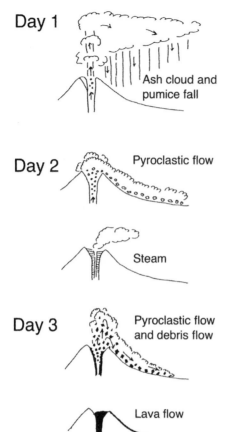

Day 1
Ash cloud and pumice fall

Day 2
Pyroclastic flow

Steam

Day 3
Pyroclastic flow and debris flow

Lava flow

FIGURE 1.1. Different eruption styles of Mount Asama, Nagano, on three successive days in 1783

SOURCE: Modified from Aramaki Shigeo, "Asama Tenmei no funka no suii to mondaiten," in *Kazanbai kōkogaku,* edited by Arai Fusao (Kokon Shoin, 1993), 105, fig. 7.

    A study of documented volcanic fatalities around the world shows that pyroclastic flows have claimed the most lives, about 29 percent of total fatalities.[10] These flows occur when a cloud of hot ash, too dense and heavy to be cast into the atmosphere, instead rolls down the volcano's flanks at high speed, carrying with it large rocks and anything else standing in the way. In 1991, the eruption of the Fugen-dake lava dome in the Unzen volcanic complex of northwest Kyushu generated a pyroclastic flow moving as fast as 200 kilometers per hour; it took fourteen lives, including those of several professional volcanologists (demonstrating the difficulty of predicting the form an eruption will take). Next in line, in the study for claiming lives, come indirect causes (famine and so on) and tsunami caused by volcanic landslides.

The previous eruption of Mount Unzen, Nagasaki Prefecture, in 1792, is one of the world's seven eruption events that have killed more than ten thousand from all causes since the year 1500.[11] It was Japan's worst documented volcanic disaster, taking fifteen thousand lives by debris avalanche and tsunami (the record has now been met by the Great East Japan Earthquake and tsunami). Another in the list is Mount Laki, whose eruption in Iceland in 1783 is notable for having produced "the largest lava flow on Earth observed in historic times, 2.9 cubic miles of lava which inundated 218 square miles."[12] It killed one-quarter of the population of Iceland and more than 60 percent of the cattle—the latter mainly by release of 0.5 billion tons of toxic gases.[13] These examples reveal that it is often not the actual eruption or earthquake that causes the most damage but the subsequent landslides, tsunami, ash dispersal, and gas emissions, which can affect more widespread areas.

Many more people live near volcanoes than ever before: three million are considered at risk when Mount Vesuvius next erupts.[14] Risk (R) calculation for geohazards depends not only on the hazard (H) itself (in this case, volcanic eruption) but also on the element (E) at risk and its vulnerability (V), as in $R = H \times E \times V$.[15] Risk entails "both physical and social components," while vulnerability can be defined as "the magnitude of loss potential."[16] Volcanoes that erupt in Hokkaido, far from any centers of habitation and whose ash clouds are blown northeast out to sea, entail very little risk because few people live nearby. Risk from volcanic hazard is thus spread unevenly throughout the Japanese islands. The eruption of Mount Fuji (west of Tokyo) or Sakurajima (in southern Kyushu) would pose a high risk of fatalities because of the high number of elements at risk and their vulnerability, as when thirty thousand people perished in the 1902 eruption of Mount Pelée over the town of St. Pierre on the island of Martinique.[17] Graphs of cumulative fatalities and of fatal eruptions both show upward curves, nearly exponential, from the fourteenth to the twentieth centuries. The authors of the aforementioned study reason that this is not because volcanic eruptions are becoming more frequent or devastating, but because more and more people are living near volcanoes and becoming vulnerable to eruptions. They note that tephra, not lava, is the cause of most deaths during volcanic eruptions, "killing mainly by collapse of ash-covered roofs … or by projectile impact."[18] They argue that improved building codes and public education to keep people away from volcanoes that pose a danger of eruption are the most efficient ways to lower the potential death rate in volcanic eruptions.

That observation brings us to consider Japan's record of historic volcanic eruptions. One of the first documented eruptions involved a collapsed house, written about in the ninth century CE and uncovered by archaeology.

◼ Past Volcanic Disasters Revealed by "Volcanic Ash Archaeology"

The Hashimure-gawa Site, Kagoshima Prefecture

At the southern tip of Kyushu, on the Satsuma Peninsula of western Kagoshima Prefecture, stands Mount Kaimondake, a volcano that has erupted at least twelve times in the postglacial period, most significantly in Late Jōmon times (4000 BP), in the Middle Yayoi period (2200 BP), in the last quarter of the seventh century CE (Late Kofun period), and in 874.[19] Nearby, the Hashimure-gawa site sits on a base layer of ash laid down by eruptions from the Ikeda caldera (5500 BP), overlain by a succession of subsequent ash fall layers from Mount Kaimondake. Thus, a sequence of time slices of the representative periods of Japanese archaeology are preserved at this site by tephra cover.

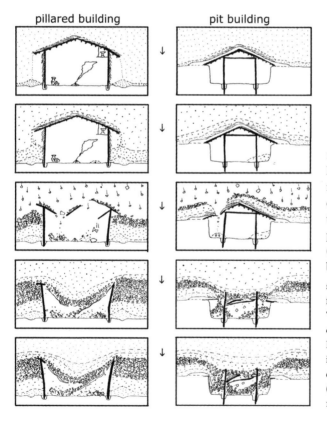

pillared building   pit building

FIGURE 1.2. Five hypothesized stages, from top to bottom, of destruction from tephra fall on a pillared building (left) and a pit building (right)

SOURCE: Modified from Soda Tsutomu, "Kofun jidai ni okotta Haruna-san Futatsudake no funka," in Kazanbai kōkogaku, edited by Arai Fusao (Kokon Shoin, 1993), 146, fig. 7.

The Hashimure-gawa site has been excavated more than twenty-five times, first in 1918–19 by Hamada Kōsaku, then professor of archaeology at Kyoto University. Designated a national historic site in 1924, it covers about 3.75 hectares and has yielded remains from the Jōmon through the Heian periods. Features recovered from under tephra layers at Hashimure-gawa include Kofun-period pit buildings, paths, canals, harrow marks, pottery dumps, and the course of the ancient river, including ramps leading up to the riverbank. The Heian-period remains include embedded-pillar buildings and ridge-and-furrow fields, cleanly covered by tephra, as well as more paths, bridge ramps, and shell dumps.

The tenth-century *Nihon sandai jitsuroku* records the destruction of the Heian-period settlement from scoria (lava cinders) and ash fall on the fourth day of the third month in 874; much of the destroyed village has been recovered in excavation.[20] One of the new features revealed by this excavation is the remains of uprooted trees. Most extraordinarily, the eyewitness account in *Nihon sandai jitsuroku* can be correlated with the sequence of the collapse of the buildings in archaeological terms (figure 1.2). The Heian-period structures destroyed in the Kaimondake tephra fall have been reconstructed at the Hashimure-gawa site in Ibusuki City, accompanied by a new site museum, the Ibusuki Archaeological Museum.

The Mount Asama Maekake Eruption in 1783, Gunma Prefecture

The 874 Mount Kaimondake eruption process may be compared with the detailed description of the progression and effects of the 1783 eruption of Mount Maekake, the latest volcanic mount of Mount Asama (As) in Gunma Prefecture.[21] Over the course of a three-month eruption, eyewitnesses itemized debris falls and flows and their destruction of houses, landscapes, and lives, revealing the horrors suffered by people living next to an active Quaternary volcano. One of the interesting issues arising from the comparison of the documentary and archaeological records of the Maekake eruption is the contradictions and omissions of the former.[22] For example, no mention is made of the final extrusion of lava; all eyes were on the incredible damage done by the pyroclastic flows and the flaming scoria falls that set fire to houses 10 kilometers distant. The succession of tephra falls and flows and their timings as documented in the field are also at variance with the eyewitness accounts.

Kuroimine and Mitsudera Sites, Gunma Prefecture

Mount Futatsudake is the most recent volcanic mount of Mount Haruna (Hr). It is believed to have erupted three times in the mid-first millennium CE. The first eruption (vent unknown) in the fifth century was relatively mild, spewing the Arima Ash (Hr-AA) in a thin layer across the landscape. The early-sixth-century eruption was more serious, resulting in the Futatsudake Ash (Hr-FA), known locally as the Shibukawa Tephra; and the mid-sixth-century eruption of Mount Haruna distributed a 2-meter-thick pumice layer—the Futatsudake Pumice (Hr-FP), known locally as the Ikaho Tephra—across the landscape to the northeast.[23]

Excavations in the winter of 2012–13 revealed the extraordinary skeletal remains of an infant and a young man wearing lamellar armor, both buried by Shibukawa Tephra in a ditch. It was the first recovery of human victims of volcanic ash fall dating to the Kofun period and a rare find of such armor outside a tomb.[24]

The later Ikaho Tephra buried a Kofun-period village, now known as Kuroimine and designated a national historic site.[25] Kuroimine was covered by more than 2 meters of ash and pumice; being so close to the source volcano, it was also bombarded by blocks of pumice as large as 30 centimeters in diameter, which broke through roofs and hastened the destruction of the houses.

The destruction by the mid-sixth-century Mount Futatsudake eruption was widespread. In addition to damage caused by the Ikaho Tephra fall, river drainages were blocked by pyroclastic flow and many locations were flooded. In particular, the Mitsudera, which sits along the bullet train (Shinkansen) route, probably sustained flood damage, including 78 centimeters of mud deposition, that led to its abandonment.[26] The first elite settlement to have been excavated, Mitsudera is extremely important in the history of Kofun-period archaeology.[27]

Preserved Paddy Fields

Some of the most eye-opening discoveries of the late twentieth century involved the excavation of intact field systems under substantial ash fall layers in the northern Kanto region of Gunma Prefecture (figure 1.3). Beginning in 1978, with the discovery of mid-fourth-century paddy fields at Hidaka, scores of field remains have been excavated from under three major protohistoric tephra falls: As-C from Asama in the late third century, Hr-FA from Haruna at the beginning of the sixth century, and Hr-FP from Haruna in the mid-sixth century; at least fifteen of these sites have yielded dry fields *(hatake)*, as opposed to fields for growing wet rice.[28]

One of the surprising aspects of these field systems is that the early paddy fields are much smaller than later, historical examples; generally rectangular and measuring about 1.5 by 2 meters, they are barely large enough to be worked by two people at a time using hand tools. Moreover, the fields are separated by slightly mounded ridges that are far less substantial than historical bunds, which served as raised paths between fields. In setting the field grid, ridges for the long sides of the rectangles were made first, dividing the land area into columns; then the length of the rectangular field was determined by making the short ridges, complete with small openings for water distribution. Irrigation water was thus passed through the columns of fields. It is thought that at some sites, the entire field area was leveled after harvest and every year the ridges and field pattern were created anew; at the very least, the horizontal ridges were dug out and the fields were redug within their columnar layout. These observations indicate that the protohistoric field system was not as permanent as the *jōri* field system—a grid of 1-hectare-square fields instituted in the Nara-Heian periods and still in evidence in some places today—and that more labor was invested on an annual basis for digging up and re-creating the field layout.

Cases have also been noted where land use has changed after a tephra fall. Phytolith analysis on samples in Gunma Prefecture taken from immediately

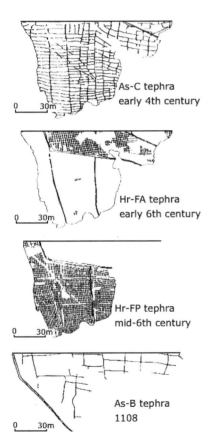

As-C tephra
early 4th century

0   30m

Hr-FA tephra
early 6th century

0   30m

Hr-FP tephra
mid-6th century

0   30m

As-B tephra
1108

0   30m

FIGURE 1.3. Stratified layers of paddy field alignments under different tephra falls in one location at the Dōdō site, Gunma

SOURCE: Modified from Hidetoshi Saitoh, "The Study of Cattle Ploughing and the Size of the Rice Fields of the Ancient Period," *Bulletin of Gunma Archaeological Research Foundation*, 17 (1999): 26, fig. 1.

above and below the three historical tephra layers—As-C (first half of the fourth century), Hr-FA (early sixth century), and As-B (dated to 1108)—indicate that rice was cultivated only from the fifth century in the Maebashi City area.[29] And yet, after paddy fields on the Maebashi Terrace were buried by the Asama B tephra (1108), the area was replanted in dry-land crops.[30]

Regional Destruction

Archaeologists have noted that populations of Paleolithic and Jōmon peoples in Japan were instantaneously exterminated when the Kikai eruption of ca. 5300 BCE deposited thick pyroclastic materials (the Akahoya tephra) over a 100-kilometer radius in southern Kyushu (figure 1.4).[31] The Kikai eruption extruded more than 170 cubic kilometers of volcanic material: up to 25 centimeters of Kikai pumice (K-KyP) was deposited on the southern tip of Ōsumi Peninsula in Kagoshima Prefecture, and up to 20 centimeters of Kikai-Akahoya tephra (K-Ah) was deposited in southeastern Kyushu, Shikoku, the San'yō coast, and southern Osaka and Wakayama prefectures.[32] From modern assessments, ash falls between 10 and 15 centimeters deep and lasting for five to seven days will kill pasture plants, crops, and soil microbes—leaving these areas sterile for up to a year; if more than 15 centimeters of ash falls, then "soil formation must begin again from this 'time zero.'"[33] Recent analyses demonstrate that the evergreen oak forest on the southern tip of Kyushu was wiped out in the area of Kikai pyroclastic flow, and other forests in the Inland Sea area were severely damaged by ash fall, floods, and landslides.

Human recolonization after volcanic eruptions is usually attested by artifact distributions. The Jōmon pottery types used prior to the Kikai eruption completely disappeared in Kyushu, to be replaced after a hiatus of nine hundred years or so by other types derived from the north and west.[34] An earlier and mightier explosive eruption ca. 28,000–30,000 BCE from the Aira caldera (now forming upper Kagoshima Bay) extruded more than 450 cubic kilometers of material: A-Ito pyroclastic flow deposit or ignimbrite from central Miyazaki Prefecture southward (figure 1.4) and its co-ignimbrite ash Aira-Tn (AT), with AT ash distributed into northern Tohoku.[35] All of Japan west of Nagoya was covered by 20 centimeters of ash. At this time, the pre-AT Paleolithic assemblage of southern Kyushu was mainly made of obsidian and consisted of knife-shaped tools, burins, and end scrapers; after the AT deposition, the material used was quartzite and included projectile points.[36] In the Kanto, the deposition of Aira-Tn (AT) ash was followed by rapid cultural change and technological diversity. Evidence from the Nogawa site, which exists as ten cultural units within thirteen natural strata of the Tachikawa loam, indicates that the shift from an early core-and-flake

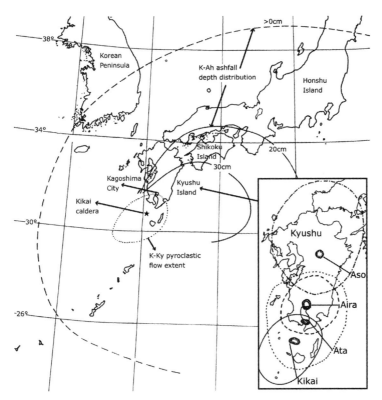

FIGURE 1.4. Tephra distribution from the Kikai caldera eruption, ca. 5330 BCE, in southern Kyushu

SOURCE: Compiled from Machida Hiroshi and Arai Fusao, *Kazanbai atorasu—Nihon rettō to sono shūhen* (Tōkyō Daigaku Shuppankai, 1992), 56, fig. 2.1–2; and Machida Hiroshi, Sugiyama Shinji, and Moriwaki Hiroshi, *Chisō no chishiki: Daiyonki o saguru* (Tōkyō Bijutsu, 1986), 80, fig. 19.

technology (Phase I) to projectile point and knife-shaped tools (Phase II) began just after the deposition of the AT ash; moreover, a change is also noted in the site structure at that time.[37] Not only do these findings enable the comparison of Paleolithic assemblages throughout Japan from below and above the AT ash layer, they also indicate that volcanic eruptions may have been severe enough to truncate certain regional lifestyles, which were subsequently replaced with other cultural forms.[38] However, Shimoyama cautions that the correlation of volcanic activity and cultural change should not be mistaken for volcanic *causation* of cultural change.[39]

Environmental change due to volcanic eruptions is thus thought to have affected both settlement patterns and subsistence regimes. The change in

Paleolithic technology over the entire Japanese islands following the AT deposition seems explainable only as the readjustment of tool technologies to new environments. More Jōmon reactions can be seen to the Akahoya deposition. One case study of shell mounds around Ise Bay has documented a decrease in ark shell *(haigai)* utilization immediately following the Kikai eruption; it is thought that erosion of Akahoya ash and its secondary deposition into the bay destroyed the productive shellfish beds on which the coastal communities had relied.[40] With the removal of this food source, the local population abandoned the area and presumably relocated. Considerable movement of populations, seen in the intrusion of Tōkai region pottery into the Kanto, or Kansai pottery into the Izu Islands, seems to have characterized the terminal Initial Jōmon period after the Kikai eruption.[41] These people, too, were probably in search of new food resources following the dumping of 20 centimeters of tephra over southwestern Japan.

At the Hashimure-gawa site, introduced above, a 20-centimeter ash fall destroyed fields but not houses; the former were abandoned but the latter were cleaned out and reused. But after a 30-centimeter ash fall and much damage to village structures, the occupants apparently left voluntarily.[42] In Gunma, the human response to tephra, except for instances of exceptionally thick cover, was to bravely recolonize the land and continue previous activities.[43] This is especially apparent where several stratigraphic layers of paddy fields, successively covered by tephra, have been excavated. In some cases where the tephra fall was light enough, farmers reused surviving paddy bunds to re-create field areas. The time lapse between field destruction and revitalization has been estimated as minimal, and even in areas of heavy tephra fall, such as the 2-meter-thick pumice cover, house pits were dug directly into the upper pumice surface before any soil formation could take place.[44] That the area was immediately reoccupied as soon as tephra deposition ceased may indicate that Kofun technology was more able to mitigate the environmental consequences than the technologies available to the Paleolithic and Jōmon peoples as seen above. Another, later example comes from Gunma Prefecture, where the 1783 eruption of Asama caused mudflows that destroyed 70 percent of crops and 30 percent of houses in paddy areas; afterward, parallel channels were dug in the earth and filled with the debris, then backfilled with the excavated dirt to remake the topsoil.[45] At the lowest level of settlement and subsistence, then, it might be possible to say that as we know from elsewhere, humans are incredibly tenacious and territorial in their habits, when not exterminated by the natural disaster itself.

▓ Volcanic Hazards in Modern Japanese Life

Scientific Monitoring of Volcanic Activity

It was not until 1973 that a Special Measures Law for Active Volcanoes (Kazan tokubetsu sochi hō) was introduced, and from the next year (1974), five-year plans for predicting volcanic eruptions began at the same time as the Committee for Predicting Volcanic Activity was formed of university specialists and government workers in hazard prevention.[46] The first list of active volcanoes made by this committee was issued in 1974, with 66 volcanoes recorded as active within the last thousand years; the list has been revised in 1999 (active within two thousand years), and in 2003 (active within ten thousand years).[47] Currently, 108 volcanoes are listed as active.[48]

In 1989, a research project on the prediction of a volcanic eruption was started within the National Research Institute for Earth Science and Disaster Prevention, now overseen by the Ministry of Education, Culture, Sports, Science and Technology. From 1994, a series of twenty-four-hour monitoring stations at twenty-six active volcanoes has gathered information on their activity, feeding these into four volcanic observation and information centers (VOICs) administered by the Japan Meteorological Agency (JMA). These centers are placed in Sapporo, Sendai, Tokyo, and Fukuoka—near the main clusters of active volcanic chains but with regional responsibilities that cover areas at risk. The monitoring stations are mainly seismic, recording earth tremors or small earthquakes that indicate the movement of magma within the volcano, but GPS, infrasonic microphones, tiltmeter, and high-sensitive visual cameras are also deployed.[49]

Satellite GPS can monitor changes in ground-level movement, tracking expansion or contraction of the volcano. For example, tiltmeter readings revealed that during the eruption of Mount Miyakejima in the Izu Islands, magma moved to the northwest, and Miyakejima Island itself contracted.[50] This is a sophisticated replacement for the visual tracking of volcano expansion, of which the most famous incident occurred in 1944–45. Then, the Usu volcano in Hokkaido rose 300 meters in twenty months before it erupted, as documented by an office worker tracing its increasingly large outline on a piece of paper taped to a window.[51]

Three other new technologies are being used in Japan.[52] Laser profiling was employed for the first time on the Usu and Miyakejima volcanoes and has been used in mapping sections of Mount Fuji that are densely forested. And, in addition to a correlation spectrometer (known as COSPEC), a large heavy instrument traditionally employed in measuring sulfur dioxide (SO_2) levels in volcanic ash clouds, a newer, smaller, lightweight instrument called Mini-DOAS (differential optical absorption spectroscopy) is making it possible to monitor emissions other than SO_2. Finally, improvements in radar have also allowed volcanoes to be "seen" through ash clouds. The technology, synthetic aperture radar (SAR), is mounted on aircraft, satellites, or space vehicles. An antenna picks up variable signals from an object as the craft moves over it; an image is then synthesized (or more accurately, simulated) from the successive signals.

The Tokyo Volcanic Ash Advisory Center (VAAC) is one of nine worldwide regional centers established in 1993 under the framework for the International Airways Volcano Watch for the express purpose of notifying airline companies, civil aviation authorities, and the like about volcanic ash cloud hazards. Any ash cloud detected by satellite imagery is monitored at intervals of six hours unless more rapid change demands closer tracking. The Tokyo VAAC, administered by JMA, cooperates with the Anchorage VAAC to the northeast, the Washington VAAC covering the South Pacific, and the Darwin VAAC to the south in monitoring volcanic activity along the western Pacific Rim.

Twenty-four-hour advisories are published on the Tokyo VAAC homepage for volcanoes located from Kamchatka in the north to the northern half of the Philippines. For any eruption of ash within this area, data are provided on the eruption time and height of the ash plume, and forecasts in graphic format and satellite images show movement of the volcanic ash cloud at six-hour intervals at three altitudes (from the surface up to 55,000 feet); however, the cloud's effect on aviation (via the aviation color code) is not specified in the Tokyo VAAC presentations. The online database goes back three months or so, and the online archive begins from 2006. Monthly reports describe volcanic eruptions and unrest (including gas and steam plumes, seismicity and number of tremors, and ground movements determined by GPS measurements).

Mitigating Public Damage

In 2003, the Committee for Predicting Volcanic Eruptions (Kazan Funka Yochi Renrakukai), under the auspices of JMA, undertook its latest ranking of Japan's active volcanoes for their likelihood of eruption. Three ranks were established (table 1.1). "Especially High Possibility" of eruption (A) includes volcanoes that had erupted more than five times in the past hundred years or more than ten times in the past ten thousand years. "High Possibility" of eruption (B) includes volcanoes that had erupted more than once in the past hundred years, or more than seven times in the past ten thousand years. Finally, "Low Probability" of eruption (C) refers to others among the active volcano list and includes volcanoes in the northern Kurile chain or submarine volcanoes.

A = very high possibility, B = high possibility, C = low probability

Region	A	B	C	Exclusions	Regional totals
Hokkaido	4	6	8	—	18
Kuriles	—	—	—	11	11
Tohoku	—	11	7	—	18
Kanto, Chubu	1	9	9	—	19
Itō, Ogasawara	3	4	4	10	21
Chugoku	—	—	2	—	2
Kyushu	3	3	4	1	11
Nanseitō	2	3	1	1	7
Rank totals	13	36	35	23	107

SOURCE: Compiled from map data in Japan Meteorological Agency, "Nihon no katsu-kazan bunpu," http://www.jma.go.jp/jma/kishou/intro/gyomu/index95zu.html.

Volcanic eruptions are potentially very likely in Hokkaido, the Izu Islands, and Kyushu but less so in Tohoku and Kanto. Indeed, there have been recent volcanic eruptions of Miyakejima in the Izu Islands and Shinmoedake and Sakurajima in Kyushu. Yamasato notes that data on eruptions of undersea volcanoes or those in the Kuriles are as yet insufficient to justify ranking of the risks.[53]

For informing the public of potential eruptions, JMA runs a volcanic warning and forecast system with three levels of warning or forecast and five levels of alert based on the feed from the volcanic observation and information centers (table 1.2).[54] According to Aizawa, volcanic alerts are issued "when volcanic activities are extremely intensified and urgent countermeasures to prevent/mitigate damage to human lives are required"; in contrast, advisories are issued "when unusual volcanic phenomena are observed and preparation for disaster prevention/mitigation is required."[55] These communications are distributed by the local municipality, but current volcanic activity throughout the islands is tabulated on the JMA website.[56] Nevertheless, complaints have been received about the warning levels—for example, that the statements themselves are hard to understand and that there are no cutoff dates. Hazard prevention experts have urged JMA to adopt the internationally used color code, but to no avail.[57]

The prediction network was instrumental in dealing with the eruptions of both Usu in Hokkaido and Miyakejima in the Izu Islands, in 2000. Nakata reports that for the first time, the real-time movement of magma in the chambers could be monitored, allowing eruption predictions, but that the system was

TABLE 1.2. Alert levels for volcanic eruptions in Japan

Warning level	Action	Volcanic activity	Eruption
5 Alert: very large eruption	Evacuate residential areas around volcano	Very large eruption or possibility thereof	Possibility of lava flow and/or pyroclastic flow reaching area distant from mountain
4 Alert: medium-large eruption	Prepare to evacuate residential areas around volcano (early evacuation of disabled persons)	Large eruption or possibility thereof; big explosion affects town	Volcanic bombs reach >3 km; small pyroclastic flow, air shock affects houses in town
3 Advisory: small eruption caution	Do not approach volcano (residents stand by for evacuation, prepare to evacuate disabled persons)	Explosive eruption at summit crater, or possibility thereof; precursors include large flux of volcanic gas, seismic swarm, glowing	Explosive eruption; volcanic bombs reach 3 km; very small pyroclastic flow
2 Active	Do not approach volcano crater (for hikers and sightseers)	Volcano is rather active: volcanic gas increases, volcanic earthquakes	Very small eruption might occur occasionally
1 Calm	Volcano is normal and calm and may be approached; however, there may be eruptions of ash, etc.; "threat to life is possible"	Volcanic activity (smoke, quakes) is low	Low possibility of eruption
0 Dormant		Dormant	No possibility of eruption

SOURCE: Compiled from Aizawa, "Present Situation."

still deficient in predicting how the magma would erupt (cf. figure 1.1) and how long the eruption would last.[58] He is optimistic that especially with volcanoes such as these, which erupt every twenty to fifty years, close scrutiny of multiple eruptions will help define and refine predictions of where and when they might erupt again. On the other hand, he warns that it is extremely unwise to try to predict *how* the volcano might erupt—to predict its "eruption scenario," as he calls it (cf. figure 1.1)—since so many variables can produce an enormous range of results, making a wrong prediction quite dangerous.

The September 1, 2004, eruption of Asama further illustrates the working of the observation network.[59] A month before eruption, a volcanic glow was captured by high-sensitive video camera, and on August 31, the hourly number of earth tremors increased dramatically. The next day, the observation level was increased from a Level 3 (advisory) to a Level 2 (alert), and the volcano erupted at 8:02 that evening. Volcanic bombs were thrown 2 kilometers from the crater. Eruptions continued in late September, and another large eruption occurred in February 2009, when again the observation level was increased from Level 3 to Level 2.

Directions of ash fall deposited in lobate forms depend on the prevailing winds, which tend to blow from the west across Japan. In the Asama eruption in Gunma in 2009, the ash fall extended 140 kilometers to the southeast (including Tokyo), whereas in the 2004 eruption, it was blown 250 kilometers northeast, close to Sendai. Each volcano has its historical set of lobate ash fall distributions of varying directions through various eruptions.[60] At least people downwind can be warned of impending ash fall, but surprisingly, there is not much public information on how this can be mitigated. Kamakura City, for example, explains on its website what to do only in case of an earthquake or typhoon, even though it is located approximately 75 kilometers directly east of Mount Fuji. The volcano last erupted in 1707, but more than seventy-five eruptions have been documented for the past twenty-two hundred years.[61] Perhaps the citizenry are under the illusion that "it can't happen here."

In southern Kyushu, Sakurajima, across the bay from Kagoshima, has been in "a period of frequent intermittent moderate explosive eruptions" since 1955.[62] Fukushima and Ishihara note a measure of complacency among the citizens of Kagoshima, attributable to the increasing sophistication of volcanic eruption prediction technology.[63] They observe that the improvements tend

> to decrease disaster prevention awareness of local people … residents may become complacent, feeling as though they live in an entirely safe environment and forgetting about previous disasters. In the long term, and with the disasters being distant memories, stories of disasters are not passed on to younger generations.

Their suggested solution is to establish "eco-museums" in areas affected by natural disasters, to bring the local environment to people's attention and keep them aware of the possibilities of reoccurrence. Fukushima and Ishihara tested their ideas during a postcard exhibition of the 1914 eruption of Sakurajima, which was viewed by about thirty-five hundred people; 80 percent of their surveyed viewers said their awareness for disaster prevention had increased even though many were not interested in the subject.

By 2008, sixty volcanic hazard maps had been published, "but no local governments have yet conducted assessments of volcanic risk analyses"; what is really needed, proponents say, is "real-time hazard maps with probability tree algorithms for forecasting volcanic events."[64] The second edition of *Volcanic Hazard Maps of Japan* was made available in 2013, documenting the development of both preanalysis and real-time analysis of volcanic hazards as the next generation of hazard maps.[65] Moreover, the announcement that "the preparation and release of volcanic hazard maps and disaster management maps for our country's major active volcanoes are almost completed" is greatly welcomed.[66]

▨ Conclusions

The Quaternary volcanic setting of the Japanese islands has greatly influenced human habitation. Periodic ash falls have not only cost lives but also interrupted settlement and cropping patterns. These changes are under intense scrutiny as scientists analyze human responses through the centuries. For both the ninth-century Kaimondake eruption in Kyushu and the eighteenth-century Maekake eruption in northern Kanto, the reactions of bureaucrats in the central government are being studied by historians for insights to their political concerns and duties, including the extension of welfare or tax relief to survivors and regional inhabitants.[67]

In recent years, much effort has been expended on monitoring volcanic eruptions because the risk factor rises with denser human occupation near Japan's 108 active volcanoes. Despite considerable scientific resources devoted to the problem, however, public education and acknowledgment of risk lag behind. People living close to constantly erupting volcanoes, such as Sakurajima, are blasé, and those surprised by the sudden eruption of, for example, Shinmoedake resume their lives close to the volcano when eruption "ceases." In Japan, where so many active volcanoes are close to prime flatlands for occupation, agriculture, and industry, the risk to person and property is somehow perceived as less than earthquake or landslide risk: people willingly coexist with volcanoes. As was reported after the Great East Japan Earthquake of 2011, "The fact remains ... that when it comes to the danger of natural disasters, complacency is the norm."[68]

This chapter is dedicated to those unfortunate enough to have been caught unawares by the sudden eruption of Mount Ontake in Nagano Prefecture on September 27, 2014, demonstrating that predicting volcanic eruptions is as yet an inexact science.

NOTES

*   I am grateful for comments and criticisms from two anonymous reviewers, from Soda Tsutomu and Machida Hiroshi concerning tephroarchaeology, and from Tina Niemi concerning earthquakes and faulting. Permission for reproducing modified versions of published illustrations were granted by Soda Tsutomu, Machida Hiroshi, and Saitō Hidetoshi. Thanks also go to Saitō Hidetoshi and Joseph Ryan for information concerning the new armor finds in Gunma.

1   "So Much for the Myth of Safety," Nikkei Weekly Special Magazine Issue (Summer, 2011): 5.

2   Tephroarchaeology is a term preferred by Soda Tsutomu (personal communication, August 17, 2013); see also Soda Tsutomu, "Evidence of Earthquakes and Tsunamis Recovered at Archaeological Sites in Japan," Kōkogaku jānaru 557 (2008): 21–26 (in Japanese with English title).

3   Gina L. Barnes, "Earthquake Archaeology in Japan: An Introduction," in Ancient Earthquakes, ed. Manuel Sintubin, Iain Stewart, Tina Niemi, and Erhan Altunel (Boulder, Colo.: Geological Society of America, 2010), 81–96.

4   California Institute of Technology Tectonics Observatory, "What Happened during the 2004 Sumatra Earthquake?" http://www.tectonics.caltech.edu/outreach/highlights/sumatra/what.html (accessed June 25, 2011).

5   USGS Earthquake Hazards Laboratory, "Magnitude 9.0—Near the East Coast of Honshu, Japan," http://earthquake.usgs.gov/earthquakes/recenteqsww/Quakes/usc0001xgp.php#tsunami (accessed June 25, 2011).

6   According to National Police Agency figures, September 2011. Kuniyoshi Takeuchi, "Disaster Management and Sustainability: Challenges of IRDR," International Conference on Science and Technology for Sustainability 2011—Building up Regional to Global Sustainability: Asian Vision—September 14-16, 2011, Kyoto, Japan, http://www.scj.go.jp/ja/int/kaisai/jizoku2011/pdf/presentation/prese20.pdf (accessed December 3, 2011). IRDR stands for integrated research on disaster risk.

7   Richard A. Warrick, "Volcanoes as Hazard: An Overview," in Volcanic Activity and Human Ecology, ed. Payson D. Sheets and Donald K. Grayson (New York: Academic Press, 1979), 161–94.

8   Katsuhiko Ishibashi, "Status of Historical Seismology in Japan," Annals of Geophysics 47, no. 2/3 (2004): 339–68.

9   Volcanic ash is not carbonized ash as from a fire; it consists of glass bubble fragments from the frothing magma and individual crystals of different minerals. It is termed "ash" because its grain size falls in the size range of carbon ash.

10   Tom Simkin, Lee Siebert, and Russell Blong, "Volcano Fatalities—Lessons from the Historical Record," Science 291, no. 5502 (2001): 255.

11   Simkin et al., "Volcano Fatalities," 255.

12   USGS Cascades Volcano Observatory, "Iceland Volcanoes and Volcanics: Laki—1783 Eruption,"
      http://vulcan.wr.usgs.gov/Volcanoes/Iceland/description_iceland_volcanics.html (accessed
      June 24, 2011).

13   Jan Kozák and Vladimir Cermák, *The Illustrated History of Natural Disasters* (Dordrecht:
      Springer, 2010), 79.

14   Christian Fraser, "Vesuvius Escape Plan 'Insufficient,'" BBC News, January 10, 2007,
      http://news.bbc.co.uk/1/hi/world/europe/6247573.stm (accessed October 15, 2013).

15   Martin Culshaw, "Geological Hazards: How Safe Is Britain?" Shell Lecture at Geological
      Society of London, December 8, 2010.

16   Warrick, "Volcanoes as Hazard," 174, 175.

17   Fred M. Bullard, "Volcanoes and Their Activity," in *Volcanic Activity and Human Ecology*, ed.
      Payson D. Sheets and Donald K. Grayson (New York: Academic Press, 1979), 9–48.

18   Simkin et al., "Volcano Fatalities," 255.

19   Fujino Naoki and Kobayashi Kobayashi, "Eruptive History of Kaimondake Volcano, Southern
      Kyushu, Japan," *Kazan* 42, no. 3 (1997): 195–211 (in Japanese with English title and abstract).

20   The date according to the lunar calendar; the corresponding solar calendar date was March 25,
      874.

21   Aramaki Shigeo, "Asama Tenmei no funka no suii to mondaiten," in *Kazanbai kōkogaku*, ed.
      Arai Fusao (Kokon Shoin, 1993), 83–110.

22   Aramaki, "Asama," 91, 105.

23   Soda Tsutomu, "Gunma-ken no shizen to fūdo," in *Gunma kenshi 1: Genshi-kodai 1*, ed. Gunma
      Kenshi Hensan Iinkai (Maebashi: Gunma-ken, 1990), 39–129.

24   Gunma Prefecture, "Kofun jidai no yoroi chakushō jinkotsu ni tsuite" (2013), http://www.pref
      .gunma.jp/03/x4500038.html (accessed October 29, 2013).

25   Tsude Hiroshi, "Kuroimine," in *Ancient Japan*, ed. Richard Pearson (Washington, D.C.: George
      Braziller and Arthur Sackler Gallery, Smithsonian Institution, 1992), 223–25.

26   Soda Tsutomu, "Kofun jidai ni okotta Haruna-san Futatsudake no funka," in *Kazanbai
      kokogaku*, edited by Arai Fusao (Kokon Shoin, 1993), 149.

27   Shiraishi Taichirō, "Mitsudera," in *Ancient Japan*, ed. Pearson, 220–22.

28   Soda, "Gunma-ken," 112, 116; the date for As-C has been revised from the original publication
      stating mid-fourth century to the late third century (Soda, personal communication, August
      17, 2013).

29   Masami Inoue and Hajime Sakaguchi, "Estimating the Withers Height of the Ancient Japanese
      Horse from Hoof Prints," *Anthropozoologica* 25/26 (1996): 119–30.

30   Soda, "Gunma-ken," 58.

31   Hiroshi Machida and Shinji Sugiyama, "The Impact of the Kikai-Akahoya Explosive Eruptions
      on Human Societies," in *Natural Disasters and Cultural Change*, ed. Robin Torrence and John
      Grattan (London: Routledge, 2002), 313–25.

32   Machida Hiroshi and Arai Fusao, *Kazanbai atorasu—Nihon retto to sono shūhen* (Tōkyō
      Daigaku Shuppankai, 1992), 47, 56–57.

33   USGS Volcano Hazard Program, "Volcanic Ash: Effects and Mitigation Strategies," section on
      "Effects on Pasture," http://volcanoes.usgs.gov/ash/agric/index.html#pasture (accessed June
      24, 2011).

34   Machida and Sugiyama, "Kikai-Akahoya," 322–23.

35   Machida and Arai, "Kazanbai atorasu," 60–65.

36   Oda Shizuo, "Kyūsekki jidai to Jōmon jidai no kazan higai," in *Kazanbai kōkogaku*, ed. Arai
      Fusao (Kokon Shoin, 1993), 207–24; Soda Tsutomu, Izuho Masami, and Sato Hiroyuki,

"Human Adaptation to the Environmental Change Caused by the Gigantic AT Eruption (28–30ka) of the Ito Caldera in Southern Kyushu, Japan" (paper presented at the International Field Conference and Workshop on Tephrochronology, Volcanism and Human Activity, Kirishima City, Kagoshima, Japan, May 9–17, 2012).

37 Oda, "Kyūsekki jidai," 210.

38 Ibid., 222.

39 Satoru Shimoyama, "Volcanic Disasters and Archaeological Sites in Southern Kyushu, Japan," in *Natural Disasters and Cultural Change*, ed. Robin Torrence and John Grattan (London: Routledge, 2002), 326–41.

40 Oda, "Kyūsekki jidai," 221.

41 Ibid., 222.

42 Shimoyama, "Volcanic Disasters," 336, 339.

43 Soda, "Gunma-ken," 111.

44 Site seen by the author in February 2001.

45 Saitō Satoshi, "Omowanu tokoro ni kodai shūraku," *Maibun Gunma* 45 (2006): 4–5.

46 Nakata Setsuya, "Kazan bōsai sofuto gijutsu no saizensen," in *Kazan funka ni sonaete*, ed. Doboku Gakkaishi (Doboku Gakkai, 2005), 131.

47 Yamasato Hitoshi, "Katsu-kazan no bunrui to kazan katsudōdo reberu no dōnyū," in *Kazan funka ni sonaete*, ed. Doboku Gakkaishi (Doboku Gakkai, 2005), 138.

48 Koji Aizawa, "Present Situation of Monitoring, Prediction and Information for Volcanic Disaster Mitigation in Japan" (paper presented at the Sixth Joint Meeting of the UJNR Panel on Earthquake Research, Tokushima, Japan, November 8–11, 2006, http://cais.gsi.go.jp /UJNR/6th/orally/O08_Present.pdf, accessed June 24, 2011).

49 Aizawa, "Present Situation."

50 Motoo Ukawa, Eisuke Fujita, Eiji Yamamoto, Yoshimitsu Okada, and Masae Kikuchi, "The 2000 Miyakejima Eruption: Crustal Deformation and Earthquakes Observed," *Earth Planets Space* 52, no. 8 (2000): 19–26.

51 Patrick D. Nunn, *Oceanic Islands* (Oxford: Blackwell, 1994), 96, fig. 3.15.

52 Nakata, "Kazan bōsai," 135.

53 Yamasato, "Katsu-kazan," 141.

54 Japan Meteorological Agency, "Weather and Earthquakes: Volcanic Warnings and Volcanic Alert Levels," http://www.data.jma.go.jp/svd/vois/data/tokyo/STOCK/kaisetsu/English/level .html (accessed December 16, 2014).

55 Aizawa, "Present Situation."

56 Japan Meteorological Agency, "Volcanic Warnings," http://www.jma.go.jp/en/volcano/ (accessed December 16, 2014).

57 Yamasato, "Katsu-kazan," 143.

58 Nakata, "Kazan bōsai," 132–33.

59 Aizawa, "Present Situation."

60 Machida and Arai, *Kazanbai atorasu*.

61 David Wolman, "Mount Fuji Overdue for Eruption, Experts Warn," *National Geographic News*, July 17, 2006, http://news.nationalgeographic.com/news/2006/07/060717-mount-fuji.html (accessed June 24, 2011).

62 See the section on Sakurajima at the website of the Global Volcanism Program of the Smithsonian National Museum of Natural History, http://www.volcano.si.edu/world/volcano .cfm?vnum=0802-08=&volpage=photos (accessed June 24, 2011).

63 Daisuke Fukushima and Kazuhiro Ishihara, "Volcanic Disaster Prevention and Community Development: How to Convert the Volcano into a Museum," *Annuals of Disaster Prevention*

Research Institute, Kyoto University 48C (2005), http://www.dpri.kyoto-u.ac.jp/nenpo /nenpo_e.html (accessed December 16, 2014).

64 Yoichi Nakamura, Kazuyoshi Fukushima, Xinghai Jin, Motoo Ukawa, Teruko Sato, and Yayoi Hotta, "Mitigation Systems by Hazard Maps, Mitigation Plans, and Risk Analyses Regarding Volcanic Disasters in Japan," *Journal of Disaster Research* 3, no. 4 (2008): 297–304.

65 National Research Institute for Earth Science and Disaster Prevention (NIED), "Contents of Volcanic Hazard Maps, Second Edition," Database on Volcanic Hazard Maps and Reference Material (1983–), http://vivaweb2.bosai.go.jp/v-hazard/articles-e.html (accessed October 28, 2013).

66 Nobuo Anyoji, "Technical Efforts to Prepare Volcanic Hazard Maps," *Technical Note of the National Research Institute for Earth Science and Disaster Prevention*, no. 380 (July, 2013), 127.

67 Cf. Aramaki, "Asama Tenmei."

68 "So Much for the Myth of Safety," 5.

2

Settlement Patterns and Environment of Heijō-kyō, an Ancient Capital City Site in Japan

TATSUNORI KAWASUMI

Like Gina Barnes's essay in this volume, this chapter employs geological and archaeological data—here used in combination with geographic information system (GIS) methodologies to help us understand how eighth-century Japanese chose the locations and layout of imperial capitals, and how people and institutions were ordered over urban space. The study is thus inherently interdisciplinary. It makes use of data on geological layers uncovered as archaeologists excavated more than seven hundred sites in the area of modern-day Nara. Putting these data into a GIS database permits us to re-create changes in the landscape over more than a thousand years. This innovative approach yields new insights into how humans interacted with their natural environment, a topic largely beyond the reach of extant written records.

In contrast to most past studies by Japanese scholars, my work here focuses on Japanese interactions with nature in an urban rather than a rural context. Unlike rural areas, where lightly built wooden buildings predominate, urban areas—and particularly capital cities—have higher concentrations of large buildings with stone foundations and tile roofs and floors. The new research methods alluded to above allow us to explore the degree to which ancient architects planned urban sites to take that characteristic into consideration as they considered appropriate foundations for monumental structures. Further, by definition, urban areas concentrate human populations and the activities that support them—transportation, trade, and manufacture. The use of GIS allows us to explore both how fresh water was supplied to support human activities and also the degree to which those activities influenced groundwater, and through it, human health.

▨ Contexts: The Founding of Heijō-kyō

Until the end of the seventh century CE, Japan lacked significant urban centers. Royal palaces existed but were moved frequently. Sometimes this took place upon the death of one monarch and the ascension of a new one. On other occasions, the palace was moved when the fortunes of a reign dimmed or seers recommended a more auspicious site. In any case, unlike their Chinese counterparts, Japanese palaces were not associated with cities (indeed, early Japan had no urban centers at all) despite the efforts of Japanese political leaders to adopt many elements of the political, religious, and architectural culture of Tang China, the epitome of high political and cultural accomplishment in East Asia at the time. These administrative efforts intensified with the Taika reforms of 645 and thereafter.

The establishment of Heijō-kyō (modern Nara) in 710 marked the first conscious effort to create a permanent imperial capital. (A short-lived predecessor was Fujiwara-kyō, 684–710.) The effort was not entirely successful: Emperor Shōmu temporarily abandoned Heijō-kyō in the mid-eighth century; another short-lived capital, Nagaoka-kyō, was constructed starting in 784; and the capital was moved permanently to Heian-kyō (modern Kyoto) in 794. Nonetheless, Heijō-kyō was Japan's first relatively long-lived imperial capital, and it set the pattern for what was to follow.

The layout of Heijō-kyō's streets and major buildings reflected Chinese influence. The street grid was modeled on that of the Tang dynasty capital at Chang'an (modern Xi'an). Streets were laid out in regular rectangular fashion, with important buildings located at the north-central part of the city. Siting of the city was also based partly on *feng shui* principles, which attempted to locate buildings and towns in such a way as to harmonize with the forces of both Heaven and Earth.

Although the general facts of Heijō-kyō's siting and construction are well known, the spatial distribution of its population within the city grid is not. Practices like *feng shui* and models such as Chang'an provide important information about the site and orientation of an urban center, and even tell us a bit about the location of imperial residences and administration, but they tell us next to nothing about how upper and lower orders of society were distributed throughout urban space. They also tell us little about how urban centers were situated in their environments to fill other needs—water supply, sewage handling, and siting of monumental architecture for both governmental and religious purposes. The effort here is part of a series of excursions into Japan's early environmental history that explores heretofore unexamined facets of urban culture.

Urban studies relating to the environment have typically focused on current issues, addressing urban problems arising from overpopulation and giving center stage to policy concerns rather than history. Historically oriented studies have analyzed the interrelationships between human activities and the natural environment in primitive urban areas. This helped establish the idea of harmonious relations between society and its urban environment, which provided a stimulus to address modern environmental problems. As a result, studies in a variety of fields, including geography, history, archaeology, and ethnology, have examined historical relations between human activities and the natural environment in order to clarify their modern ramifications.[1] In contrast to this approach, Miyataki Kōji has gone further and argued that environmental history should be undertaken on its own terms as one element of research into ancient history in Japan, regardless of its potential to resolve modern environmental problems.[2]

While urban problems loom large on the agendas of those studying modern environmental issues, Japanese historical geographical research on ancient environmental history has largely focused on agricultural communities.[3] Even in this realm, however, the number of studies is limited. Based on existing research, a generalized model of the spatial structure of villages has emerged, although scholars recognize wide variation in practice. This model posits that residences were located on relatively higher ground, with paddy fields on lower ground and dry fields in the remaining sections of the village.

Research on rural Japan poses one important question for scholars of ancient urban geography: does the model developed for agricultural communities have an urban parallel? Although urban residents often cultivated some vegetables and the like in their own gardens, these efforts supplemented other work that formed the core of their economic activities. Did this lead to different patterns of settlement than we see in rural areas?

To explore these and other issues, my colleagues and I have adopted a geo-archaeological approach for reconstructing landform changes in historical times based on observations and records of the surface geology in Heian-kyō and Fujiwara-kyō.[4] This research is somewhat unusual because few other studies of ancient history have focused on landforms, which can change every several centuries. Using a geo-archaeological approach has allowed us to analyze the relationship between ancient landforms and land use; we have discovered that landform and landform changes were closely related to the siting and development or transformation of ancient capital cities. Further, we have explored the landform's influence on water supply, groundwater quality, and other urban planning issues. Such studies require a truly long-term perspective, one we have utilized.

The relationships between the natural environment and human activities in Heijō-kyō (710–784) have yet to be sufficiently investigated, a gap this study hopes to partially fill. I focus on historical interactions between landforms and human activities in an attempt to explore the influence of the environment on the location and development of this ancient city. Specifically, I identify and reconstruct a series of landform changes that occurred in historical times and attempt to correlate them with changes in land use in the same location over time. This study also examines the characteristics of environmental utilization in Heijō-kyō in relation to other ancient capitals (figure 2.1, figure 2.2).

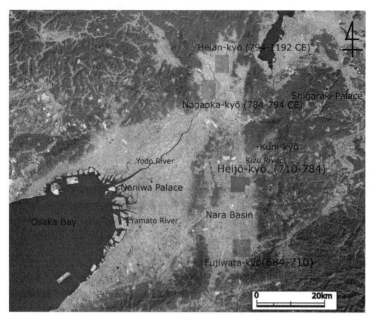

FIGURE 2.1. Capital cities of ancient Japan

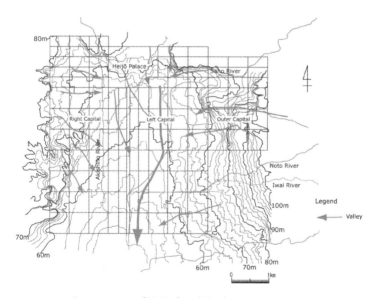

FIGURE 2.2. Contour map of Heijō-kyō (Nara) area
Contour lines are at 2-meter intervals.

SOURCE: Base map traced from Hachiga Susumu, "Kodai tojō no senchi ni tsuite:
Sono chikei teki kankyō," *Gakusō* 1 (1979): 32, fig. 2.

■ Heijō-kyō's Site and Climate

Heijō-kyō was established as Japan's capital in the north of the Nara basin in 710, following abandonment of the previous capital farther south in the basin, Fujiwara-kyō. Apart from a brief hiatus during the 740s, Nara functioned as Japan's capital until 784, when the capital was moved to Nagaoka-kyō in the Kyoto basin. From 740 to 744, three temporary courts were set up—at Naniwa Palace in the Osaka area, Shigaraki Palace in the Shiga region, and Kuni-kyō in the southern Kyoto district—but ultimately, only Heijō-kyō and Naniwa Palace, which was treated as a subsidiary imperial center, remained.[5] In ancient Japan, as noted above, movement of palaces and capitals was frequent and occurred against a varied background of political, religious, intellectual, and environmental factors.

Heijō-kyō was laid out in a grid pattern, called *jōbōsei*, in which streets formed regular blocks laid out along north-south and east-west axes. This grid pattern was modeled on that of China's Tang dynasty capital, Chang'an. The *jōbōsei* arrangement created an urban space that reproduced the distribution of dry field and paddy on agricultural land. In recent years many excavations of historical remains have uncovered the remnants of this *jōbōsei*, providing an evidentiary basis for reconstructing the original design of the city.

The north-south axis of Heijō-kyō was formed by Suzaku Avenue, the city's main street. At the southern end of Suzaku Avenue stood the main gate to the city, the Rajōmon. At the north end of the street stood the main gate to the Heijō Palace grounds, site of the central political and administrative facilities as well as the emperor's residence. Within this compound was the Audience Hall (Daigoku-den), a large central structure where imperial ceremonies took place. The Audience Hall is emblematic of the monumental structures of Heijō-kyō. Based on archaeological findings, it is estimated to have been 44 meters wide, 20 meters long, and 27 meters high.

The area in Heijō-kyō outside the palace was called *kyō*, the capital; the section of the city to the west of Suzaku Avenue was called the Right Capital *(ukyō),* and that to the east was called the Left Capital *(sakyō)*. An area connected to the northwest part of the Left Capital was called the Outer Capital *(gaikyō)*. The existence of an Outer Capital differentiated Heijō-kyō from other ancient Japanese imperial cities.

Residences of aristocrats and commoners occupied much of the capital's area, but in its southern portions were two commercial centers, the Eastern Market and the Western Market, with foodstuffs and other goods collected as taxes from Heijō-kyō and the country as a whole. Within the city, taxes were levied not on paddy but on small dry fields in the residential areas. In addition to the palace, administrative offices, residences, and markets, the city also featured many

Buddhist temples. Seven main temples, called the Seven Great Temples of Nara, were built, three in the Right Capital and four in the Left.

The road network of Heijō-kyō was planned to link with the existing main roads of the Nara basin. Suzaku Avenue connected to the main north-south road of the basin, called the Shimotsumichi. Higashi Kyōgoku Avenue connected to another main road, the Nakatsumichi. The Shimotsumichi and Nakatsumichi both ran to the former capital of Fujiwara-kyō. In addition, the capital was linked to the road to the Naniwa Palace, the Tōkaido Highway, and to the main roads connecting Heijō-kyō to nearby provinces.

Waterborne transport was also important. The Yamato River and the Kizu River were of particular significance, even though neither was deep enough to accommodate large boats. Through Heijō-kyō's location in the north of the Nara basin flowed the various tributaries of the Yamato River, which empties into Osaka Bay. The largest of these tributaries (by size of the drainage basin, 128 square kilometers) was the Saho River. Its source was the Yamato plateau at the northeast of the Nara basin. The Saho River flowed from the northeast of the basin through Heijō-kyō, and then to the south through the Left Capital district. Flowing from the northeast through the Left Capital area was the Iwai River (with a drainage basin of 13 square kilometers), a branch of the Saho. The Higashihori River, a man-made stream, flowed parallel to the Saho in the Left Capital area. Smaller streams originating in the piedmont area of the Yamato plateau joined the Higashihori River. Alongside the Saho River, to its west, was the Komo River, a small stream that flowed through the lowest part of the basin. What is now called the Akishino River (with a drainage basin of 23 square kilometers) ran from its source in the Saho Hills through the Right Capital section of Heijō-kyō. This river corresponds to the man-made Nara-era Nishihori River. It flowed roughly parallel to but on the opposite side of Suzaku Avenue from the Higashihori River, whose path has been determined by archaeological excavations. Smaller streams flow into the Akishino River from the Nishinokyō Hills.

The special characteristic of the rivers in the northern part of the Nara basin lay in their small watersheds and the small amounts of water they ordinarily transported. This circumstance made for distinct advantages as well as problems. On the one hand, limited volumes of water made it possible to construct water control facilities for irrigation and other purposes. Reflecting their low water volume, the streams flowed through Heijō-kyō in straight lines. Perhaps as an outcome of civil engineering on these streams, documents from the Nara era record only one flood.[6] On the other hand, small catchment size and limited water supply posed real problems for both human consumption and productive agriculture. Despite the irrigation facilities, droughts were likely, and finding a secure and reliable source of water for daily use was problematic. This issue was partly solved by digging wells, but as we shall see, that solution, in combination with limits on existing excavation technologies, brought its own complications.

The limited volume of water reduced the risk of flooding but raised the risk of shortages. And in fact, Nara-era documents record frequent droughts. Archaeological excavations of Heijō-kyō have led to the discovery of many wells, which drew the groundwater of relatively shallow aquifers.

Furthermore, the limited supply of water was also easily polluted by wastewater from daily business and household activities. Today, because of the high population density of Osaka and environs, the modern Yamato River network has very poor water quality, with Japan's second-highest levels of biochemical oxygen demand, indicative of high levels of organic contamination. In the Nara era, the estimated population of Heijō-kyō was between fifty thousand and one hundred thousand people, which would indicate a population density of two thousand to five thousand people per square kilometer, a relatively high level that, paralleling the experience of modern Osaka, placed a heavy burden on the environment.[7] Toilet facilities in Heijō-kyō uncovered by archaeologists took advantage of nearby small streams to naturally flush sewage directly into streets and ditches.[8] Garbage disposal was a problem, too, and excavations suggest that it was simply thrown into pits dug within the confines of residential lands, creating additional sources of pollution in residential compounds.

What do we know of the climate of Heijō-kyō? According to the empirical research of Yoshino Masatoshi, the Sinitic government of ancient Japan was established in an era of warming.[9] The original city of Heijō-kyō thus functioned under warm conditions at the start of the Medieval Warm Period. This trend partly explains why droughts were relatively common in Heijō-kyō but flooding was rare.

Heijō-kyō, surrounded on three sides by mountains or hills, had access to significant forest resources within a relatively short distance. Based on identification of the plant and pollen remains in Nara-era sites, it is clear that secondary forests had spread throughout the hillsides around the Nara basin.[10] Forests supplied construction materials, although not always in the ways we might expect. In the Narayama Hills, located to the north end of Heijō-kyō, were distributed several kilns for firing tiles used to roof buildings in the capital; they were fed by fuel from the area's woods and forests. However, the timbers used to construct many buildings came from regions relatively far away, outside Yamato Province; for example, supplies from Ōmi and Iga provinces have been documented.[11] In addition, when the capital was moved from Fujiwara-kyō, we can verify the reuse of timbers from buildings in the old capital that were torn down and transported to Heijō-kyō.[12]

◼ Historical Landform Changes in Heijō-kyō

As already noted, Heijō-kyō is situated in the north of the Nara basin, surrounded by hills and plateaus to the east, west, and north. These highlands and craggy cliffs are distributed along an active fault. The Saho, Akishino, Iwai, Noto, and other tributaries of the Yamato flow through the Heijō-kyō area and empty into Osaka Bay. A distinctive characteristic of these rivers is their very small drainage area as they flow into Heijō-kyō. Consequently, the area of floods, when they occurred, was relatively restricted, and their influence in creating landforms was rather limited.

The Saho River has its origins in the Yamato plateau and has the largest drainage basin of any of the rivers that flow through Heijō-kyō (plate 2). Its floodplain extends from the alluvial fan near the modern city hall to a region of natural levees. The Holocene (the present back to about twelve thousand years ago) alluvial fan created a terrace some 2 meters high along the shallow valley of the river. The Iwai River, which like the Saho has its origins in the Yamato plateau, and the Noto River also gave rise to alluvial deposits to create a fan where the rivers join before flowing into the Saho River. The Akishino River begins in the northwest section of the uplands and created a gentle sedimentary belt during the Holocene epoch.

Despite those constraints, Heijō-kyō's riverine environment underwent important changes during the city's history. To reconstruct historical landform changes, I surveyed the strata uncovered by excavations at archaeological sites. Other information on landforms and geological strata contained in excavation reports was used to supplement the observations that formed the core database. As a result, it was possible to classify historical landform changes in the Saho River basin within the city of Heijō-kyō. Five clear stages are evident (figure 2.3).

Stage 1 (eighth–early ninth centuries). It was not possible to confirm any landform changes in this period. Surveys conducted at various excavation sites have failed to reveal any significant flood deposits from the Saho River during this stage. There appear to have been only a few flood overflows, resulting in limited landform changes. This finding was somewhat surprising for sites located so close to a river. Circumstances changed in later eras, when soil deposition and flooding were more common. The absence of major flooding during Stage 1 was not the result of landforms that differed from those of later ages—the broad topographical contours of the Heijō-kyō area remained essentially the same. (Riverbed changes are noted below.)

Stage 2 (mid-ninth–early fourteenth centuries). Between the mid-ninth century and the twelfth century, the floodplain dropped approximately 2 meters, and lateral erosion occurred in the upstream channel of the Saho River as it

Stage 5: Mid-Twentieth Century-Present

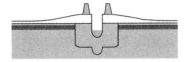

Stage 4: Sixteenth Century - Early Twentieth Century

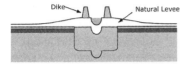

Stage 3: ca. the Fifteenth Century

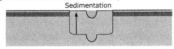

Stage 2: Mid-Ninth Century - Early Fourteenth Century

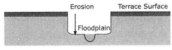

Stage 1: Eighth Century - Early Ninth Century

FIGURE 2.3. Model of landform changes in Saho River floodplain (Nara) in historical times. Stages are placed with the earliest at the bottom, and the most recent at the top.

passed through the northeastern parts of Heijō-kyō. This suggests more frequent and more significant flooding than in Stage 1. Although archaeological excavations reveal major floods in the Heian era, thereafter channel erosion deepened the riverbed and the area exposed to flooding actually decreased. This riverbed erosion essentially created a sharp drop-off from the original terrace, forming a cliff from the terrace down to the river. The Saho River basin was thus divided into two landform surfaces: a floodplain up to about a hundred meters wide, bounded on both sides by cliffs leading up to the original terrace surface. Consequently, the flood overflows were confined to the floodplain at the base of the terrace cliffs, and in this stage, the Holocene terrace surface continued to be largely free of flood overflows.

Stage 3 (ca. fifteenth century). As the Stage 2 floodplain was continuously subjected to deposition of upstream erosional soils, both the plain and the river bottom rose in elevation. On the terraces, too, flooding and accompanying sedimentary processes reshaped the surface of the Holocene terraces. In this stage, flood overflows influenced the Holocene terrace surface more than in the preceding stages. This once again suggests more flood activity and severity in this part of Nara than during Stages 1 and 2.

Stage 4 (sixteenth–early twentieth centuries). In this stage, the frequency and scale of flood overflows increased still further. Additional natural levees were formed and the riverbed rose in elevation. The rise in the bed of the Saho River also raised the local groundwater level, causing waterlogging of nearby land-forms. On the Saho River, dikes were constructed to prevent flooding, but as occurred on the Tenjō and other rivers in the area, soil deposition raised the riverbed gradually. Ultimately, when dikes were breached, the resulting floods were more severe than those before dike construction.

Stage 5 (mid-twentieth century–present). During modern times the bed of the Saho River has been artificially dredged and the river channel deepened to control flooding. These efforts have been particularly intense since the end of World War II, and the riverbed is now at a lower level than in the previous stage.

Just why flooding increased in Stage 3, and especially in Stage 4, remains to be investigated. However, we can point to several possible explanations. First, beginning in the fifteenth century, the Little Ice Age changed the climate in some parts of Japan, causing an increase in precipitation. In addition, growth in population during this time led to extensive deforestation and the expansion of arable land. These developments probably increased both the risk of flooding and its upstream erosional effects.

As indicated by the historical landform changes in the Saho River basin, noted above, other landforms in the Nara period, too, differed from those of today. To reconstruct the landforms of the Nara period, it is necessary to remove later ones from contemporary landform classification maps (plate 2). Specifically, our GIS model allows us to reconstruct Nara-period landform distribution patterns in this period by eliminating from the map those that resulted from channel erosion in Stage 2 and natural levee formation in Stages 3 and 4 (plate 3).

In the Nara period, Heijō-kyō was located in an area that witnessed few flood overflows, yet surface geology surveys of the basins of the Akishino, Iwai, and Noto rivers show no evidence of flood deposits from the eighth century. We can therefore assume that flooding was infrequent in the city of Heijō-kyō. Further, extant historical documents show that the area was hit by only one flood disaster during the Nara period, in 728 (the fifth year of Jinki).[13]

Most of Heijō-kyō's archaeological remains are buried deep below the current surface level. To reconstruct the uneven ground surface of Heijō-kyō as it existed in the Nara period, I obtained data from as many excavation sites as possible indicating how far below surface level the ancient remains lay. Identifying the spatial distribution and depth of such Nara-era remnants provides a foundation on which we can reconstruct the shape of the ground surface in that period. I generated a digital elevation model (DEM) that represents the undulations in the ground surface by subtracting the depth of Heijō-kyō's archaeological remains from a DEM of the present ground surface, and then, from the many known points and depths, used algorithms to create a model of Nara-period topography for the study area of the city.

Plate 4 illustrates the spatial distribution of the detected depth of the surface of Heijō-kyō site based on this methodology. The depth increases from the end of the Saho River's alluvial fan through the upstream section of the natural levee. Excavation surveys have proved that most of this sediment accumulated from the sixteenth century onward.

The Relationship between Landforms and Land Use in Heijō-kyō

Not only can we reconstruct the topography of Heijō-kyō, we can also create a map of land-use patterns in the Nara era. For purposes of this research I combined data for land use and landforms using GIS. To display land use on a block-by-block basis according to the old *jōbōsei* divisions, I created block-level polygons based on previous reconstructions of the city. Combining information on land use compiled from previous research on Heijō-kyō with the urban polygon data made it possible to display land use on the scale of one *jōbōsei* block on the GIS platform.[14] Finally, data from the most recent archaeological excavations were added.

The excavated building remains of Heijō-kyō exhibit differences from the beginning of the period through the end of the era. We can imagine that there were also changes in land use over the course of the Nara era, but at present there are no research results that can be structured into a database and subdivided by time period. Thus, for purposes of the present research, if any archaeological or written evidence demonstrated that at some time a unit of area was residential land during the Nara era, that land was treated as urban land.

A comparison of Heijō-kyō's landforms and the reconstructed map of land use shows a close correspondence between the two. In this section, I discuss my research results regarding the location of Heijō Palace, the large temples, urban areas occupied by ordinary people, and residences of the nobility (plate 5).

To begin, Heijō Palace was built atop a Pleistocene terrace and various incised valleys at the northern end of the Nara basin. The first and second audience halls (Daigokuden) were built on the terrace surface. These very large structures required firm foundations, and the terrace surface provided exactly the kind of foundation on which such magnificent edifices could be erected.

For the same reason, large temples were also constructed on Pleistocene terrace surfaces. Prominent examples include Tōdaiji, Kōfukuji, Daianji, and Saidaiji. Tōshōdaiji and Yakushiji temples in the Nishinokyō area include some flat alluvial land within their precincts, but the main buildings were all on terraces.

In commoner and aristocratic residential areas the land-use pattern suggests that when selecting urban areas, planners gave priority to geographical features that allowed easy access to groundwater. For example, Heijō-kyō's common people mostly lived on the floodplain of the Saho River, incised valleys within the Pleistocene terraces, valleys in the river's alluvial fan, and around the border of the alluvial fan and the natural levee zone. These geographical features were at high risk of flood overflows and had a relatively high groundwater table. In these areas archaeological excavations reveal numerous wells near the residences of commoners.

The location of aristocratic residences likewise suggests the importance of water availability for city residents. The residences of the nobility in Heijō-kyō were mostly located in incised valleys in Nishinokyō near the alluvial lowland, near the border of the alluvial fan and the natural levee zone of the Saho and Akishino rivers, and on the natural levee zone (floodplain) of the Saho River. The residences were located on floodplains near buried paleochannels and incised valleys.

In general, these are Heijō-kyō landforms with a high groundwater table, once again indicating that land where water was easily accessible was chosen for constructing residences. For example, surveys of land plots *(tsubo)* 15 and 16 of the block defined by the fifth east-west avenue *(jō)* and sixth north-south boulevard *(bō)* of the Left Capital, where residences of the nobility have been excavated, uncovered sand-filled paleochannels of the Jōmon period.[15] These paleochannels are likely to have been passages for underground water because this area had higher water permeability than the surrounding areas. Wells were dug along the sand-filled paleochannels. There are many examples of nobles' residences in Heijō-kyō where wells were dug into such paleochannels.

High population densities had deleterious effects on the subsurface water. There were places where good-quality groundwater could be obtained, but it is also easy to imagine that population concentration polluted the resource. As noted previously, in the aristocratic residential areas of Heijō-kyō, excavations have uncovered the remains of toilets whose waste ran through the compounds directly into the streets and roadside ditches. However, residential compounds also had wells, and this combination led to contamination of drinking water and the spread of disease. In the remains of the toilets, researchers found fossil eggs of intestinal worms, fecal worms, oriental liver flukes, and other parasites, informing us of the poor hygienic environment of the age. Also, documentary evidence reveals the reasons for public officials' absences from official duties, and frequent mentions of diarrhea and dysentery provide further indications of the unhygienic conditions of the time.[16] Further, excavations of these compounds have revealed many pits that appear to have been dug for the purpose of waste disposal, creating further sites of pollution close to human habitation.

The apparent close relationship between land use and landforms leads me to conclude that land use was determined by taking into account the different landforms and their characteristics. However, we also see a tendency to select construction sites that sometimes incorporated topographical features and conditions that would be regarded as unsatisfactory by present-day land evaluation standards. For example, the water table in Heijō-kyō's aristocratic residential district was high, and thus, in principle, the area was at risk of serious flooding. Presumably, given the technological constraints of the age, aristocrats had a preference for sites where it was easy to draw groundwater from wells.

The river waters of Heijō-kyō were not abundant and were not deemed appropriate for human consumption. Accordingly, it is widely believed that well water was used for drinking and cooking. Of course, we can also imagine that this water was used for washing kitchen implements and clothing. At the time, there was apparently no custom of daily bathing: no remains that can be associated with baths have been found in the residences of either commoners or aristocrats.

Compared with land use in Heian-kyō in the first half of the Heian period, Heijō-kyō exhibits a more conspicuous expansion of the urban area into the floodplain. It appears that this type of land use was closely related to the scale of river floods. In other words, it is likely that settlement on floodplains near Heian-kyō was impeded by flooding from the Kamo and Katsura rivers. Both rivers have basins much larger than those of the Saho and Akishino rivers, and in fact, floods are well documented historically.[17] This situation contrasts sharply with the evidence from archaeological excavations of the area of modern-day Nara, as noted above.

Conclusions

In summary, two important factors appear to have operated in the choice of Heijō-kyō as an imperial capital and in the distribution of the court buildings, religious institutions, and residences.

The first factor was the nature of the ground surface. The siting of large buildings such as audience halls and temples had less to do with water availability or susceptibility to flooding than with the sites' ability to support monumental structures. These buildings required stable, solid foundations, a condition that was met by Heijō-kyō's Pleistocene terraces. Of course, sites on high ground also offered attractive views, which may also have been a factor in their selection. The second factor was water availability. The city's small water catchment reduced the risk of flood damage but also created a significant seasonal risk of water shortages. As summer dry spells and heat reduced water flows, people had to find reliable sources of water. The relatively high water table of areas near the rivers afforded just such a circumstance. Therefore it is likely that the highest priority in the determination of land use in Heijō-kyō was obtaining water for everyday use. However, the concentration of population in the city gradually caused the water quality of both rivers and underground water to deteriorate, and diarrhea, dysentery, and other parasitic illnesses likely increased. Unlike contemporary urban areas that we typically associate with diminished plant and animal life, here we have an example of human activity that fostered the growth of animals—parasites—harmful to human life.

NOTES

1   Geography: Yoshikoshi Akihisa, "Toshi no rekishiteki suimon kankyō," in *Toshi no suimon kankyō*, ed. Arai Tadashi, Shindō Shizuo, Ichikawa Arata, and Yoshikosi Akihisa (Kyōritsu Shuppan, 1987), 201–52. History: Kobayashi Takehiro, *Kindai Nihon to kōshū eisei: Toshi shakaishi no kokoromi*, Yūzankaku Shuppan, 2001. Archaeology: Matsui Akira, *Kankyō kōkogaku*, Nihon no bijutsu 423 (Shibundō, 2001); Edo Iseki Kenkyūkai, *Saigai to Edo jidai* (Yoshikawa Kōbunkan, 2009). Ethnology: Ono Yoshirō, *Mizu no kankyōshi: "Kyō no meisui" wa naze ushinawaretaka* (PHP Kenkyūjo, 2001).

2   Miyataki Kōji, "'Kankyōshi,' saigaishi ni fumidashita Nihon kodaishi kenkyū," *Rekishi hyōron* 626 (2002): 60–65; Miyataki, "Ima naze kankyōshi, saigaishi no shiten ka: Nihon kodaishi no tachiba kara," *Atarashii rekishigaku no tame ni* 259 (2005): 32–35.

3   Takahashi Manabu, "Kodai matsu ikō ni okeru chikei kankyō no henbō to tochi kaihatsu," *Nihonshi kenkyū* 380 (1994), 38–48; Takahashi, "Kodai matsu ikō ni okeru rinkai heiya no chikei kankyō to tochi kaihatsu: Kawachi heiya no shimabatake kaihatsu o chūshin ni," *Rekishi chirigaku* 36:1 (1994) 1–15; Kinda Akihiro, *Bichikei to chūsei sonraku* (Yoshikawa Kōbunkan, 1993), 284.

4   Kawasumi Tatsunori, "Heian-kyō ni okeru chikei kankyō henka to toshiteki tochi riyō no hensen," *Kōkogaku to shizen kagaku*, 42 (2001): 35–54; Kawasumi, "Rekishi jidai ni okeru Kyōto no kōzui to hanrangen no chikei henka: Iseki ni kirokusareta saigai jōhō o mochiita suigaishi no saikōchiku," *Kyōto rekishi saigai kenkyū* 1 (2004): 13–23; Kawasumi Tatsunori, Harasawa Ryōta, and Yoshikoshi Akihisa, "Chūsei Kyōto no chikei kankyō henka," in *Chūsei no naka no "Kyōto" (Chūsei toshi kenkyū)*, ed. Takahashi Yasuo and Chūsei Toshi Kenkyūkai (Shinjinbutsu Ōraisha, 2006), 151–79.

5   Ogasawara Yoshihiko, "Kuni-kyō, Shigaraki-kyō, Naniwa-kyō," in *Heijō-kyō no jidai*, vol. 2, *Kodai no miyako*, ed. Tanabe Ikuo and Satō Makoto (Yoshikawa Kōbunkan, 2010), 216–37.

6   Aoki Shigekazu, *Nara-ken kishō saigaishi* (Yōtokusha, 1956), 452.

7   Tanabe Ikuo and Satō Makoto, eds., *Kodai no miyako 2: Heijō-kyō no jidai* (Yoshikawa Kōbunkan), 2010.

8   Kurosaki Tadashi, *Suisen toire wa kodai ni mo atta: Toire kōkogaku nyūmon* (Yoshikawa Kōbunkan, 2009), 252.

9   Yoshino Masatoshi, *Kōdai Nihon no kikō to hitobito* (Gakuseisha, 2011).

10 Kanehara Masaaki, *Chūsei kōkogaku to shizen kagaku ni okeru gaku yūgō no kanōsei: Chūsei sōgō shiryōgaku no kanōsei* (Shinjinbutsu Ōraisha, 2004).

11 Hachiga Susumu, "Tojō no zōei gijutsu," in *Tojō no seitai*, vol. 9, *Nihon no kodai*, ed. Kishi Toshio (Chūō Kōronsha, 1996), 204–205.

12 Shimada Toshio, "Kenchiku shizai no risaikuru," in *Koto hakkutsu*, ed. Tanaka Migaku, 156–58. Iwanami shinsho 468 (Iwanami Shoten, 1996).

13 Aoki, *Nara-ken kishō saigaishi*, 452.

14 Tanaka Migaku, *Heijō-kyō* (Iwanami Shoten, 1984); Nakai Isao, "Dai kibō takuchi to sono ruikei," in *Kodai toshi no kōzō to tenkai*, Kodai Tojōsei Kenkyūkai dai 3 kai hōkokushū, ed. Kodai Tojōsei Kenkyū Shūkai Jikkō Iinkai (Nara: Nara Kokuritsu Bunkazai Kenkyūsho, 1998), 185–216; Nara Kokuritsu Bunkazai Kenkyūsho and Asahi Shinbunsha Ōsaka Honsha Kikakubu, *Heijō-kyō ten* (Osaka: Asahi Shinbunsha Ōsaka Honsha Kikakubu, 1989), 26–27.

15 Nara Kenritsu Kashihara Kōkogaku Kenkyūjo, *Heijō-kyō sakyō gojō nibō jūgo, jūroku tsubo* (Nara: Nara-ken Kyōiku Iinkai, 2006), 245.

16 Kishi Toshio, *Tojō no seitai*, vol. 9, *Nihon no kodai* (Chūō Kōronsha, 1987).

17 Nakajima Chōtarō, "Kamogawa suigaishi (1)," *Kyōto daigaku bōsai kenkyūjo nenpō* 26 (B-2) (1983): 75–92.

3

Earthquakes as Social Drama in the Tokugawa Period

GREGORY SMITS

On June 16, 1662, Kyoto suddenly shook amid a thunderous roar. On realizing it was an earthquake, people on the streets shouted, *Yonaoshi, yonaoshi!* ("Oh, God!").[1] Buildings began to sway. Confusion reigned as townspeople fled into the main avenues. Aristocratic women, their sashes untied, robes and hair in disarray, "forgetting all shame," fled screaming into the streets. Thus began the estimated magnitude (M) 7.25–7.6 Kanbun Earthquake, described in Asai Ryōi's *Kaname'ishi (Foundation Stone,* published in 1662).[2]

Ryōi's three-volume account was Japan's earliest book-length work of popular literature about an earthquake. Its illustrated pages surveyed the extent of damage, with specific examples sometimes described in lurid detail. The work also dealt with rumors, earthquake lore, past earthquakes, and academic theories about earthquakes. It was a comprehensive and entertaining treatment of the event. Ryōi's brush transformed a destructive earthquake into social drama. Blurring the line between journalistic reporting and fiction, *Kaname'ishi* became the basic model for later earthquake books, such as the 1856 *Ansei kenmonroku (Ansei Record)* and *Ansei kenmonshi (Ansei Chronicle),* as well as lesser-known works from other Tokugawa-period earthquakes.

Beginning with *Kaname'ishi,* I examine several representative works of early modern earthquake literature in chronological order. All of these works use the extreme circumstances inherent in major earthquakes to heighten a sense of drama; however, I argue that they also promoted social resilience by normalizing earthquakes. Major seismic events in any given location were typically so infrequent that when they occurred, they seemed "unprecedented" *(zendai mimon* or *mizō)* because there was no living social memory of a previous comparable event. Much of the literature following major earthquakes functioned both to entertain and to reassure. These functions offered some measure of solace to survivors worried about the future amid aftershocks, death, and destruction. There were two complementary approaches to reassurance. One was to normalize earthquakes by historicizing them: earthquakes have occurred frequently in the past, both in Japan and elsewhere, and are thus to be expected. The other approach was to explain the mechanisms that cause earthquakes: as in most societies, early modern Japanese explanations of earthquakes often included both religious and mechanical components. Over time, however, there was a trend toward emphasizing rational, mechanical explanations. Relatively high literacy rates in early modern Japan made this literature accessible to wide audiences.[3]

Kaname'ishi, 1662

The first volume of *Kaname'ishi* takes the reader through Kyoto and is divided into sections, each of which ends with two lines of verse to segue into the next topic. "Destruction to houses in Kyoto" explains that all houses, shrines, and temples are constructed in the same basic manner. The integrity of these structures depends on the main roof beam *(munagi)*. When it breaks, dislodges, or bends, the house becomes a danger zone, especially for anyone on the second story. Many people died from being crushed beneath falling beams, and the cries of the injured rang out—a situation never previously encountered *(zendai-minon)* in Kyoto.[4] Being pinned under a beam was the most common dramatic scenario in earthquake literature, creating situations in which one must choose who lives or dies, or amputate one's own body parts to survive, or be crushed to death, whether for a good reason (divine retribution) or not (moral ambiguity).

In the next section, the author describes the deaths of two children on the grounds of a shrine. Crushed by a falling stone lantern, "from their heads to their arms and legs, there was nothing connecting any of these parts, so mangled were their bodies." The text briefly speculates that such a result could have occurred because of the karmic situation from a past life, but the main point is to dwell on the tragic nature of the situation.[5] That a falling lantern could kill both children seems farfetched, but such details surely enhanced the morbid appeal of the work.

Next Ryōi takes readers to the Muromachi district for the death of the wife of a leading citizen. She was seventeen and pregnant. Along with three household servants, she fled toward an open area in the back of the house. A collapsing storehouse, however, killed them all. In passing, the text suggests the possibility of fate in the current lifetime shared by two or more people because of something they did in a past life *(ichigō shokan)*. Here too, the emphasis is on the tragic nature of the events, with no serious attempt to explain causal factors. A gruesome description of "twisted entrails" and the five-month-old fetus falling from the mother's body as she was removed from the rubble again adds morbid appeal.[6] The top-heavy construction of earth-plastered storehouses designed to withstand fire apparently made them especially susceptible to collapse in earthquakes.[7] Moreover, large, heavy storehouses were symbols of worldly wealth and its ephemeral nature.

The fate of shrines and temples was a common topic in earthquake literature. Kyoto's Daibutsu (Great Buddha) was undergoing major repairs at the time of the Kanbun Earthquake, and its head had been detached. Workers at first assumed the shaking was a divine punishment for having removed the head. One hundred or so workers cried out, "Praise to the Buddha Shakyamuni" and apologized, begging for their lives. It was only after they had fled the building that they realized an earthquake had occurred.

Next was Toyotomi Hideyoshi's (1536–1598) mound of ears. *Kaname'ishi* first explains the origins of this macabre monument, saying that Hideyoshi had felt sorry for those whose ears had been shipped to him from battlefields in Korea. He ordered them buried in a mound with full Buddhist ceremonies and a small pagoda monument *(gorintō)* placed on top of it. The earthquake knocked the top off the monument, punching a hole in the mound. Somebody looked in the hole and asked, in verse, whether the ears had heard the earthquake. Supposedly, a rumbling sound came out, answering that it took an earthquake so powerful that it probably shook land in China to open the mound of ears.[8]

Other landmarks enter the story. At the Ishibashi stone bridge, two people fell through the collapsing boards. One hit his head on a stone and died, and the other escaped with a minor knee injury. At the Kiyomizu Temple a stone pagoda collapsed, and at the south gate of the Gion Shrine a stone shrine gateway collapsed. The shaking caused liquefaction of the ground, which led to a scene of confusion. Parents abandoned children, elder brothers forgot their siblings, and a man who thought a teahouse attendant was his wife took her hand and began to flee. A woman thought that a large jar was her child and stumbled along, carrying it.[9]

Because most of the population could not recall a previous major earthquake, phenomena such as aftershocks or mud oozing from fissures in the earth were especially terrifying. Ryōi pointed out, however, that precisely the same kinds of things had taken place during the Keichō-Fushimi Earthquake of 1596, and before then, there were instances of fire coming out of the earth.

Ryōi portrayed the current earthquake as a social leveler, in that people of high and low status alike dwelled in simple huts to ride out the aftershocks and thousands of huts could be found in every open area. Heavy rains and other inconveniences made everyone suffer. A noblewoman, for example, composed a verse about feeling helpless when people happened to be hanging around the hut while she was urinating.[10]

The final section of the first volume talks about a shining object *(hikarimono)* seen moving across the sky while the earth was shaking.[11] Other sources connected with this earthquake mention something similar, and *Konoe Diary* even included a small line drawing of a shining object resembling a spoon, ladle, or possibly a comet.[12] From this time onward, flashes, objects, or pillars of light became a standard component in Japanese descriptions of earthquakes. One likely reason was the dominant theory of the shaking mechanism: yang energy trapped within the earth accumulated and sought to escape upward. This same process, aboveground, created thunder and lightning. Similarly, earthquakes and rainstorms also became associated in the popular imagination.[13]

The linkage of light flashes, atmospheric phenomena, and earthquakes persisted well into the twentieth century. Pioneering seismologist Musha Kinkichi, for example, urged students of earthquakes to take these descriptions of light emanations seriously.[14] Gregory Clancey points out that the Meiji seismologists of the late nineteenth century had inherited a rich lore of folk wisdom about these relationships and were obliged seriously to consider the possibility of a connection between seismicity and meteorology.[15]

The second volume of *Kaname'ishi* is similar to the first, describing tales from the regions surrounding Kyoto, but the third volume begins with a detailed survey of past earthquakes. Ryōi explains that there was an earthquake as early as the reign of the twentieth human emperor, Ingyō, in 416. A great earthquake in 598 caused mountains to collapse everywhere and prompted official prayers. An earthquake in 684 was "unprecedented," causing many deaths, and in 855, an earthquake beheaded the Great Buddha at Nara's Tōdaiji and killed tens of thousands; homes, mountains, and famous places collapsed. The march of earthquakes continued right up until "the current prosperous reign of the 112th emperor above." "Surely," readers are reminded, "severe earthquakes have occurred throughout the past."[16]

Moreover, "even in China" we find the same dramatic history of violent shaking. Ryōi provides only a few examples, but their details resonate with accounts of Japanese earthquakes. A great earthquake in 1291, for example, toppled many buildings and killed seven thousand. Great earthquakes occurred in 1303 and 1306, killing more than five thousand, "starting with palace ladies and officials of state." In short:

> Whether in foreign countries or in Japan *(honchō)*, there is nowhere without a
> history of major earthquakes. Because earthquakes occur periodically and there have
> been so many past examples, is it contrary to reason *(kotowari narazu ya)* for high
> and low alike to be taken by surprise?[17]

Though phrased in the manner of a typical rhetorical question, the answer is, "Yes, it is contrary to reason."

Ryōi next briefly examines theories of earthquake causality. He begins with Buddhist theory, as explained in *Great Wisdom Discourse (Daichi doron,* attributed to Nāgārjuna), that earthquakes are caused by the movement of one of four entities: the fire or hearth deity, the dragon deity, the golden-winged bird, and the lord of heaven. Next is basic Buddhist cosmology with respect to the composition of the earth, starting from the surface of the earth and moving downward in the following sequence of layers: earth, gold, water, and wind. The basic earthquake mechanism was the movement of the deepest layer, wind. This agitated the water layer, which transmitted its agitation to the earth via the gold layer.[18]

Next Ryōi explains Chinese earthquake theory, whereby yang energy is trapped within the earth and seeks to rise. Yin energy suppresses it, but eventually the situation becomes unstable, causing the earth to shake. He likens this mechanism to certain diseases whereby yang energy causes trembling of one's body. Moreover, this yang energy in the air is called thunder if it makes a sound and lightning if it is silent. When accumulated energy moves under the earth, it is called an earthquake. He concludes that there is no way to prevent an earthquake.[19]

Following this discussion of earthquake mechanics, the author describes at length the connection of shrines and shine rituals to earthquakes. Ryōi does not assert that the deities cause earthquakes. His ultimate point seems to be that earthquakes have the socially beneficial side effect of causing people throughout society to attend shrine services and purify themselves.[20]

The main purpose of the final section of *Kaname'ishi* is reassurance. Ryōi cites a poem by Su Dongpo (Su Shi) referring to earthquake divination. Depending on which month an earthquake occurs, it is a harbinger of a subsequent development, often bad, but sometimes beneficial. With no apparent evidence, Ryōi says that the timing of the current earthquake signals an abundant harvest. He ends on a further positive note that explains the title of his book:

> In popular lore there is the notion that the world is supported by a dragon king whose anger is the cause of earthquakes. The Kashima deity suppresses this dragon king whose head and tail are twisted and overlap at one point. Because the foundation stone *(kaname'ishi)* is located above this point, no matter how violently the dragon king shakes, it will not destroy human society. An old saying goes, "The *kaname'ishi* will not be thrown off no matter how great the shaking, as long as the Kashima deity is present." Therefore, I name this record of the earthquake "Kaname'ishi."[21]

Thus ends a skillful attempt to present a major earthquake as social drama.

Around seven a.m. on December 18, 1828, the earth around Sanjō in Echigo (Niigata Prefecture) began to shake. The estimated M6.9 inland, shallow-focus earthquake caused 1,443 deaths and destroyed more than ten thousand structures. Fires were a major cause of death and destruction. The Sanjō Earthquake inspired Koizumi Kinmei's *Chōshin hiroku (Account of Chastisement and Shaking)*. Like *Kaname'ishi*, this narrative includes extensive illustrations and is a comprehensive account of the earthquake.

Chōshin hiroku begins with a detailed description of the geography of "our Echigo," then relates harbingers of the shaking. For example, the sun became very powerful, melting snow even in the high mountains, and the warm energy *(ki)* caused various plants to bud and azalea flowers to open.[22] Here is another instance of the tendency to posit links between atmospheric phenomena and earthquakes.[23]

Next, several sections describe the earthquake and its dramatic effects. For example, landslides in the mountains blocked or diverted the paths of several rivers, causing flooding and disrupting village life. In some places the force of the shaking expelled water from wells; in others, "fire *ki*" emanated from the earth, causing lanterns to flare.[24]

Although the discussion of earthquake-related phenomena is detailed, the focus of *Chōshin hiroku* is the human drama, with a strong sense of moral didacticism. Sections with titles like "Greed Is Difficult to Stop," "Honest to a Fault," "Things beyond One's Strength," "Chastity," "An Impressive Person," "Power of the Deities," "Loyalty and Courage," and "Cowardice" use the social drama the earthquake created to make moral points. The first sections feature cases of merchants so attached to their wealth that they perish in awful ways amid the conflagration. "Even tens of millions of coins" cannot purchase one's life: that is the baldly stated moral of a tale of a man who, intent on retrieving his money boxes, was trapped under a beam, had to amputate his foot to escape, and died of his wounds.[25] Some tales highlight positive examples, such as the husband who bravely came to the aid of his wife and child only to be surrounded by flames on all sides. He could have saved himself but refused to abandon his family.[26]

In a tale about the power of the deities, a bride's household received a home altar *(kamidana)* from the Ise Shrine. She became completely devoted to the deity in the months before the earthquake. Although their house was devastated, everyone survived because of the wife's courage and piety.[27] "A Heartwarming Person" tells of the farmer Jirōbei from Nakasai Village, who had been honest, devoted, hardworking, and helpful ever since he was young. He was buried under the collapsed roof of his house, but when neighbors tore a hole in the roof, there was Jirōbei, calmly seated with a grandchild on each knee. He had whisked them into a safe corner of the house, away from beams, and was a model of composure.[28]

The extreme circumstances of the earthquake amplified and brought to the surface the internal qualities of those caught up in nature's fury. *Chōshin hiroku* ends with a discussion of earthquake lore and the causes of earthquakes, more detailed but similar to that in *Kaname'ishi*.

▓ *Jishinkō* and *Honchō jishinki*, 1830

In the afternoon of August 19, 1830, an earthquake of approximately M6.5 shook Kyoto. The death toll was between two hundred and three hundred, and serious shaking was limited to the city itself. The earthquake was a major event, to be sure, but press reports described the destruction in wildly exaggerated terms. In some accounts the entire city lay in smoldering ruins. Moreover, the tabloid press reported that areas far from Kyoto suffered massive damage, in some cases from a tsunami that never actually occurred. In a detailed study, historical seismologist Miki Haruo speculates that rumors of sinking villages came directly from literature published 130 to 170 years earlier about severe floods and other disasters in that area, including the Kanbun Earthquake.[29] Miki concludes that in 1830, the mass media propagated rumors by exaggerating the destruction.[30]

Several factors exacerbated the tendency to sensationalize. First, Kyoto was the imperial capital, and 1830 was a year of special religious significance that featured mass pilgrimages to the Ise Shrine *(okage-mairi, nuke-mairi)*. Moreover, serious aftershocks continued for several months, keeping the general population in a state of agitation. For these reasons, the Kyoto Earthquake became a social drama that played out all across Japan, with stories distributed through information networks rooted in the major cities.

This situation engendered a counterreaction in the form of two books, both of which sought to explain the earthquake in rational terms and place it in scientific and historical perspective. One was *Honchō jishinki (Record of Earthquakes in Japan),* which started with an entry "earthquake followed by rain" for 645.[31] Hashimoto Manpei regards *Honchō jishinki* as the earliest earthquake history in Japan.[32] Although *Honchō jishinki* was comprehensive, by 1830 it had become common practice for writers of both scientific and popular literature to include discussions of past earthquakes, especially when writing in the wake of a major seismic event. The purpose of these historical accounts was to calm fears that the recent earthquake was an unprecedented catastrophe and possible sign of even more serious problems ahead by pointing out the frequency with which even more severe earthquakes had occurred.

More influential than *Honchō jishinki* was *Jishinkō (Thoughts on Earthquakes)*, written by Kojima Tōzan and one of his students. Its explicit purpose was to calm fears by explaining earthquakes in detail. Discussion includes the basic mechanism of shaking, precursors, aftershocks (leftover yang energy), and earthquake lore.[33] Especially noteworthy is the idea of an earthquake center, explained in part via a diagram featuring a circle representing the earth. *Jishin* ("earth + center"), a homophone for earthquakes, indicated the center of the earth. Directly above it, a small circle at the surface marked the epicenter of an earthquake. Two small dots on each side of the epicenter showed the range of shaking.[34] *Jishinkō* became the most influential work on earthquakes for the rest of the Tokugawa period. Literature in the wake of the 1855 Ansei Edo Earthquake, even catfish prints, frequently cited *Jishinkō* or appropriated some of its content. The exaggerated drama of the Kyoto Earthquake had the effect of advancing and disseminating academic knowledge of earthquakes.

■ "Shinshū jishin ōezu" and Other Works, 1847

Around ten p.m. on May 8, 1847, disaster struck areas in present-day Nagano Prefecture in the form of an estimated M7.4 shallow-focus earthquake. Called the Zenkōji Earthquake, it brought pronounced devastation because seven thousand to eight thousand pilgrims from all parts of Japan were lodged in cramped quarters around the major temple from which its moniker was derived.[35] The pilgrims were there for a special public display of an image of the Amida Buddha with two attendants, a viewing that takes place just once every seven years. The main shock took place at night, and it dislodged the many lanterns hung around the temple, igniting an inferno.

Fire was just one dramatic feature of the Zenkōji Earthquake; water was another. Mount Iwakura (also known as Mount Kokuzō) collapsed, blocking the Sai River. For twenty days, the resulting lake swelled, eventually stretching 40 kilometers and engulfing some thirty villages.[36] On the evening of May 27, the earth dam formed by the landslide broke. A deluge burst onto the Zenkōji Plain. At Koichi, a town along the river at the edge of the plain, the water level briefly reached 20 meters. The wild rush of water lasted four hours, washing away some eight hundred dwellings and burying another two thousand with sand and mud. Only about a hundred people perished, however, thanks to the Matsushiro Domain's emergency measures, including evacuation downstream of the temporary lake.[37]

The Zenkōji Earthquake produced hundreds of accounts of human drama. One tale, reminiscent of the pious young wife in *Chōshin hiroku*, involved a twenty-year-old woman who survived for some twenty days in a house buried under mud. Fortuitous location of furniture provided air space and access to food, and she extracted water from the mud. According to the tale, the main reason she survived, however, was her constant intoning of the names of the *kami* and Buddhas.[38] Although the earthquake did not produce a comprehensive account for mass consumption like *Kaname'ishi*, it did inspire dramatic prints, including early versions of catfish prints, much like those that became popular in 1855.[39]

Perhaps the most striking cultural product of the Zenkōji Earthquake was "Shinshū jishin ōezu" ("Large Illustration of the Shinshū Earthquake"), a complex, two-page print by Hara Masakoto.[40] It depicted a wide territory, from the Matsumoto Domain to the Iiyama Domain.[41] The bakufu's censors, the Gakumonjo, approved its publication, and Masakoto offered it for sale in many places in Japan. He even sold the work to baronial daimyo and other samurai households. "Shinshū jishin ōezu" was the first disaster map sold to a nationwide audience, and Masakato envisioned the project not only as a commercial venture but also as a way of memorializing the victims.[42]

Because so many pilgrims from throughout Japan failed to return to their homes, the Zenkōji Earthquake became widely known nationwide. Nevertheless, in Edo in 1847, it made only a slight impression, producing little more than a tasteless satirical verse *(senryū)* lauding Zenkōji as a place that provides three kinds of funerals: fire, water, and earth.[43] But the Zenkōji Earthquake gained special prominence eight years later, after the 1855 Ansei Edo Earthquake. A dramatic print entitled "Edo namazu to Shinshū namazu" ("Edo Catfish and Shinano Catfish"), for example, features a mob attacking two giant, menacing catfish, one with "Edo" on its forehead and the other, "Shinshū" (Shinano).[44] In other words, the shakeup of Edo in 1855 focused attention on the previous Zenkōji Earthquake.

▨ *Ansei kenmonshi, Ansei kenmonroku,* and *Namazu taiheiki konzatsubanashi,* 1855

The Ansei Edo Earthquake produced public and private literature in such quantity that much still awaits systematic analysis. The event supposedly began with a light show, and numerous accounts report light flashes across the sky or emanating from the earth. Perhaps most dramatic, the spire of Sensōji (the Asakusa Temple) allegedly bent after being hit by a laserlike beam of "white *ki*" that shot out of the earth as the shaking began.[45] By 1855, people expected to see such light displays at the start of an earthquake, and sure enough, flashes of light were precisely what many accounts reported, albeit with widely varying details.

In the aftermath of the earthquake, much of the literature was a straightforward reporting of damage, casualties, fires, and other basic details. Two days after the main shock, however, a distinctive type of literature began to emerge. Kanagaki Robun and Kawanabe Kyōsai most likely produced the first catfish print of the 1855 quake. Known as "Oinamazu" ("Old Catfish"), it was in part a parody of the kabuki drama *Oimatsu (Old Pine Tree)*. According to a biography of Robun, an unnamed printer went to his house early in the morning after the earthquake. Standing in the ruins of his house, Robun wrote on the topic of *namazu no oimatsu* ("old pine tree catfish"). Kyōsai next created an image to go with the text, and the print sold several thousand copies. As a result, Robun's work was suddenly in demand. In the space of five or six days, he had written forty to fifty manuscripts for prints. They all sold well, and "because of the catfish" Robun was able to realize a windfall profit.[46] By 1855, a giant catfish had become a well-known metaphor for earthquakes.[47]

The catfish, either as a giant monster symbolizing the violent forces of nature or as an anthropomorphized symbol of social forces, was ideal for printmakers and writers of popular literature. For example, Kanagaki Robun (as Daidō Sanjin) wrote *Namazu Taiheiki konzatsubanashi (Tumultuous Catfish Taiheiki)*, which took the form of a military struggle between Kashima and other deities versus the earthquake catfish and its nefarious supporters. The sudden night attack of Namazu Nurakurō (Catfish Slimealot) and his allies, such as Jumping Fire, was devastating. The attackers faced the "cowardly warriors" (*ōkubyō musha*) of Edo, whose names included Karada Chijimaru (Shriveled Body) and Menotama Detarō (Protruding Eyeballs). The cowardly warriors fled, causing Amaterasu to convene the assembled deities and appoint Kashima as commander and Atago Gongen as vice-commander to lead them in subduing the rebellion. In the end, the forces of order prevailed and the slain catfish were sold to restaurants specializing in broiled eel *(kabayaki)*.[48]

Humorous earthquake drama undoubtedly played a role in mitigating psychological trauma, but the earthquake also provided an opportunity for writers to comment on human behavior, often in the guise of journalistic reporting of unverifiable tales. Much like *Kaname'ishi*, the literature following the Ansei Edo Earthquake blurred the boundaries between fact and fiction, between moral edification and morbid curiosity. For example, although *Ansei kenmonshi* clearly appeals to a visceral fascination with disaster and strange tales, the introduction stakes out a high moral ground:

Humans possess five states of *ki* [energy] and seven emotions. Amid joy and anger, sorrow and elation, people's minds are apt to become disordered and they lose their ordinary presence of mind. If we deepen the scope of our contemplation during ordinary times, then even at times of extreme danger or ill fortune we will be able to act without forgetting our social obligations and righteousness. Thus we present detailed exemplary tales that will inspire even ordinary women and children.[49]

This description suggests that the earthquake served as a catalyst for moral clarity, and indeed most of the social drama from this earthquake, as in the past, reinforced a message of righteous behavior leading to a relatively good outcome.

Although the literature produced in the wake of the Ansei Edo Earthquake owed much to earlier material, a close reading of the major works from 1855 and 1856 reveals several new trends. Ansei Edo literature was sometimes skeptical of received wisdom and explanations. Moreover, it sometimes acknowledged gaps in academic knowledge and moral ambiguities. Some texts acknowledged that the cosmic forces and the conventions of social morality might operate independently. Of the major works, *Ansei kenmonroku* best exemplifies these trends.

For example, one widespread rumor had it that falling strands of hair from the sacred horse of Ise saved lives in the earthquake. The *Kenmonroku* takes a skeptical stance, noting that it might have happened, but we cannot know for sure without a thorough investigation. Moreover, in 1836, strands of hair were reported in Edo. In this case, an investigator consulted Western science books and examined the material under a microscope. He thereby confirmed that the material was not hair and printed his findings as a broadside. The *Kenmonroku* author had a copy at hand and reproduced its content. The 1836 "hair" turned out to have been small worms produced by strange atmospheric conditions. The worms were dispersed by the wind and fed on plants, thus contributing to the crop failures of the time.[50] The *Kenmonroku* also relates a tale of a daughter of exemplary filial virtue and overall character. Nonetheless, in the earthquake a falling beam crushes her to death. The account concludes that despite the adage that the workings of the heavenly way *(tentō)* reward goodness and visit calamities on evildoers, this woman's case is an exception. The text speculates that what Buddhists call "residual karma" might be at work, but it takes no firm stance on this explanation.[51] Skepticism regarding supernatural phenomena and a tendency not to speculate beyond verified facts are typical of the *Kenmonroku*.

Other accounts from 1855 reflect a similarly mechanical view of the earthquake, minimizing the role of religious forces. A good example is Jōtō Sanjin's *Yabure mado no ki (Account of Broken Windows)*. He observed that the main buildings of many temples survived the shaking with little or no damage. "Although people commonly attribute it to the intervention of deities *(shinbutsu)*," Sanjin rejected such explanations in favor of rational principles *(kotowari)*. He explained that the four eaves of the support beams in the temples "naturally served to balance the structures" such that despite the shaking, they remained in equilibrium. For the same reason Sanjin also rejected the explanation that intact bridges enjoyed divine protection.[52]

■ Conclusion

By 1855, the view that earthquakes could be explained and predicted by means of rational principles (*kotowari, ri, dōri,* and other terms) had become commonplace. Texts such as the *Kenmonshi, Kenmonroku,* and *Fujiokaya nikki (Fujiokaya Diary)* abound with tales of earthquake precursors, "recognized" in retrospect. The notion that vigorous swimming of catfish is a predictor of earthquakes, for example, first appeared in 1855. It consumed considerable research resources in modern times without producing useful results.[53]

Earthquake literature in early modern Japan helped society deal with temblors' death, destruction, fear, and disruption in several ways. First, whether ostensibly journalistic reporting or fiction, these works chronicled the event. Some, such as the "Shinshū jishin ōezu" print, functioned in part to memorialize the event and its victims. Second, earthquake literature often addressed a wide range of emotions and fears, with the intent of edifying the reader and promoting social resilience. In a manner that could be entertaining to at least some readers, earthquake literature reassured a worried society that when viewed in a longer temporal perspective, the recent example was not, in fact, an unprecedented disaster, but part of the ordinary process of nature. Moreover, earthquakes could be explained in terms of mechanical causes, and a better understanding of these causes might lead to mitigation of future destruction through prediction or possibly better building construction.

NOTES

1   In this context, *Yonaoshi* functions as talismanic exclamation of surprise, roughly similar to the automatic "Bless you!" one often hears in English after someone sneezes. This *yonaoshi* is related to but different from the same term functioning as a noun meaning "world renewal." For details on this matter, see Gregory Smits, *Seismic Japan: The Long History and Continuing Legacy of the Ansei Edo Earthquake* (Honolulu: University of Hawai'i Press, 2013), 26.

2   Asai Ryōi, *Kaname'ishi,* in *Kanazōshishū,* ed. and trans. Taniwaki Masachika, Oka Masahiko, and Inoue Kazuhito (Shōgakukan, 1999), 14–15.

3   See Richard Rubinger, *Popular Literacy in Early Modern Japan* (Honolulu: University of Hawai'i Press, 2007).

4   Asai, *Kaname'ishi,* 16–17.

5   Ibid., 18–21.

6   Ibid., 21–22.

7   Ibid., 23.

8   Ibid., 24–26.

9   Ibid., 26–30.

10   Ibid., 32–37.

11   Ibid., 37–39.

12   "Konoe nikki," in *"Nihon no rekishi jishin shiryō" shūi: Tenmu Tennō 13-nen yori Shōwa 58-nen ni itaru,* ed. Usami Tatsuo (Nihon Denki Kyōkai, 1998), 149.

13   For a detailed discussion of early modern theories of earthquakes, see Smits, *Seismic Japan,*
     37–70.

14   Musha Kinkichi, *Jishin namazu* (Meiseki Shoten, 1995 [1957]), 52–104. Musha argued that such
     accounts "should not be carelessly dismissed as absurd. It is hardly the case that people of the
     past purposely wrote lies. … Accounts of things such as earthquake light could not have been
     written by the imagination" (52). He then analyzed the matter in more than fifty pages. Musha
     was writing prior to the general acceptance of plate tectonics theory in Japan. A generation
     later, in a book-length study of the 1830 Kyoto Earthquake, Miki Haruo gave only passing
     mention to light flashes in his chapter on precursors and concluded, "The identity of the shin-
     ing objects *(hikarimono)* has been unknown from then to now." Later, he concluded that there
     was no statistical relationship between earthquakes and rainstorms. See Miki Haruo, *Kyōto
     daijishin* (Shibunkaku Shuppan, 1979), 60–66.

15   Gregory Clancey, *Earthquake Nation: The Cultural Politics of Japanese Seismicity, 1868–1930*
     (Berkeley: University of California Press, 2006), esp. 152–53.

16   Asai, *Kaname'ishi,* 66–69.

17   Ibid., 69–70.

18   Ibid., 70.

19   Ibid., 70–71.

20   Ibid., 71–77.

21   Ibid., 78–81.

22   "Chōshin hiroku," in *Saikō 2-nen yori Shōwa 21-nen ni itaru,* vol. 3 in "*Nihon no rekishi jishin
     shiryō" shūi,* ed. Usami Tatsuo (Watanabe Tansa Gijutsu Kenkyūjo, 2005), 212–28.

23   By the Meiji period, unseasonably warm weather had become widely accepted as a precursor
     of earthquakes. In another incident, although most likely because of aftershocks connected
     with the Nōbi earthquake (Gifu Prefecture), a November 4 newspaper article noted that some
     residents of Tokyo reported slight shaking in early November 1891, explaining that although
     the weather was nice, many residents of the capital were on edge, because they thought the air
     was too warm. "Jishin o kizukau mono ari," in *Yomiuru shinbun,* November 4, 1891, special ed.,
     2.

24   "Chōshin hiroku," 213–18.

25   Ibid., 220–21.

26   Ibid., 226–27.

27   Ibid., 227–28.

28   Ibid., 230–31.

29   Miki, *Kyōto daijishin,* 75–78. See 79–86 for the details of other rumors. A good example of
     exaggerated damage reports appears in the first sentences of "Jishin kidan miyako manzair-
     aku," in *Dai-Nihon jishin shiryō,* ed. Shinsai Yobō Chōsakai (Shibunkaku, 1973), vol. 1 (kō), 558.

30   Miki, *Kyōto daijishin,* 87–88. For further analysis of press exaggeration of the Kyoto
     Earthquake, see Kitahara Itoko, *Jishin no shakaishi: Ansei daijishin to minshū* (Kōdansha,
     2000), 92–93.

31   Toyo Tokinari, *Honchō jishinki,* in Edo jidai josei bunko, vol. 49 (Ōzorasha, 1994). Page faces
     7–25 from the start of the main text (including the illustrations) constitute a survey of earth-
     quakes up to 1830. See also the comprehensive listing of major historical earthquakes in the
     jishin section of International Research Center for Japanese Studies, *Kojiruien* (Dictionary of
     Historical Terms) Database (Kyoto: International Research Center for Japanese Studies, 2007),
     http://www.nichibun.ac.jp/graphicversion/dbase/kojirui_e.html, under the subheading "*Jishin
     rei,*" 1366–1375 (accessed December 21, 2011).

32 Hashimoto Manpei, *Jishingaku kotohajime: Kaitakusha Sekiya Seikei no shōgai* (Asahi Shinbunsha, 1983), 26.

33 Kojima Tōzan and Tōrōan-shujin, *Jishinkō* (Kyoto: Saiseikan, 1830). See also "Jishinkō," in *Dai-Nihon jishin shiryō*, ed. Shinsai Yobō Chōsakai (Shibunkaku, 1973), vol. 1 (kō), 589–94.

34 Kojima, *Jishinkō*, page face 3 in the second section. To view the diagram, see http://archive.wul .waseda.ac.jp/kosho/w001/w001_03628/w001_03628_0002/w001_03628_0002_p0022.jpg.

35 For a detailed report on this earthquake, see Chūō Bōsai Kaigi, *1847 Zenkōji jishin hōkokusho* (Nihon Shisutemu Kaihatsu Kenkyūjo, 2007).

36 Usami Tatsuo, ed., *Seimu Tennō 3-nen yori Shōwa 39-nen ni itaru*, vol. 2 in "*Nihon no rekishi jishin shiryō*" *shūi* (Yamato Tansa Gijutsu Kabushikigaisha, 2002), 239–40, presents a hand-drawn map illustrating the collapsed mountain and affected villages. For another version and one similar to it, see Usami, ed., *Saikō 2-nen yori Shōwa 21-nen ni itaru*, 354–57. For yet another map, see http://www.um.u-tokyo.ac.jp/publish_db/1999news/02/images/030_01.jpg.

37 "Kenshūroku," in *Dai-Nihon jishin shiryō*, ed. Shinsai Yobō Chōsakai (Shibunkaku, 1973), vol. 2 (otsu), 107, 109–11; Itō Kazuaki, *Jishin to funka no Nihonshi* (Iwanami Shoten, 2002), 139–41; and Shinano Mainichi Shinbunsha Kaihatsukyoku Shuppanbu, *Zenkōji daijishin* (Nagano-shi: Shinano Mainichi Shinbunsha, 1977), 150–80.

38 "Shinano bukō, chōshin hikan," in *Saikō 2-nen yori Shōwa 21-nen ni itaru*, 334.

39 For a typical broadside describing the earthquake, see http://www.iii.u-tokyo.ac.jp/archives /digital_archives/ono_collection/image/big/52008_00.jpg. For catfish prints, see prints 1 and 2 in Miyata Noboru and Takada Mamoru, eds., *Namazue: shinsai to Nihon bunka* (Ribun Shuppan, 1995), 240–41.

40 *Shinshū jishin ōezu* can be viewed in Kitahara Itoko, "saigai to jōhō," in *Nihon Saigaishi*, ed. Kitahara Itoko (Yoshikawa Kōbunkan, 2006), 248; Chūō Bōsai Kaigi, *1847 Zenkōji jishin,* x (*kuchie* 13); Akahane Sadayuki and Kitahara Itoko, ed., *Zenkōji jishin ni manabu* (Nagano-shi: Shinano Mainichi Shinbunsha, 2003), 4–5; and online at http://www.bousai.go.jp/kouhou /h20/07/imgs/ph34.jpg.

41 Harada Kazuhiko, "Matsushiro-han de sakusei sareta jishinzuerui ni tsuite," in *1847 Zenkōji jishin,* ed. Chūō Bōsai Kaigi (Nihon Shisutemu Kaihatsu Kenkyūjo, 2007), 122–23.

42 Kitahara, "Saigai to jōhō," 247–52, and Furihata Hiroki, "Zenkōji jishin to saigai jōhō," in *Zenkōji jishin ni manabu,* ed. Akahane and Kitahara (Nagano-shi: Shinano Mainichi Shinbunsha, 2003), 156–61.

43 Noguchi Takehiko, *Ansei Edo jishin: saigai to seiji kenryōku* (Chikuma Shobō, 1997), 18.

44 Print 45, Miyata and Takada, *Namazue*, 6–7, 266–67. This print is also known as "Mizugami no tsuge." To view the relevant half of it, see http://metro2.tokyo.opac.jp/tml/tpic/imagedata /toritsu/ukiyoe/oC/0277-C004(02).jpg.

45 Anonymous, *Edo Ōjishin matsudai hanashi no tane*, 1855, 11, reproduced at http://archive.wul .waseda.ac.jp/kosho/w001/w001_03639/w001_03639_p0012.jpg. See also Musha, *Jishin namazu*, 59, and Abe Yasunari, "Jishin to hitobito no sōzōryoku," in *1855 Ansei Edo jishin hōkokusho,* ed. Chūō Bōsai Kaigi (Fuji Sōgō Kenkyūjo, 2004), 131. For a comprehensive study of this earthquake, see Smits, *Seismic Japan,* esp. 103–38.

46 Print 150 in Miyata and Takada, *Namazue*, 332–33. Regarding Kanagaki's biography, see Wakamizu Suguru, *Edokko kishitsu to namazue* (Kadokawa Gakugei Shuppan, 2007), 7–8; Takada Mamoru, "Namazue no chosakutachi: chosha, gakō o meguru bakumastu bunka jōkyō" in *Namazue,* ed. Miyata and Takada, 38; and Kitani Makoto, *Namazue shinkō: Saigai no kosumorojī* (Tsuchiura-shi: Tsukuba Shorin, 1984), 6–13. For print 150, see http://metro2.tokyo .opac.jp/tml/tpic/imagedata/toritsu/ukiyoe/oC/0277-C020.jpg.

47 Regarding the origins of Japan's earthquake catfish, see Gregory Smits, "Conduits of Power: What the Origins of Japan's Earthquake Catfish Reveal about Religious Geography," in *Japan Review* 24 (2012): 41–65, http://shinku.nichibun.ac.jp/jpub/pdf/jr/JN2402.pdf.

48 Suzuki Tōzō and Koike Shōtarō, eds., *Fujiokaya nikki*, Kinsei shomin seikatsu shiryō (San'ichi Shobō, 1995), vol. 15, 519–20. See also Kitahara, *Jishin no shakaishi*, 136–37.

49 Illustrations by Utagawa Kuniyoshi et al., author(s) and publisher unknown, *Ansei kenmon-shi* (1856), vol. 1, 1–2 (jō no ichi, jō no ni). See also Arakawa Hidetoshi, ed. *Jitsuroku, Ō-Edo kaimetsu no hi: Ansei kenmonroku, Ansei kenmonshi, Ansei fūbunshū* (Kyōikusha, 1982), 18.

50 Hattori Yasunari (text) and Utagawa Yoshiharu (illustrations), *Ansei kenmonroku* (publisher unknown, 1856), vol. 3, 13–16 (ge no jūsan–ge no jūroku). See also Arakawa *Jitsuroku, Ō-Edo kaimetsu no hi*, 90–94 and Kitahara, *Jishin no shakaishi*, 188.

51 Hattori, *Ansei kenmonroku*, vol. 1, 4–7 (jō no yon – jō no shichi). See also Arakawa *Jitsuroku, Ō-Edo kaimetsu no hi*, 29–34.

52 "Yabure mado no ki," in *Dai-Nihon jishin shiryō*, vol. 2 (otsu), 555–56.

53 For a detailed examination of aspects of the influence of early modern earthquake lore in the development of modern seismology and Japan's postwar program of earthquake prediction, see Smits, *Seismic Japan*, 181–91, and Gregory Smits, *When the Earth Roars: Lessons from the History of Earthquakes in Japan* (Lanham, Md.: Rowman and Littlefield, 2014).

PART II

Water:
Oceans, Rivers, Lakes

4

Traditional Use of Resources and Management of Littoral Environments at Lake Biwa*

SHIZUYO SANO

In recent years researchers in world environmental history and historical ecology have focused attention on cases where humans actively altered their natural environment throughout history. For example, Fairhead and Leach showed that patches of forest in the Guinea savanna were not relics of pristine woodland but were instead produced by human activities.[1] This was an unexpected example of farmers creating forests rather than destroying them. Such situations raise the interesting issue of whether human activities should be considered artificial disturbances of nature or as "natural" in themselves, since they are food production activities of the human animal that parallel the admittedly simpler patterns of other animals, such as bees.

Regardless of whether we conceive of human management of the environment as artificial or natural, findings similar to those of Fairhead and Leach can be found in research on Japan's environmental history—examples of flexibility and resilience in the face of multiple socioeconomic changes. For example, diverse secondary forests (called *satoyama*), used to obtain wood, charcoal, and fertilizer, were maintained through human agency that involved a moderate amount of disturbance. Evaluating spaces that maintained high biodiversity and showed similarly constructive (from a social perspective) human engagement continues to be a focus of Japanese scholarly efforts.[2]

Consideration of positive human effects on nature, not simply activities that inflict damage, is reflected in many studies of regions as diverse as savannas and forests; however, there has been little research applying such an approach to the study of wetlands. Human activities throughout history have caused obvious damage to the ecosystems of water areas. Examples include the destruction of littoral environments, such as wetlands in Europe, through land reclamation and the overfishing and overharvesting of marine resources. However, through studies such as those noted above, slash-and-burn agriculture, previously considered only in terms of its harm to forests, has now been found to maintain some vegetation. Thus it is rash to assert that all human activities are harmful. We must reexamine the historical effects of human activities in wetlands.

The present study examines actual effects of selected past human activities on the ecology of water areas via a case study of a freshwater lake, Lake Biwa. I explore whether the presence of human beings—through fishing, agriculture, and urbanization—resulted in environmental load and depleted resources in the lake's ecosystem or contributed to sustaining a diverse ecological arrangement. Today, as it has for centuries, Lake Biwa plays a critical role by providing drinking water to the Kyoto-Osaka region. But even more than a thousand years ago, it had a history of varied human involvement because of its proximity to the imperial capital, Kyoto, and the city's demands for agricultural and lake products. Over this time there certainly have been changes in the ecology of the lake's basin, but in core respects—the provision of clean water, fish resources, and materials for daily life—there has been a fascinating continuity, reflecting an important level of positive coexistence and resilient interconnectedness of both aquatic and human life.

A relatively large collection of historical materials on the use of this lake exists, especially from medieval times. The practices described here continued from ancient times (ca. seventh century) to the 1940s. The historical data on which this analysis is based come from medieval and early modern times (ca. fourteenth–nineteenth centuries), but some critical data are from the early twentieth century. Use of twentieth-century prewar data is justified by the continuous use of the practices described through that time.

In the terminology of limnology, Lake Biwa is an "ancient lake." In general, as sediments accumulate over time, lakes become land and are fated to disappear after several tens of thousands of years. In exceptional cases lakes continue to exist for more than a million years, but Lake Biwa formed roughly four million years ago. Ancient lakes are marked by extremely high biodiversity and unique endemic species that have evolved throughout their long histories.[3] Unlike Lakes Victoria and Malawi—ancient lakes that remained in a primitive natural state until the modern era—towns and villages bounded Lake Biwa for many centuries, and it has been an integral part of human activities.[4] Most importantly for this study, despite human involvement, Lake Biwa's water quality and high biodiversity were maintained until the 1940s.

The catchment area of Lake Biwa was heavily populated for several hundreds of years, and by the nineteenth century, the population had reached six hundred thousand people. Nonetheless, the water remained potable into the 1940s. Also, many fish species endemic to Lake Biwa's water system had been caught since ancient times and were transported to Kyoto, feeding the population of the capital. These species survived until the 1950s and supported fishing techniques and a food culture unique to Lake Biwa.

Several factors explain Lake Biwa's ability to maintain its biodiversity and water quality; our critical theme here is to elucidate the enduring framework of coexistence between humans and littoral environments. Of all the ancient lakes in the world, only Lake Biwa's deep connection to the daily lives of people resulted in extensive written materials describing its use since ancient times. Analyzing these materials provides valuable insights into the relationship between human beings and lakes and the environmental history of Lake Biwa.

◼ The Lake Biwa Region and Past Research

History of Lake Biwa and Its Use

Situated in the center of Shiga Prefecture, Lake Biwa's surface area is 670 square kilometers, and its maximum depth reaches 103.58 meters (figure 4.1). A tectonic lake, it originates from faults, and about 460 rivers and streams empty into its catchment area. This basin has a surface area of about 3,174 square kilometers, almost commensurate with the boundary of Shiga Prefecture. Currently, the volume of the reservoir is 27.5 billion cubic meters. The only natural river to flow out of Lake Biwa is the Seta River, which becomes the Uji River and then the Yodo River as it leaves Shiga Prefecture and flows through Kyoto and Osaka

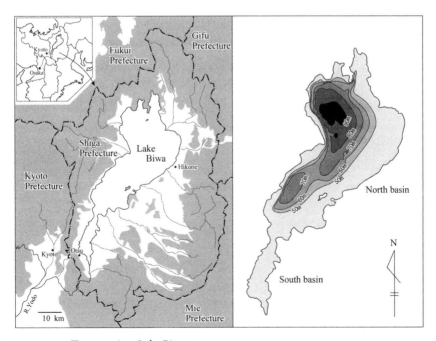

FIGURE 4.1. Target region: Lake Biwa

before emptying into Osaka Bay, providing drinking water and hydroelectric power along the way.

We have a good understanding of the activities of the people who used Lake Biwa from ancient times until the early twentieth century. Fishing is the oldest human activity: large shell mounds, such as the Awazu archeological site at the bottom of the lake, have been discovered, indicating that five thousand years ago, settlers depended on the fishery resource. The Yōrō Code, a historical document from the eighth century, mentioned crucian carp brought as tribute from the shores of Lake Biwa to the imperial court. During Japan's medieval era, aristocrats' diaries further recorded many instances of fish from Lake Biwa presented as tribute. Records tell of fishermen selling fish from the lake in shops in Kyoto. Fishing during the medieval era provided the population of Kyoto, including commoners, with a source of protein.[5] Thus settlers had established a fishing industry that exceeded self-sufficiency and provided fish for trade.

Another use of Lake Biwa during Japan's early history was water transportation. In the seventh century, the throne of Emperor Tenji was located in what is today Ōtsu City, and from the eighth century, Japan's capital was located in Nara and then Kyoto. Transportation by water along Lake Biwa became valued as the shortest route for delivering tax and tribute from the Sea of Japan region and eastern Japan to these capitals. Through the medieval era, shipping networks crossing the lake transported rice as tribute from regional manors to the aristocrats, temples, and shrines in Kyoto. Concomitant with the establishment of these routes was the establishment of many port towns on the shores of Lake Biwa, such as Katata, Shiotsu, Katsuno, and Sugaura. Lake Biwa continued to serve as a major transportation artery until the beginning of the Meiji era.[6]

Population Growth and Agriculture in the Catchment Area

Urban settlements in the catchment area of Lake Biwa included not only the residence of Emperor Tenji in the seventh century but also later settlements around provincial offices, established as the bases for regional control. In port towns, such as Katata and Sugaura, houses were built on narrow, rectangular plots during the medieval era, indicating a dense urban landscape.[7]

Lake Biwa's catchment area population grew dramatically during the sixteenth century, when the construction of castle towns hit full stride. During this, the Warring States period, Shiga, sitting at the junction of important routes, was a militarily valuable region. Daimyo castle towns were built on the shores of Lake Biwa, one after another, drawing the warrior class and merchant class residences together. Many of these castle towns, such as Ōtsu, Hikone, Nagahama, and Ōmihachiman, developed into modern, economically important cities, and their populations grew dramatically. For example, in the seventeenth century, it is estimated that the population of the Hikone castle town had already reached more than thirty thousand people.[8] Because warriors and merchants were not involved in primary industries, castle towns were large-scale consumer centers, but for our purposes it is important to recognize that they produced large amounts of human sewage that flowed through the catchment area and finally into Lake Biwa.

Agriculture in the catchment area constitutes another important human activity that affected Biwa's water quality. Close to Kyoto, the lake district served as home to outlying agricultural communities that supplied food to the capital for more than a millennium. Paddy fields occupied a large proportion of the land used in the catchment area.

Many of the streams and rivers that flow through the catchment area formed fertile plains. Since ancient times, these rivers and streams had fed well-maintained irrigation systems. At the end of the Heian period, the imperial court and aristocrats had already built many important manors and water conservation works that accompanied the development of paddy fields.[9]

As the population jumped sharply after the sixteenth century, agricultural production to support it grew in importance, and closely managed, advanced irrigation systems watered increasingly productive paddy fields. For example, the amount of rice produced overall in the area of the prefecture during the seventeenth century, the beginning of the early modern era, reached more than 830,000 *koku*.[10] After being discharged from agricultural lands, large amounts of irrigation water flowed through rivers and streams into Lake Biwa. Thus, our study must consider the effects of agricultural wastewater on the water quality of Lake Biwa.

The Influence of Topology and Vegetation

Although fishing and water transportation before the early modern period have been well studied, previous research on the occupations of people who used Lake Biwa has ignored the perspectives of topography and vegetation. Both the fishing and the transportation industries were largely determined by the topology and resultant vegetation particular to Lake Biwa. Of course, both topology and vegetation had a significant role in shaping human use of the region, too, and as I will show below, this involved not only maintenance of some natural processes but also the slowing or elimination of others.

A topological feature that greatly affected both fishing and transportation activities was the "inner lake." Because of its size, Lake Biwa experiences strong seasonal winds, and the resulting waves and currents toss sand and gravel in the same manner as seen in the formation of beach ridges and coastal sandbars. The water bodies that formed inland behind the sandbars were freshwater lagoons, called inner lakes.[11]

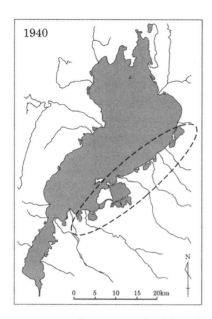
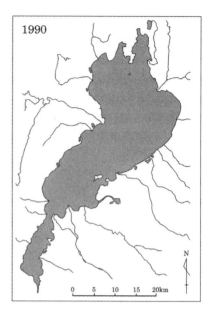

FIGURE 4.2. Changes in satellite lakes of Lake Biwa

By the 1940s, Lake Biwa's inner lakes were distributed in more than forty locations (figure 4.2). Since inner lakes were calm water, protected from the winds and waves of Lake Biwa, they had been used as naturally favorable bays since ancient times. Many of the traditional port towns on the shores of Lake Biwa, such as Katata, Shiotsu, and Katsuno, were built on inner lakes. These towns were also economic and military bases and by the end of the medieval period had become the sites of castles.[12] Hikone, Nagahama, and Ōmihachiman, the representative castle towns mentioned above, faced inner lakes and were designed to encompass them. Such towns had important inter-actions with their inner lakes.

In addition to providing ports, the shallow, calm inner lakes hosted commu-nities of reeds and submerged plants, creating favorable spawning grounds that supported abundant fish populations.[13] Many carp, including endemic species, lived in Lake Biwa's water system. Carp swimming upstream to spawn and lay their eggs among the reeds and waterweeds were a mainstay of Lake Biwa's catch, destined for Kyoto. Thus the topography of the inner lakes, together with communities of reeds and aquatic plants that grew there, had a major influence on the establishment of human occupations at Lake Biwa.

▨ The Uses of Aquatic Plants and Their Significance

Aquatic Plant Communities as Living Habitats

The importance of aquatic vegetation to Lake Biwa's wildlife ecosystem is common knowledge among biologists. However, previous historical research has completely overlooked the significant influence that Lake Biwa's aquatic vegetation exerted on the lives of people. Here I examine the relationship between human activities and Lake Biwa's ecosystem, using the history of human use of the lake's aquatic plants as a point of entry.

Lake Biwa's powerful winds and waves account for the sandy beaches and rocky shores of its littoral regions, with their meager plant life. However, the calm inner lakes and bays are characterized by muddy sediments, which allow aquatic plant beds, from underwater vegetation to wetland forests, to develop (figure 4.3). Because these plants are vulnerable to winds and waves, the areas where they can grow are limited: they do not flourish where the waves are strong even if the lake is shallow. The inner lakes, however, support large communities of aquatic plants.

Submerged and emerged aquatic plants are spawning grounds for Lake Biwa's carp. From spring until early summer, a great mass of carp stream from Lake Biwa into the inner lakes and bays. *Nigorobuna (Carassius auratus grandoculis)* and *gengorobuna (Carassius cuvieri,* Japanese crucian carp), endemic to Lake Biwa's water system, served as important catches since ancient times. Many were caught as they entered the reed beds during their spawning season, through a method of stationary fishing unique to Lake Biwa. Fishermen placed large traps

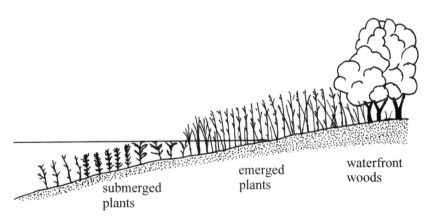

FIGURE 4.3. Aquatic plant communities on the shore of Lake Biwa

(eri) near the entrances to inner lakes and bays and waited for approaching fish (plate 6). Relics of *eri* from the Kofun period (fourth–sixth centuries) have been discovered, and mention of *eri* has been found in tenth-century historical documents. This ancient fishing technique for lakeshore villages remained dominant in such villages; the method used in deep offshore waters was not generally adopted until the twentieth century.[14]

Aquatic plant communities in these areas also had significance for winter migratory waterfowl, such as mallards. Their diet consists of waterweeds and ears of *Poaceae* grass, so emerged and submerged aquatic plant communities constitute indispensable feeding grounds. Thus the inner lakes and bays were important places for hunting waterfowl. Records from the fifteenth century include descriptions of tundra swans, bean geese, and mallards as tribute to the Muromachi shoguns. Thus we see that aquatic plant communities supported the lives of people over the centuries by providing important sources of protein.

Aquatic Plants as Materials for Daily Life

In addition to providing habitat for marine creatures, aquatic plants also had great value as material for the daily lives of people. Submerged plants, emerged plants, and waterfront woods, such as willow and Japanese alder, were all used as textile fiber materials or fuel (figure 4.3) in a variety of daily applications. Records of their importance date back to ancient times.

Japan's oldest collection of *waka* poetry, the *Man'yoshū* (eighth century), depicts the cutting of sedges on the shores of Lake Biwa. A component of reed beds, sedges are ideal for woven straw hats and raincoats, and they were valued highly. A thirteenth-century document that recorded privileges of manorial *(shōen)* rights to Biwa's lakeshore resources includes formal agreements between overlords to allow admission into grasslands that produced sedges; the joint use of sedges demonstrates awareness of this plant as a valuable resource. Reeds are still in demand today as material for screens and thatched roofs; in the past, the sprouts were widely used by villages as feed for livestock and horses. A description of this use was recorded in a well-known fourteenth-century collection of documents, called "Ōshima-Okitsushima jinja monjo," describing medieval villages. The high value of reeds as a resource can be seen in the vivid depictions of reed-growing regions in pictorial maps of the domains of medieval lakeside manors[15] and in descriptions of aggressive, sometimes lethal disputes over reeds between neighboring villages during the medieval and early modern eras.[16]

For lakeside villages that lacked access to mountain forests, reeds and waterfront woods were indispensable resources for fuel. Villagers logged willow and Japanese alder for firewood and used them to fashion items of daily life, such as wooden clogs *(geta)*. In the same way, submerged plants were important in the littoral area of the lake as fertilizer, replacing compost obtained from forests. These examples show the great importance of such plants. Submerged plants were pulled up into boats, and many images of the Lake Biwa area show farm boats loaded with these plants passing by one another during spring and summer. By the beginning of the early modern period, collecting aquatic plants for agricultural use as fertilizer was already well documented.

Thus aquatic plants were irreplaceable resources for villages on the shores of Lake Biwa; conversely, however, residents' efforts had a major influence on maintaining this vegetation over many years. Take the example of reed communities. In their natural state, emerged plant communities gradually accumulate plant remains, providing soil conditions for succession to the next stage of vegetation, such as scrub forest, on more solid land. By removing plant remains through cutting and gathering, human beings stop the succession to scrub forests, and reed communities continue into the next year. Thus, instead of eradicating aquatic vegetation, human beings maintained the vegetation through the moderate "disturbances" of traditional cutting and gathering.

Similarly, the woods established behind reed communities were maintained. Left undisturbed, willows and Japanese alders grow luxuriantly, gradually intruding into reed communities and creating scrub forests. By logging a portion of the woods on the shores of Lake Biwa for fuel and household materials, humans prevented the woods from colonizing the reed communities, maintaining the vegetation in and around Lake Biwa in its existing state.[17] Because human activities interrupted the natural succession of vegetation, we should reevaluate the role that traditional occupational activities played in the preservation of important reed communities.

Aquatic Plants and Sediment: Recovery of Nutrients

Water Quality and the Use of Aquatic Plants and Sediment

At the core of Lake Biwa's biocultural order, we see humans actively interfering with natural cycles of plant life, in a way that over the long run preserved the existing ecology. Nowhere is this better demonstrated than in the human interaction with plants and water: humans harvested nutrients that under normal circumstances would have been returned to the environment directly. Other interventions maintained the quality of the lake water, which otherwise would have been subject to deterioration.

Aquatic plants, including reeds, have the ability to absorb nutrients, such as phosphate phosphorus and nitrogen, from water (a feature gaining notice for its usefulness in purifying water).[18] Once these plants wither in the winter, their nutrients are eluted into the lake waters. Thus, for the nutrients to be recovered for use on land, the plants must be cut and gathered. What allowed the recovery of nutrients at the shores of Lake Biwa was the cutting and gathering of aquatic plants by residents for their daily needs. Their daily occupations not only preserved reed communities but also were responsible for purifying the water of Lake Biwa, maintaing its quality, and transferring lake nutrients to the land.

Because submerged and emerged plant communities form only in locations without strong winds and waves, the system of purifying water by aquatic vegetation functioned mainly in inner lakes and bays. These locations were also the site of the many lakeside castle towns, described above. For example, an inner lake existed in the northern part of the castle town Hikone. Sewage from the thirty thousand people who lived there during the seventeenth century flowed through the inner lake before emptying into Lake Biwa, along with agricultural wastewater from paddy fields that surrounded the inner lake. There the plants helped purify the water.

The relationship between effluent and the inner lakes has been observed in other castle towns and settlements on Lake Biwa.[19] After the temporary accumulation of foul water in the inner lakes and the precipitation of solids, only clear, supernatant water flowed into the main body of Lake Biwa. The inner lakes thus functioned as sedimentation septic tanks, with aquatic plants acting as the filters. Cutting and gathering these nutrient-absorbing plants for human use then transferred the nutrients to the land.

In addition to using emerged plants as textile fiber materials and submerged plants as fertilizer, humans also used the sediment of inner lakes. Sediment was dredged for fertilizer, a practice confirmed from the beginning of the early modern era.[20] Thus, a portion of the nutrients that flowed from the towns and arable lands of Lake Biwa's catchment area and settled at the bottom of inner lakes was transported directly to the land. Such activities—using aquatic plants and sediment on a regular basis—in effect created a recovery system for phosphorus and nitrogen, greatly reducing the pollution load on Lake Biwa and preserving it as a resource for the tens of thousands of people who lived in the catchment area.[21]

How much nutrient was recovered from the lake in the gathered aquatic plants and dredged sediment? To answer this question, it is necessary to examine statistical data from the past.

Statistical Analysis of Submerged Plants and Sediment Used

Shiratori and Hiratsuka estimated the amount of nutrients recovered from other lakes where aquatic plants were used for fertilizer.[22] However, the subject of their studies was submerged plants only; the nutrients in emerged plants and sediment were not considered.

A search for statistical data yielded no records on the amount of emerged plants, such as reeds, collected in the past, but information for submerged plants and sediment from the early twentieth century could be used to estimate the amount of nutrients recovered. Before presenting the data, however, I want to explain the utility of this data for understanding earlier human-ecological interactions on Lake Biwa. First, although farmers increasingly employed commercial fertilizers in the form of fish meal and other substances, the load imposed on the littoral areas of the lake did not overwhelm the system's ability to purify water, support plant life, and foster large schools of fish. As already noted, the lake water remained directly potable into the 1940s, and fish catches remained plentiful. Second, although the specific volume of lake materials collected and the proportion of all nutrients they contributed to agriculture may have changed over the years, the important point derived from early-twentieth-century data remains valid for earlier eras: the lake was a significant source of nutrients for agriculture.

The amount of submerged plants collected for use as fertilizer was tabulated as fishery statistics by government agencies in Shiga Prefecture. The oldest records, dating from 1905 to 1921, recorded the total value of submerged plants collected in the entire prefecture, but not their volume.[23] However, the total value greatly exceeded the value of sales of shellfish, which were important catches at the time, so I estimate that the volume of submerged plants collected exceeded the equivalent amount of shellfish.

The volume of submerged plants collected was recorded by towns and villages in the *Tōkeizensho* (*Consolidated Statistics*) issued by Shiga Prefecture from 1930 to 1938. In the columns for fishery statistics, the amounts and prices of *mizumo* (submerged plants) were recorded. These figures represent sales only; the amount actually gathered included plants gathered for subsistence and thus certainly exceeded the recorded figures.[24] These figures provide basic data for calculating the amount of nutrients recovered from submerged plants.

What is important about the *Tōkeizensho* is that *doromo* (sediment or detritus) was recorded as fertilizer for self-use. Villagers along the shores of Lake Biwa distinguished *doromo* from *mizumo*. The gathering periods, methods of use, and purposes of these two items differed.[25] Thus, only submerged plants were recorded in the *mizumo* columns, and mainly sediment was recorded in the *doromo* columns in the *Tōkeizensho*. However, previous researchers have not recognized these differences.

The *Tōkeizensho* described the total amount of *doromo* for the entire prefecture, making regional analysis difficult, so I reviewed the original materials to find records of the amount of *doromo* collected for 1933 at the county level.[26] With this record of *doromo* and the record of *mizumo* in the *Tōkeizensho* for Shōwa 8 (1933), I compared the amount of the two items collected by each county. These amounts are shown in figure 4.4.

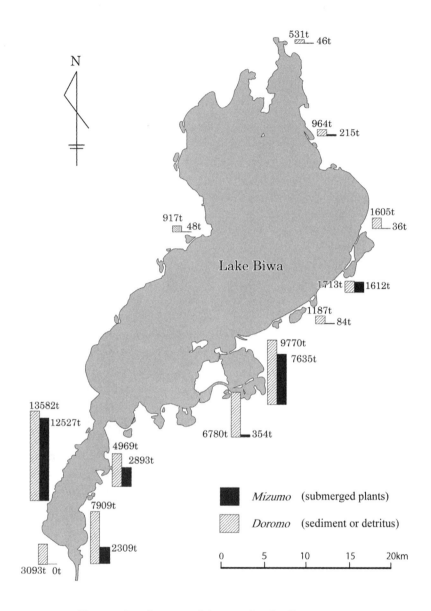

FIGURE 4.4. The quantity of waterweeds harvested and sediment gathered by county, Lake Biwa, 1933

From figure 4.4, we see that the amount of sediment collected greatly exceeded the amount of submerged plants collected, suggesting that dredging played a larger role in recovering nutrients from the lake. Also, volumes were especially high for counties in the vicinity of the inner lakes, suggesting that a system of water purification by emerged aquatic plants and a nutrient recovery system based on gathering the submerged plants was in operation in these areas.

Mizumo and *doromo* were both gathered in vast amounts in the waters in the southern basin of Lake Biwa. The southern basin is defined by the Yasu River delta, which creates a neck 1.35 kilometers wide at its narrowest point. Compared with the northern basin (average depth, 44 meters), the southern basin is shallow (average depth, 3.5 meters). It is calm water, with winds and waves from the northern lake blocked by the Yasu River delta, so aquatic plants grow luxuriantly. Waterweeds in the southern basin had been gathered for use as fertilizer since ancient times and were highly valued: in the seventeenth century, major conflicts between villages over the right to gather waterweeds were brought before the courts of the shogunate.

The surface area of the southern basin is just 58 square kilometers, around which a large population has lived since ancient times. In the 1940s, the population of Ōtsu, the biggest city in Shiga Prefecture, reached about two hundred thousand people. At that time, a large-scale sewage treatment facility had yet to be built. Despite the lack of modern water treatment, the water of the southern basin remained potable.

How much effect did submerged plants and sediment dredging have on maintaining the water quality of the southern basin? Components of sediment reflect the land uses of the catchment area and vary depending on the region and the time period. For the southern basin, we have a document that recorded the makeup of sediment and submerged plants of the same period—information that enables us to calculate the amount of nutrients recovered at the southern basin.

Amount of Nutrients Recovered at the Southern Basin

Around 1939, about thirty species of submerged plants were observed in the southern basin. Of these, three were used as fertilizer: hydrilla *(Hydrilla verticillata)*, *Vallisneria denseserrulata*, and curly-leaf pondweed *(Potamogeton crispus)*. According to the *Tōkeizensho*, the total amount of these three *mizumo* collected overall in the southern basin in Shōwa 8 (1933) was 17,730 tons (wet weight), and the total amount of *doromo* collected was 29,553 tons (wet weight).

In 1933 and 1938, the prefectural agricultural experiment station analyzed the components of sediment and submerged plants in samples from various locations in the southern basin.[27] The percentages of phosphate present in the standing crop of hydrilla, *Vallisneria denseserrulata*, and curly-leaf pondweed were 0.82 percent, 0.76 percent, and 1.01 percent, respectively.[28] These values were ratios from dry weight. Having no data on the proportions of each species, I used an average of the three values (i.e., 0.86 percent) to calculate the total amount of phosphate of the *mizumo* collected in the southern basin. The weight before drying was noted as ten times the dry weight, so the percentage of phosphate present in live plants would be 0.086 percent. Therefore, the total amount of phosphate in 17,730 tons of *mizumo* collected from the southern basin was roughly 15 tons (wet weight). Converting the amount of phosphate into phosphate phosphorus, I obtain a value of 5 tons.

The percentages of nitrogen of the standing crop (dry weight) of hydrilla, *Vallisneria denseserrulata*, and curly-leaf pondweed were 2.32 percent, 2.41 percent, and 3.26 percent, respectively. As above, the average value for live plants would be 0.266 percent. Thus, for the 17,730 tons of *mizumo* collected from the southern basin, the amount of nitrogen in the plants was roughly 47 tons (wet weight).

Doromo, or lake sediment, also contains phosphate phosphorus and nitrogen, and I determined the amounts using similar calculations. Because *doromo* is usually composted for half a year before being used as fertilizer, the agricultural experiment station analyzed the composted material and found 0.11 percent phosphate. After half a year of drying, however, the sediment was 56 percent of its original weight, or 16,549 tons. Thus, the amount of phosphate was estimated as 18 tons. Converting the value of phosphate into phosphate phosphorus, I obtain 6 tons. Using the same method, I calculate that the amount of nitrogen in the composted *doromo* was 0.28 percent of 16,459 tons, or 46 tons.

In sum, the 17,730 tons of *mizumo* and 29,553 tons of *doromo* yielded large amounts of phosphate phosphorus and nitrogen, 11 tons and 93 tons, respectively, which were then returned to the land. The figures reflect the nutrients only in submerged plants and sediment; if emerged plants, such as reeds, were added, the actual amount of nutrients recovered would be even higher.

Although the above calculations are but rough estimates, it is clear that the amounts are substantial. The reason for the pristine quality of Lake Biwa's water in 1933 was not that people's activities were minimal or that their pollution fell within the tolerable range of nature's purifying capacity. Instead, through their plant-gathering and sediment-harvesting activities, humans created a system that maintained water quality. Lake Biwa did not represent primeval nature after all, but was clearly a "nature" created in partnership with human hands.

A Human-Generated and Sustained Littoral Ecosystem

Links between Cities and Villages

The use of aquatic plants and sediment was central to the system of maintaining Lake Biwa's water quality and its ability to support a population of several tens of thousands in its catchment area. Human intervention to maintain Lake Biwa's water quality applied also to municipal wastewater.

Human waste produced in urban catchment areas often constitutes a major problem because of its damage to water quality. However, the city of Ōtsu, with its population of about two hundred thousand, did not install a sewage treatment facility until 1950: for centuries, the city's human waste was purchased by surrounding villages, transported by boat, fermented, and ultimately used to fertilize paddy fields. This practice prevented waste from towns and settlements in the catchment area from flowing directly into the lake and contaminating it.

In short, factors that maintained Lake Biwa's water quality were its multiple waste disposal and nutrient circulation systems. Figure 4.5 provides a schematic of the systems. It shows that people's active use of Lake Biwa's aquatic plants and sediment both preserved the water quality of Lake Biwa and preserved aquatic plants as habitats for food animals. Unlike other ancient lakes in their primitive natural state, Lake Biwa's ecosystem had a socioecological balance that included constructive, moderate human activities.

This interactive pattern established by human involvement with nature brings to mind *satoyama*, which were diverse secondary forests that supplied charcoal

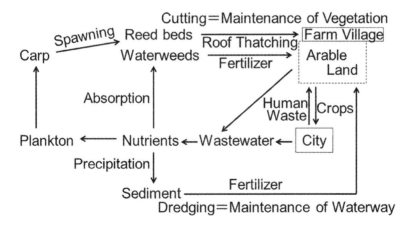

FIGURE 4.5. The littoral zone resource cycle of Lake Biwa

fuel and fertilizer. The Japanese consciously maintained those forests by preventing succession to subsequent vegetation through a moderate amount of human disturbance for their own ends. Similarly, the inner lakes and bays of Lake Biwa were "actively preserved nature," precluded from "natural" change into the sediment-filled swamps they would have become without human intervention. This type of ecosystem existed not only in areas near mountains in Japan but also at the edge of water bodies.

Communal Resource Management by Villages

To establish a littoral ecosystem that accommodated human beings, one precondition must not be overlooked: a social system that carried out the collection of aquatic plants and sediment in appropriate amounts and at appropriate times. In other words, villages had to have established a communal resource management system.

Emerged plants used as thatching materials and textile fiber and submerged plants and sediment used for fertilizer were strictly managed as resources by villages. For example, by the sixteenth century, villages along the shores of Lake Biwa sought to prevent exhausting this resource by prohibiting the cutting of young reeds by individuals.[29] For the majority of villages, reeds were considered communal property. Expenses incurred in preserving reeds and earnings produced from this preservation customarily were shared equally among the residents.

Moreover, because aquatic plant communities were critical as spawning grounds for carp, the villages strictly managed the gathering periods to protect spawning throughout the breeding season. The periods for dredging sediment were also managed in similar ways. Long-term efforts to protect spawning grounds have been confirmed in multiple areas.[30]

Resource management by villages also applied to fishery resources. In spring, many *eri* traps lined the inner lakes and bays to capture schools of crucian carp swimming upstream to spawning grounds. Because *eri* placed at the entrance of the inner lakes can catch a particularly great haul of fish, fishing at those locations was most profitable. Many cases have been found in which the *eri* at the inner lake entrances were treated as the communal property of villages, and the fish caught were distributed equally to residents of the villages.[31] The size of the traps and periods for their use fell under the supervision of the villages. Selfish acts that consumed fishery resources were not accepted. In this way, fishery resources were protected and maintained over the years even as the carp population was subjected to constant fishing.

Previous studies have focused attention on mountainside environments as communal resources, such as water for irrigation and forests for fertilizer, wood, and charcoal.[32] This study shows that associations for using littoral resources had also been established. Such village resource management systems eliminated selfish exploitation of goods by individuals and preserved the sustainability of resources.

The systems of village resource management described above were tied to shrine rituals of villages. In many cases, *eri* traps belonged to the community, and reed beds were treated as properties of the village shrine. Many examples of these shrines come from the early modern period, and many date back to the medieval era.[33] The relationship between the communal properties of a village and its shrine is deep. Treating communal properties as "belonging to the gods" may indicate that villages wanted to guarantee the public nature of these properties and their use by affording them the support of transcendent, divine powers. Ritualistic or religious norms may have encouraged the smooth functioning of village resource management systems.

◼ The Disintegration of Lake Biwa's Littoral Ecosystem

The human-inhabited littoral ecosystem in Lake Biwa deteriorated when a policy of lowering Lake Biwa's water level was implemented in the 1940s. Alteration of lake water levels began in 1905, when the Nango Dam on the Seta River, the only river to flow out of Lake Biwa, made it possible to adjust the amount of water released from the lake. The amount of water discharged from Lake Biwa immediately jumped to accommodate demand from the Kyoto-Osaka region downstream.

During World War II, the water level of Lake Biwa was further lowered as part of the First-Phase Control of Rivers and Streams policy, carried out by the government in 1942 to supply hydroelectric power and water for industrial use. At this time, in part to provide compensation for the lowered water level of Lake Biwa, reclamation of inner lakes was formally permitted in Shiga Prefecture. Afterward, because of the food shortages during and after the war, many other inner lakes and bays were converted to paddy fields.

The reclamation and filling of inner lakes and bays continued into the 1950s, shrinking the aquatic plant communities and decreasing the surface area for natural growth. The economic development philosophy underlying the policies of lowering Lake Biwa's water level and developing landfills continued into the 1970s in the form of the national Comprehensive Development of Lake Biwa policy. The development of water resources and the destruction of aquatic plant communities intensified.

From the latter half of the 1940s, agriculture underwent a major change as inexpensive chemical fertilizers became available. The use of waterweeds, which required intensive labor, and human waste for fertilizer plummeted. Submerged plants and accumulated sediment lost their value and were neglected. The cycle of returning nutrients to the land through agriculture was interrupted. Vast amounts of agricultural wastewater, now with chemical fertilizers and agricultural chemicals, flowed into Lake Biwa, while inner lakes, inner bays, and aquatic plants continued to be landfilled. The former lakeshore systems of water purification and nutrient recovery were abandoned.

In addition, from the 1950s, the delivery of drinking water through piped distribution systems spread rapidly to all areas in the prefecture, and accompanying this, the amount of wastewater from households increased dramatically, to the detriment of Lake Biwa's water quality. Eutrophication proceeded. The lake could not process the increased amounts of agricultural and household wastewater. Thus the historic system for purifying Lake Biwa's water, established through the active participation of human beings, completely disintegrated. People began playing a very different, less constructive role in the littoral ecosystem.

▓ Conclusion

This chapter has explored Lake Biwa, an ancient lake closely linked to human society with large populations in its catchment area. I have elucidated, from historical materials, the effects of human activities on the ecosystem of the water region over the past thousand years. How people used Lake Biwa—specifically, the use of aquatic plants—has generally been slighted in previous historical research; I analyzed the system of returning nutrients to the land by harvesting aquatic plants and sediment.

From the southern basin of Lake Biwa, vast amounts of nutrients—11 tons of phosphate phosphorus and 93 tons of nitrogen—were recovered in 1933. It was not the lack of human activity that maintained the lake's pristine water quality but human intervention, on a grand scale. The aquatic vegetation was communally managed in a manner comparable to that of woods and forests by village associations. Social regulations supported the sustainable use of resources. This is a major reason why, of all ancient lakes, Lake Biwa alone maintained its biodiversity and high water quality, despite its large surrounding human population.

This history of Lake Biwa illustrates a pattern emblematic of both resilience theory and adaptive cycles: it is a story of a socioenvironmental system with a long-term capacity to maintain a balance of critical processes essential to maintaining water quality, marine resources, and agriculture. As external demands on the system mounted in the mid-twentieth century, these patterns were compromised, and interrelated systems lost resilience. In the language of adaptive cycles, they were kicked from a conservation mode to a release phase.

The type of "natural" ecology seen in the Lake Biwa basin, established by the moderate participation of human beings but flexible and adaptive to changing conditions over centuries, is also found in the case of *satoyama* near Japan's mountain lands, where wood, charcoal, and fertilizer resources were sustainably managed. Recognizing that, like *satoyama*, the littoral regions of Lake Biwa were actively preserved through conscious activities designed to serve human beings' own ends can guide our thinking today about how to preserve the lake. Maintaining a specific ecological system does not require removing human involvement and leaving nature as is, but realizing that a moderate degree of human intervention is essential to sustaining it and preventing an otherwise entropic deterioration. Standards for determining how—and how much—humans should be involved and what use and management strategies are appropriate can be found in past occupational activities in littoral regions, as in *satoyama*. Research on the past can suggest means to preserve and regenerate littoral regions in the future. Environmental historical research that sheds light on the relationship between human beings and lakes has the potential for great practical significance.

NOTES

*   I would like to express my deep appreciation to Philip Brown for his advice and help with the preparation of this chapter.

1   James Fairhead and Melissa Leach, *Misreading the African Landscape* (Cambridge: Cambridge University Press, 1996).

2   Washitani Izumi and Yahara Tetsukazu, *Hozen seitaigaku nyūmon* (Bun'ichi Sōgō Shuppan, 1996). A collection of essays introducing multiple perspectives on this practice is K. Takeuchi, R. D. Brown, I. Washitani, A. Tsunekawa, and M. Yokohari, eds., *Satoyama: The Traditional Rural Landscape of Japan* (Tokyo: Springer, 2003).

3   Hiroya Kawanabe, "Biological and Cultural Diversities in Lake Biwa," in *Ancient Lakes: Their Culture and Biological Diversity*, ed. Hiroya Kawanabe, G. W. Coulter, and Anna Curtenius Roosevelt (Ghent: Kenobi Productions, 1999), 17–41.

4   Ibid.; A. Rossiter, "Lake Biwa as a Topical Ancient Lake," in *Ancient Lakes: Biodiversity, Ecology and Evolution*, ed. A. Rossiter and Hiroya Kawanabe, *Advances in Ecological Research,* vol. 31, 571–98 (London: Academic Press, 2000).

5   Iga Toshirō, ed., *Shiga-ken gyogyōshi*, vol. 1 (Ōtsu: Shiga-ken Gyogyō Kyōdō Kumiai Rengōkai, 1954); Biwako Hakubutsukan, *Nihon chūsei gyokairui shōhi no kenkyū* (Kusatsu: Biwako Hakubutsukan, 2010).

6   Kitamura Toshio, *Ōmi keizaishi ronkō* (Taigadō, 1946).

7   Amino Yoshihiko, *Nihon chūsei toshi no sekai* (Chikuma Shobō, 1996).

8   Shiga-ken, *Shiga kenshi*, vol. 3, *Chūsei-kinsei* (Ōtsu: Shiga-ken,1928).

9   Sano Shizuyo, *Chū-kinsei no sonraku to mizube no kankyōshi* (Yoshikawa Kobunkan, 2008).

10 *Koku* is a traditional Japanese unit of volume equal to approximately 180 liters.

11 Nishino Machiko and Hamabata Etsuji, *Naiko kara no messēji—Biwako shūhen no shitchi saisei to seibutsu tayōsei hozen* (Sunrise Shuppan, 2005).

12 Sano, *Chū-kinsei no sonraku to mizube no kankyōshi.*

13   Nishino and Hamabata, *Naiko kara no messēji—Biwako shūhen no shitchi saisei to seibutsu tayōsei hozen.*

14   This is examined in detail in Sano, *Chū-kinsei no sonraku to mizube no kankyōshi.*

15   From the fifteenth-century map "Ōmi no kuni Hira no shō ezu," reproduced as a supplement to Shiga Chōshi Henshū Iinkai, *Shiga chōshi*, vol. 2 (Shiga [Shiga Prefecture]: Shiga Chōshi Henshū Iinkai, 1999).

16   From the fifteenth-century document "Iroirochō," listed as Chūsei document #1 in Chūzu-chō Kyōiku Iinkai, *Ōmi no kuni Yasu gun Awaji Kuyū monjo mokuroku* (Chūzu [Shiga Prefecture]: Chūzu-chō Kyōiku Iinkai, 1995).

17   From observations made by members of the 2009 Shiga Prefectural Council for the Preservation of Reed Communities (Shiga-ken Yoshigunraku Hozen Shingikai).

18   In recent years, Shiga Prefecture has implemented regulations to preserve reed beds. Preservation of aquatic vegetation, such as reed communities, has benefits in addition to enhancing habitats and natural scenery: the prefecture expects the aquatic plants' ability to absorb nutrients will purify the water.

19   Sano, *Chū-kinsei no sonraku to mizube no kankyōshi.*

20   Ibid.

21   If offshore sediment had been dredged from Lake Biwa in summer, when the water is warm, the nutrients stirred up from the lake bottom might have promoted eutrophication. In practice, however, dredging was carried out in winter and early spring so that the sediment could be applied to paddy fields before the start of cultivation. Also, most sediment was collected from creeks near paddy fields rather than from offshore, thus avoiding the risk of eutrophication.

22   Shiratori Kōji, "Inbanuma ni okeru 'mokutori' no jittai," *Shizen to bunka* 3 (1996): 35–40; Hiratsuka Jun'ichi, "1960 nen izen no nakaumi ni okeru hiryōmo saishūgyō no jittai," *Eco-Sophia (Ekosofia)* 13 (2004): 97–112.

23   Shiga-ken Suisan Shikenjō, *Biwako suisan zōshoku jigyō seiseki hōkoku*, vol. 1 (Ōtsu: Shiga-ken Suisan Shikenjō, 1923).

24   Because a boatload was used as the unit (wet weight) for the amount of submerged plants sold in the prefecture, the volume recorded is regarded as the weight of live plants, not dry weight.

25   Aquatic plants pinched between two thin bamboo poles and hoisted up during the summer were called *mizumo*. They were placed in fields to fertilize vegetables. Sediment dredged during the winter and early spring and used as dressing and fertilizer for paddy fields was called *doromo*. It was plowed into paddy fields after having been composted for half a year. (From Shiga Kenritsu Nōji Shikenjō, *Biwako engan mizumo no riyō to sono hikō* [Ōtsu: Shiga Kenritsu Nōji Shikenjō, 1939] and the author's interviews at various villages.)

26   Recorded in "Agricultural Materials," submitted by the commission on Lake Biwa measures in Shōwa 10 (1935).

27   Shiga Kenritsu Nōji Shikenjō, *Biwako engan mizumo no riyō to sono hikō.* This is a report of the prefectural agricultural experiment station on the use of submerged plants of Lake Biwa as fertilizer.

28   The prefectural agricultural experiment station used the method of N. v. Lorenz to estimate the percentage of phosphate of the standing crop of the plants (see L. Gisiger, "Die titrimetrisehe Bestimmung der Phosphorsäure auf der Grundlage der Methode von N. v. Lorenz unter Anwendung der Tauchfiltration," *Zeitschrift für analytische Chemie* 115 (1938): 15–29. Here, I assume the amount of phosphate to be that of PO_4.

29   From "*Awajimura yoshiokite*," Chūsei document #69 in Chūzu-chō Kyōiku Iinkai, *Ōmi no kuni Yasu gun Awaji Kuyū monjo mokuroku.*

30   Sano, *Chū-kinsei no sonraku to mizube no kankyōshi.*

31   Details were reported in a series of reports on regional folk culture assets compiled by the Shiga Prefecture Board of Education: Shiga-ken Kyōiku Iinkai, *Biwako sōgō kaihatsu chiiki minzoku bunkazai tokubetsu chōsa hōkokusho*, vols. 1–5 (Ōtsu: Shiga-ken Kyōiku Iinkai, 1979–1983).

32   Hagiwara Tatsuo, *Chūsei saishi soshiki no kenkyū* (Yoshikawa Kōbunkan, 1965); Harada Toshimaru, *Kinsei sonraku no keizai to shakai* (Yamakawa Shuppansha, 1983).

33   Sano, *Chū-kinsei no sonraku to mizube no kankyōshi.*

5

Floods, Drainage, and River Projects in Early Modern Japan

Civil Engineering and the Foundations of Resilience

PHILIP C. BROWN

For large-scale societal contractions and reconfiguration, the concept of "resilience" figures prominently in scholars' analyses of interactions among complex systems responding to a variety of pressures and challenges. As noted in the introduction to this volume, the typical framing explores erstwhile society-wide movements. When scholars think in dramatic terms, they may label a substantial societal contraction as a "collapse," a severe, broadly societal dysfunction in response to a significant change in fundamentals, especially environmental ones. For example, the best-known study of collapse, Jared Diamond's work by that title, stresses societal response to environmental change and treats Japan as a success story.[1]

Here I take a somewhat different perspective. I emphasize cases in which repeated, widespread, but smaller-scale resilience (lowercase r) is a critical component of ordinary societal functioning that must operate if a society is either to maintain itself or to grow (the latter, especially in premodern agricultural societies). The focus in this instance is on what happens in local society, not the society as a whole ("nationally" in a modern framing), in contexts where the land itself can be ruined and rendered unproductive.

I argue that Japan, not typically thought to have collapsed, actually experienced widespread social dysfunction in the sixteenth century from which it recovered largely because of repeated local "practice" at resilience in rural areas.[2] In contrast to the hunter-gatherer societies, raiding societies, and other relatively nonurban societies typically studied by archaeologists interested in Resilience (capital R) at the broad societal scale, Japan presents an example of collapse and rebound in a relatively urbanized, commercially complex society—even in the context of the late sixteenth and early seventeenth centuries.

Japanese Resilience in the face of environmental challenges is often alluded to or introduced explicitly but briefly in English-language books and texts. Discussion includes a brief geographical contextualization for history books in particular. Nonetheless, after rather quick introductory comments, the subject is seldom addressed at any length. Volcanoes and earthquakes represent dramatic examples of natural forces to which authors allude, and to be sure, Japanese society has had to accommodate such events.

Even less attention—virtually none—is devoted to the more common problems of water control, problems far more widespread than the dramatic displays of natural force embodied in volcanic eruptions and earthquakes. Water supply to improve agricultural productivity constitutes one set of such problems, but so, too, is oversupply of water—flooding—and that is the focus of this essay. Regions subject to frequent inundation offer us an opportunity to consider the response of local communities to these deadly and destructive threats. Despite the havoc created by floods, communities generally rebounded and trucked on, establishing patterns of response that arguably underlay successful reactions to broader shocks to Japanese society.

Here I take up floods and reactions to them in agricultural contexts.[3] Although water supply is important to all agriculture—which requires the right amount at the right times—a particularly important issue for Japan is not simply adequate supply of water but its dramatic oversupply and the accompanying high risk of flooding. Today, more than 50 percent of Japan's population is subject to flood risk, a proportion that was likely larger in premodern times.[4] In part, this results from many steep-sloped mountains (some 85 percent of the country) that push snowmelt and rain from relatively large drainage networks into narrow valleys and small plains. Even when appropriate ground cover anchors topsoil and soaks up rainwater, floods repeatedly strike many parts of Japan. Further, floods and landslides destroy not only crops, fields, and homes but also critical water supply networks and sluice gates. All these effects reduced domain governments' revenues, most of which derived from land taxes calculated as a proportion of putative agricultural yield *(kokudaka)*. Thus, the developments explored below were a central and constant element in Japanese agriculturalists' and rulers' interactions with the physical environment.

As a case study, I take up the region of the lower Shinano River, Japan's longest. Unlike the Kanto Plain (on which Edo, now Tokyo, is located), which is actually quite hilly, this district slopes gently and smoothly to the Sea of Japan. In particular, I examine the interactions between largely rural communities and the lower Shinano River system near its mouth during the Tokugawa era. At the start of the early modern era, these streams were constrained only by natural levies supplemented by modest human efforts. They flooded often, and though they did not have room to shift course to the same degree as China's famous Yellow River, traverse the Echigo Plain they did, destroying villages and fields that lined their banks.

Despite repeated and widespread flooding, villagers in the Echigo Plain never gave up. Instead, they displayed considerable ingenuity and resilience. Life was hard, to be sure, but their efforts to protect their communities and to promote agriculture were such that Echigo became the most populous province after Musashi, where Edo was located. This fact is all the more remarkable because the province sported no large metropolitan area like Osaka, Kyoto, or Nagoya.

A variety of mechanisms underlay both the resilience of these Echigo communities and Japan's Resilience as demonstrated by its ability to recover from a major, society-wide crisis during the late fifteenth through mid-sixteenth centuries. Among other factors, a combination of adaptive farming and civil engineering made this trajectory possible locally. Regarding the former, I simply note as an example that villagers employed special techniques for cultivation in swampy land, often using boats to do so. The latter realm constitutes the heart of this essay: villagers, townsmen, and even their military baron (daimyo) overlords planned and executed sophisticated civil engineering projects that built dikes and new rivers, rechanneled streams, and dried out marshes in efforts to limit flood damage, improve productivity of arable land, and expand acreage under the plow. The techniques employed here reflect those marshaled in many parts of Japan, and analysis of them reveals the determination and resilience exhibited in many Japanese villages and towns that confronted numerous floods and landslides. It also identifies the attitudes and flexibility that underlay recovery and, later, the expansion of agriculture from the seventeenth century through the Tokugawa era.

▓ The Sixteenth-Century Disruptions

Although Japanese communities have historically faced the need to recover from flood damage as well as landslides, earthquakes, and volcanic eruptions, no specific incident or collection of natural events created conditions for a massive, country-wide social disruption; however, political conditions could. The sixteenth century presents one striking example. Examining this major disruption highlights what is at stake when patterns of farmland maintenance, renewal, and reconstruction are disrupted.

Earlier examples in Japanese history might be characterized as "collapse," but the society's descent into widespread warfare in the late fifteenth and early sixteenth centuries must surely be the best-documented premodern case of collapse in governance. The outcome was a frequent disruption of agricultural activity. Infrastructure essential to paddy agriculture, flood amelioration efforts, and the like was not well maintained by villagers, districts, or their samurai overlords. Even when not destroyed outright by military activities, these systems tended toward entropy without regular maintenance, and conscription of rural labor for military efforts often disrupted precisely such maintenance.

It is difficult to tie this late medieval event to ecological change, although some have investigated this possibility. For example, Bruce Batten, in a recent effort to synthesize Japanese climate history, writes of the links between climate change and disorder at this time:

> Many specific uprisings can be traced to crop failures resulting from bad weather. One is therefore tempted to generalize and blame all of the sociopolitical problems of this period on the cooler climate. However that would be a mistake. Although temperatures were on average lower than during the Medieval Warm Period, overall that was probably a plus. . . . Further arguing against any simple link between climate and social conditions is the example of the subsequent stage, which saw a return to political stability despite profound climatic deterioration.[5]

Whatever influences ecological change had on the deterioration of political stability, the disruption was pervasive. Tensions among prominent regional lords (the military governors, or *shugo*) in the capital (Kyoto) commenced in 1467 and were not resolved during the next decade when this conflict, the Ōnin War, concluded. During the Ōnin War, subordinates of the military governors sought to advance their own power, pressing from below even as their fellow governors attacked them on the battlefields. In the end, the power of the military governors fractured and devolved to small military leaders.[6] Pervasive combat among them continued through the mid-sixteenth century, when a handful of these men began to develop into regional powers or to participate in regional alliances. Nonetheless, until the late 1580s, many regions of Japan continued to experience warfare that disrupted agricultural communities.

The consequences of disorder for Japan's farmers and their ability to sustain agriculture on much of the acreage cultivated up to that time are clear: field structures and water control systems for both supply and oversupply suffered. First, such structures were directly destroyed by armies sweeping back and forth across the land. Second, facing heavy tax and corvée exactions, villagers fled individually or as whole communities, abandoning their fields.[7] Finally, warlords conscripted labor, diverting it from critical seasonal tasks like maintaining dikes, irrigation channels, and water retention ponds even when not directly engaged in combat locally.

Irrigation facilities typically are built up gradually over many years, and fields likewise were conditioned over the years; rebuilding required extended, long-term efforts, not just a short, intensive burst of labor. This was particularly true for paddy, in which the pans need careful hardening and leveling to control the depth and flow of water on the field. Both irrigation systems and paddy ridges require constant upkeep.[8] Dikes, too, are subject to the wear and tear of hydraulic forces, even when not subject to the stress of high waters. The damage and exactions of war all detracted from or rendered impossible regular maintenance, much less extension of arable, irrigation, and diking.

Measuring this loss directly is impossible, but we do know that conflict bred pressures to which many daimyo responded with considerable creativity. In addition to improving military organization and technology, and using forceful revenue-raising devices, the most forward looking made investments in natural resources and expansion of agriculture. They invested in programs to expand arable under the plow, extend irrigation systems, and build dikes to limit destructive flooding. Such efforts paid dividends, starting with return of abandoned land to cultivation and restoration of flood control facilities from the mid-sixteenth century.

One significant technological innovation (in the Japanese context; the technique was certainly known elsewhere) focused on flood control. The technique was nicknamed the Shingen-tei (a.k.a. Shingen *tsutsumi,* Shingen dike) after the daimyo Takeda Shingen, who according to legend made considerable use of this design in his own territories in central Japan. These dikes were not just solid barricades designed to constrain floodwaters. Rather, at planned intervals, they allowed overflow, directing it to safe catchment areas that would drain once floodwaters receded. In effect, these catchments were zoned, permitting use for certain purposes, but not for cultivating important crops or for residences.

Such investments in land fostered significant recultivation of abandoned fields, improvements to arable (conversion of dry field to higher-calorie-yielding rice paddy and enhancement of poor-quality dry field), and expansion of cultivation into new lands. The full benefit of such developments was not manifest until after the mid-seventeenth century, but by that time, with the demands of war eliminated and defense expenditures curtailed, daimyo frequently used their resources to engage in large-scale reclamation projects, major improvements of irrigation systems, and the like. Tax burdens on villagers stabilized and daimyo curtailed the worst abuses of rural populations sufficiently to end the absconding of whole villages.[9] Daimyo proffered regular incentives to villagers in the form of land tax abatements, individual and collective, to improve land and extend cultivation.

Although difficult to measure with any precision, the gains from restoration of abandoned fields to cultivation and the extension or improvement of arable from ca. 1550 to ca. 1650 were clear: population grew at a remarkable rate for a premodern society, and the value of land subject to taxation increased significantly. Nationwide population estimates for the early seventeenth century range from twelve million to eighteen million, but whichever figure is chosen, by the early eighteenth century Japan boasted a population of about twenty-two million.[10] This figure is even more remarkable given the rapid growth of urban areas, where the natural rate of premodern population growth without immigration is typically negative. Daimyo castle towns (well over two hundred) expanded from a few thousand residents to a hundred thousand (Kanazawa) or a million (Edo) at the extreme. Given problems in land survey processes, manipulation of units of measure, and the difficulty in getting an accurate fix on cultivated land subject to taxation, estimates of area under the plow during the sixteenth and seventeenth centuries must be treated cautiously, but the population increase Japan witnessed over the seventeenth century would not have been possible without significant improvements in the amount and quality of land under the plow.[11]

Data commonly cited as evidence of the expansion of arable provide the following picture:

TABLE 5.1. Estimated total arable land in Japan, in *chō**

	Estimated total arable
ca. 1450	946,000
ca. 1600	1,635,000
ca. 1720	2,970,000
1874	3,050,000

*1 *chō* = approximately 9,917 m^2

SOURCE: From Table II, Kitajima Masamoto, ed., *Tochi seidoshi*, vol. 2, Taikei Nihonshi sōsho 7 (Yamakawa Shuppansha, 1975), 28.

Each step but the last shows considerable progress. Although the trajectory painted here has the expected direction, the magnitude of change across the first three steps is subject to question, especially for the era between 1450 and 1600. Areas of measure were not consistent over time: for example, many regions that used 360 *bu* for a *tan* prior to the 1580s switched to 300 *bu* or *tan* thereafter, a desktop change that would increase area under the plow by 20 percent in the swish of the writing brush.[12] Changes in measuring units continued in different regions at different paces after 1600 as well. Even as daimyo gained a firmer hold on their agrarian resources through surveys of land, in many instances surveys were not based on accurate, replicable measuring techniques but involved eyeball estimates and village-reported data. Much arable land remained unregistered as taxable land, to be added only with repeated efforts over the first half of the seventeenth century and beyond.[13]

The data are compiled on an inconsistent basis and incomplete; their best use is as an indication of the direction of long-term change. The lack of data for the sixteenth century, when we can reasonably expect stasis and decline under conditions of widespread internecine warfare, masks significant movements over that century. This stasis and decline can only be surmised from the descriptions of disruption in the documentary literature, plus what we know about what happens to paddy, irrigation, and premodern dikes if not regularly maintained—rapid deterioration due to floods, erosion, burrowing animals, and rot.[14]

Kozo Yamamura, the Japan Society of Civil Engineers (Nihon Doboku Gakkai), and others have stressed the role of large-scale projects in the expansion and improvement of arable; however, these projects account for but a fraction of the apparent increase in arable in the seventeenth century.[15] Understanding their contribution to increased agricultural output requires context, and even that context suggests the larger problem to which I draw attention, the role of flooding as a stimulus to action.

The largest projects were frequently intended to cultivate new sections of floodplains, convert riverine islands to arable land, or drain marshes and swamps—precisely the kinds of environments that illustrate the need for maintenance on the one hand and the high risk of flooding on the other. In each of these environments, flooding continued to be a major hazard, with civil engineering projects providing some relief from common low-level threats but not from major inundations. Given the sharp slope of most of Japan's rivers, decreased velocity of flow at the intermittent flatlands and in the coastal plains deposited significant upstream debris that forced streams into new channels. Islands and floodplain fields could be swept away in short order.[16] Consolidation of local power under early modern daimyo provided a governing framework that allowed assembly of adequate resources to tackle projects on a much larger scale than heretofore and certainly contributed to expansion of arable into new environments as well as recultivation of abandoned fields. But by their nature, such projects further caution us about the danger of ignoring the need for consistent and regular maintenance—work rendered very difficult during the sixteenth-century wars.

The Echigo Plain

Examination of developments in one particular region further illustrates the importance of routine adaptations to potential floods. For this purpose, I have chosen Echigo Province, and within it, the Echigo Plain in particular. Although thought of as rustic and out of the way, at the time of the Meiji Restoration (1868), Echigo had a very large population despite its lack of a single large castle town.[17] For most of the period, the Echigo region was divided into multiple domains of small to middling size. Governance of the province also rested in part on the shogun (e.g., the area around Niigata Town) or the daimyo who oversaw shogunal lands on his behalf (e.g., the area of Uonuma County around modern Tōkamachi). Although Niigata was important as a conduit of trade and a link to the gold and silver mines on Sado Island, it was a small town even in the 1870s.[18] Nagaoka and Takada, headquarters of eponymous domains, were larger but still of modest scale.

The Echigo Plain is one of Japan's larger plains, and unlike the Kanto Plain, it is, in fact, quite flat. Through it flows Japan's longest river, the Shinano, which begins its course high in the Japan Alps as the Chikuma River.[19] Numerous smaller streams crisscross the plain as well, all subject to regular flooding.

We often imagine Japan's physical geography as an outcome of earthquakes thrusting mountains upward and of volcanic activity. Niigata Prefecture, including parts of the upstream Shinano, sits uncomfortably at the conjunction of multiple fault lines along which earthquakes occur, and there was also volcanic activity in the area. The Echigo Plain was indirectly created by the resultant mountains, specifically by erosion. Historical geologic maps (figure 5.1) show the extensive role of sedimentation (black areas) within the Echigo region. Coastal areas figure prominently in this depiction, but notice the broad finger of sediment that strikes south from the Sea of Japan in the center of the map, one that follows the Shinano River as it winds into the mountains of central Japan.

Recent reconstructions of the changing coastline and flow of rivers in the prefecture highlight both the risk of inundation and the wandering course of streams on the Echigo Plain. These are shown in the various segments of plate 7. (The scale of the maps has been adjusted to focus attention on the differences.)

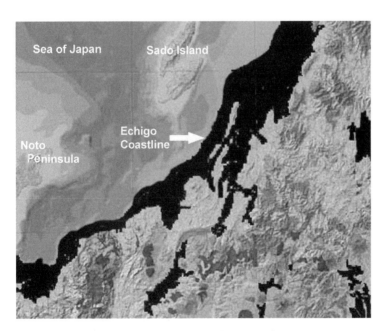

FIGURE 5.1. Echigo historical geography: Sedimentary formations

SOURCE: Based on Kensetsushō Kokudo Chiri'in, *Nihon kokusei zu* (Kensetsushō Kokudo Chiri'in, 1997), CD-ROM.

Plate 7A shows that some sixty-eight hundred years before the present, the area of the modern city of Niigata was still ocean. Note, too, that the old riverbeds (the darker streams) were considerably wider than their modern counterparts (lighter-colored streams). Plate 7B indicates that by a thousand years ago, the modern coastline had largely formed, although the old riverbeds, as well as their width, differed considerably from their course today. Plate 7C, to which we shall return shortly, shows the same coastline as 7B, but it is clear that the precursors of the modern Shinano and Agano rivers, which now empty into the Sea of Japan separately, were joined, with the Agano entering the Shinano rather than flowing into the sea directly. By the mid-nineteenth century (7D), that situation changed again. Although rivers appear to flow along paths that approximate their present-day courses, their beds are wider, and if one looks closely at plate 7D, a number of small lakes or ponds south of Niigata City are still visible.

The potential for flooding was well described by nineteenth-century English traveler Isabella Bird:

> In the freshets, which occur to a greater or less extent every year, enormous volumes of water pour over these [riverbed] wastes, carrying sand and detritus down to the mouths [of streams], which are all obstructed by bars. Of these rivers the Shinano, being the biggest, is the most refractory, and has piled up a bar at its entrance through which there is only a passage seven feet deep, which is perpetually shallowing. The minds of engineers are much exercised upon the Shinano, and the Government is most anxious to deepen the channel and give Western Japan what it has not—a harbour.... [20]

Her observations point not only to frequent flooding but also to the effect of eroded materials carried downstream, filling up both riverbed and harbor. To these observations can be added those of English engineer Henry Brunton, who calculated the Shinano's floodwaters as much greater than those of Europe's famous rivers and noted that in many places, the river's floodplain upstream from Niigata was three miles wide. [21]

In part, the meandering of the Shinano and other, smaller rivers in the area results from the very shallow slope of the lands through which they flow. Much of these lands historically lay less than a meter above sea level, even 15 kilometers south of the Shinano's mouth. Drainage was often poor through the nineteenth century; farming frequently required the use of boats, and the major form of transportation in the Niigata Town area was waterways, not footpaths. [22]

In this environment, preservation and maintenance of arable were virtually indistinguishable from efforts to improve the land. Small-scale efforts were common, of course, but tend to be poorly documented. Here I focus on several projects to suggest the lengths to which residents went to reduce the threats of flooding and improve drainage, efforts that did not wait for modern power equipment, explosives, or construction materials to push and pull the natural surroundings into a shape more commensurate with human exploitation. Well documented, they suggest the techniques employed and issues confronted in numerous smaller projects.

▨ The Matsugasaki Diversion, 1731

Plate 7C shows the Shinano and Agano rivers (the two largest streams in the map) as strikingly dissimilar to the modern geography of the area. The Agano (the northern of the two streams) now empties directly into the Sea of Japan; however, that was not the case as Japan entered the early modern age. In 1600, the Agano first entered the Shinano before its waters discharged into the ocean. (Note that in plate 7B, the two rivers have separate mouths on the Japan Sea, additional evidence of the transitory nature of streams on the Echigo Plain.) The broad bulge of the Shinano just inland from the conjunction of the two rivers helped accommodate the flow of the combined waterways. Even with the presence of that expanded riverbed, flooding was frequent on both rivers as well as between the Agano and Shiunji Lake, and along the old Kajikawa riverbed. From the early seventeenth century, villagers made repeated efforts to improve drainage, achieving less than satisfactory results. The Agano's modern path to the sea resulted from a combination of human effort to control flooding and a major flood that broke the dikes designed to accomplish that task. In the end, the product of this conversion of natural and human forces still functions today.

In 1727, with the aim of reclaiming part of Shiunji Lake (approximately the area of modern Suibara) and developing some 250 *chō* of farmland on the edges of the lake, one Takemae Shōhachirō (resident of Yonago Village, Shinano Province, well upstream the Shinano) submitted a plan to the shogunate, a bold plan that marked the start of sustained development of farmland in the area. With funds from one merchant from Edo and another from Kashiwazaki (just north of modern Jōetsu City on Niigata's Japan Sea coast), the project got under way. Using only hand labor, the Ochibori drainage construction was undertaken (north of modern Shibata City), and the smaller Sakai River was diverted to improve drainage. Shōhachirō did not live to see the completion of the full project, and in 1729 his younger brother Gonbei took charge.

To simplify administrative challenges due to the fragmented political control of this region, the shogunate turned over control of some of its territory to Shibata Domain to facilitate construction of a diversion channel for the Agano. This planned channel siphoned off some of the river's flow before it entered the Shinano and directed it into the Sea of Japan at Matsugasaki.[23] The diversion channel was completed late in 1730. (Figure 5.2A shows a dike constructed in such a way as to allow controlled runoff of floodwaters.) The channel from that point to the sea was excavated by hand and partially lined with dikes to control the overflow. However, the hopes and confidence of designers, investors, and governing officials in this structure proved misplaced.

The following year, in 1731, a flood destroyed the drainage channel facility, washed away the dikes lining the diversion channel, and widened the stream considerably (figure 5.2B). As a result, the drainage channel was transformed from its role as a release valve for high waters into the main stream of the Agano. All the river's waters now flowed straight into the Sea of Japan, not into the Shinano River. While this mishap relieved some flooding in downstream areas, it also caused a sharp drop in the water level of the Kajikawa, which at the time flowed directly into the Agano.

This reduced the utility of the Kajikawa for transport as well as for irrigation.[24] Although arguably beneficial to many, the outcome made it clear that projects of this sort could have unintended consequences—in particular, failures that interrupted river-borne commerce, irrigation, and even effective drainage. Opponents of similar projects cited these failures over the following

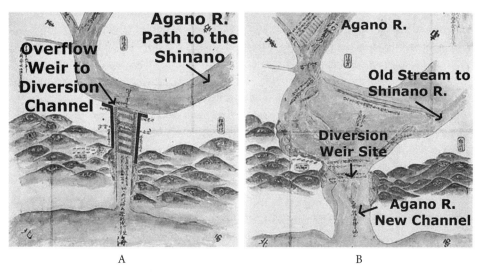

A   B

FIGURE 5.2. The Matsugasaki diversion channel project, Echigo (a) The completed diversion facility (1730) showing the overflow weir on the Agano River and the drainage channel to the Sea of Japan. The main stream emptied into the Shinano River to the south. (b) The redirected main channel, postflood (1731), showing the broken weir, the widened channel draining the main stream into the Sea of Japan, and the drastically reduced flow along the old channel that took the Agano's water into the Shinano River.

SOURCE: Fukunaga-ku Manuscript Collection, Agano City, http://mobile.pref.niigata.lg.jp /HTML_Article/559/319/P7,0.pdf.

two centuries, including when the two most famous diversion channels on the Echigo Plain, the Shinkawa and the Ōkōtsu, were created.[25]

Dikes and Drainage on the Shinano's West Bank

To the west of the Shinano, just inland from where the wide, modern Sekiya diversion channel takes overflow from the river, lies the area until recently known as Kurosaki Town (now a part of Niigata City; upper-right corner of plate 7D). To its south lies the Nakanokuchi River, where it joins the Shinano. Although it does not border the town, to the west is the Nishikawa River, which also flows into the Shinano. Kurosaki was thus faced with major sources of flooding from at least three directions. Like the area of the Matsugasaki diversion project, a comparison of a late-nineteenth-century or early-twentieth-century map of Kurosaki with its late-twentieth-century counterpart would show two very different landscapes. The modern map shows no trace of the marshes, swamps, and small lakes present up to the dawn of the twentieth century. Postwar civil engineering projects have largely removed these features from the landscape, but the process began much earlier. As in the case of Matsugasaki, the impetus was a desire to limit flooding and expand arable.

Surviving documents outline a story of repeated efforts to limit flooding. Between 1691 and 1868, a span of 177 years, the dozen villages in the district recorded at least twenty-two incidents of dikes broken by floodwaters, an average of one every eight years.[26] At best this record is partial, but it is sufficient to tell us of a long-term contest with the rivers. Broken dikes speak to the deliberate efforts of Kurosaki villagers to contain the high waters of the Shinano and Nakanokuchi. They speak further to efforts of their neighbors on the Nishikawa to maintain and extend natural levees. Dikes were typically built piecemeal, not as one large project, and repaired frequently, but even with regular maintenance, they were unequal to the forces of nature.

Of course, broken dikes tell only part of the story. Flooding can occur even without such incidents, and even if dikes were repaired promptly, floods destroyed crops already in the ground, left debris on fields, and destroyed irrigation channels and control facilities, all of which required extended efforts to repair. To handle such reconstruction, as well as to provide sustenance during rebuilding, villagers were often left on their own. In the cases of dike breaches in Kurosaki, in only four of the twenty-two incidents did villagers receive any assistance from their overlords. Further, in this region, water could stand long after inundation, failing to drain even after the flood-stage waters receded.[27]

As previously noted, poor drainage in this region stems from the Echigo Plain's shallow slope. Other than a large sand dune on the Japan Sea side of the district, with a maximum elevation of 27 meters and only a mile wide, land in this area is consistently only 1 or 2 meters above sea level.[28] The sand dune drains well but produces no rivers or streams and no large runoff during rains, although its erosion can fill the Nishikawa. The main sources of floodwaters lie upstream: waters carried from the mountains and collected from the broad Echigo lowlands during rainstorms or after snowmelt.

Thus, it is unsurprising that residents in the Kurosaki area and its environs devoted much attention to improving drainage, including emptying the numerous lakes, marshes, and swamps that dotted the countryside. These bodies of water, like rivers, generated floodwaters that damaged surrounding fields. Finding effective ways to drain them not only opened new land to cultivation, it also served to protect already cultivated fields surrounding them.

The draining of the so-called Three Lakes (Sangata) of Yoroigata, Ōgata, and Tagata is legendary in Niigata Prefecture, but the area just south of Niigata City was pocked with smaller bodies of water that were likewise the object of drainage efforts. During the Tokugawa era, progress in reclaiming these lowlands, even partially, depended on the accumulated effect of multiple, typically small efforts. One major project, the creation of the Shinkawa (New River), was a part of the Tokugawa efforts. I have treated this impressive project elsewhere, so here I will focus on the smaller endeavors.[29]

At least nine ponds were associated with the Kurosaki area, the farthest from the Shinano being Uragata.[30] In addition, some of the land closest to the Nakanokuchi River served as an overflow zone for floods, development of which was restricted.[31] Because of sedimentation from the sand dune above it, the Nishikawa's bed sometimes rose, decreasing its utility as a drain for the area and forcing water back into some of the smaller channels upstream. This situation continued until the completion of the Shinkawa in the early nineteenth century.[32] During the Tokugawa, villages throughout the area south of Niigata developed a variety of plans to add arable acreage or reduce flooding. Given the varied trade-offs of benefits and risks among the district's villages, planning projects was complicated. Some villages took irrigation water from these ponds or in other ways benefited from the existing conditions, so any plan for villagers in one location to develop land or change drainage patterns encountered doubters in others. At the least, planners had to make adjustments to avoid injury to other villages or to provide compensation for loss of utility.

Villages in Nagaoka Domain, initially intent on developing land around Shiunji Lake near the Agano, also had plans for the Kurosaki area, and in 1732 they submitted the first plan for draining the Sangata area by creating a riparian underpass under the Nishikawa (a precursor of the plan eventually adopted for the early-nineteenth-century Shinkawa).[33] Repeated failures with different versions of this plan finally caused the villagers to turn to an alternate plan in 1745, one that drained this area by creating channels that passed through two Kurosaki villages, Torihara and Kurotori, and carried water into the Shinano River directly. After affected villages were consulted, and compensation for land that would be lost to the new channel had been arranged, the bakufu granted permission in 1746. Division of responsibility for construction, limits on construction to prevent problems with other villages, responsibility for maintenance, and the like were carefully apportioned. The locations of intake and discharge gates, as well as the scale of dikes along the drainage channel, were all carefully specified. Agreements even specified the contours of flood aid to Torihara and Kurotori, with assistance to come from neighboring villages that benefited from the new channel even though they might not be directly affected by any specific flood. Over the next half-dozen years, despite flooding that destroyed part of the construction, the project proceeded, and in 1751 new lands under the plow were surveyed at 3449 *koku* (1 *koku* = about 5 bushels) of assessed yield.

Final Reflections

The preceding treatment of civil engineering and floods has focused on one important subset of their causes—the overflow of rivers and banks, especially in a broad flatland area. In the Echigo area there were other sources of floods. The eighteenth-century Echigo author Suzuki Bokushi describes localized flooding in hill and mountain villages due to the combined influence of snowmelt and the packing of snow on paths and roads where villagers tread during the winter. In these instances, snow blocked drainage paths, damming up waters from melting snow and forcing them along footpaths into the homes of townspeople and villagers. In other instances, snow blocked rivers and streams as it rolled off hillsides or as the surface ice on rivers collapsed into the stream. No civil engineering could prevent this kind of hazard.[34]

I would be remiss if I did not also mention villagers' engineering efforts to deal with another water-related hazard, landslides. We are aware that premodern peoples understood the relationship between maintenance of ground cover and prevention of landslides, a subject well explored by the historian of Japan Conrad Totman.[35] However, we typically presume that prior to the modern era, engineering efforts to ameliorate or prevent landslides were not known. This is an erroneous presupposition. To cite one Echigo case, in 1865 villagers in Nakajō-mura (in Tōkamachi City) constructed erosion control dams and made efforts to divert water that would otherwise contribute to landslides. They employed a variety of techniques not described elsewhere, suggesting a degree of inventiveness and originality.[36] These efforts may not have been on a par with modern efforts, but they clearly reveal an understanding that the key to prevention lay in control of groundwater.

Although not all flooding could be addressed through civil engineering, and its use to combat landslides was unusual in sixteenth- to nineteenth-century Japan, civil engineering was widely employed to ameliorate flood hazards. The projects discussed above are broadly representative of such activities throughout early modern Japan. In these particular cases, the interactions of two basic forces—natural processes that shape the physical environment and the human need for food to sustain life and reproduce—combined to stimulate activities that radically altered the physical environment.

The above discussion focuses on the creation of new rivers: large in the case of the Agano River and the Matsugasaki diversion channel, smaller streams in the case of the Kurosaki area of modern Niigata City. Two other large projects, the creation of the Shinkawa and Ōkōtsu diversion channel, have been mentioned in passing. The latter, while proposed in the early eighteenth century, was not realized until the twentieth century, and it was a controversial project even after its completion in 1922.

Over the long run, the creation of new streams in Japan's early modern era set a precedent for later efforts. The costs of new technologies placed constraints on these efforts before World War II, but this approach has been a preferred method for alleviating flood risk in the post–World War II era. Up to the end of World War II, three man-made rivers emptied directly into the Sea of Japan; today, there are twenty such streams. In addition, other new streams have been created and old ones reshaped as part of the effort to limit flood damage and improve irrigation.

New construction materials and techniques now limit entropy in fields, flood control, and irrigation systems, but the problem remains. The challenge appears quite visibly in the constant efforts to keep debris from clogging dams and irrigation gates. Today the most noticeable detritus often consists of paper, plastic bottles, and similar artifacts of Japan's modern consumer-oriented economy, but this should not obscure the continued buildup of siltation and other natural debris that historically contributed to the flooding that challenged residents of the Echigo Plain. Nor should the reshaped landscape and the resultant expansion of urban and suburban space distract our attention from the cold, hard facts: flooding persists and the financial loss associated with each incident continues to grow.

Finally, the premodern predecessors of modern civil engineering discussed here illustrate not only the efforts that built and routinely sustained Japanese paddy agriculture, but also those essential to rebuilding the agricultural foundations of Japan after the late-fifteenth- and early-sixteenth-century political crisis. Such efforts extended beyond the day-to-day maintenance of paddy pans, irrigation channels, and water control gates but for much of Japan represented common endeavors. Once the destruction, distractions, and disruptions of rampant internecine warfare had ceased, routine practice with resilience provided the foundation for Japan's late-sixteenth- and early-seventeenth-century recovery. In this way, Japanese endurance of frequent floods and landslides, combined with long-standing efforts to expand and improve land under the plow, provided a clear foundation for long-term, large-scale Resilience.

NOTES

1   In dealing with Japan, Diamond argues that a successful response to countrywide deforestation was grounded in both elite and subject conceptions of long-term rather than short-term self-interest, even in environmentally robust conditions. Jared Diamond, *Collapse: How Societies Choose to Fail or Succeed* (New York: Viking, 2005), especially 294–306.

2   Diamond stresses landowners' private long-term landownership interests (305) as a key to avoiding crisis, but as I discuss in *Cultivating Commons: Joint Ownership of Arable Land in Early Modern Japan* (Honolulu: University of Hawai'i Press, 2011), arable land in much of Japan was not held privately but by village corporations, which also made investments in land like those noted here.

3   Quite apart from agriculture, the development of Edo (modern Tokyo) out of lowland, with its associated risk of flooding, is a well-known story—even the subject of Japanese television shows. Indeed, much of the early modern city was built on landfill, a pattern that continues today. Part of that construction involved building new rivers. The redirection of the Tone River is perhaps most famous: it was diverted from its original path—through the city into what is now Tokyo Bay—to its current course, which traverses Chiba Prefecture to empty into the Pacific. Further construction of waterways is likewise narrated as part of the story of building a new urban space, providing water, and improving transportation networks. In English, part of this story is told in Marcia Yonemoto, *Mapping Early Modern Japan: Space, Place, and Culture in the Tokugawa Period, 1603–1868*, Asia-Local Studies/Global Themes (Berkeley: University of California Press, 2003), chap 1.

4   World Commission on Dams, *Dams and Development: A New Framework for Decision-Making* (London and Sterling, Virginia: Earthscan Publications, 2000).

5   Bruce L. Batten, "Climate Change in Japanese History and Prehistory: A Comparative Overview," Occasional Papers in Japanese Studies Number 2009-01 (Cambridge, Mass.: Edwin O. Reischauer Institute of Japanese Studies, Harvard University, 2009), http://rijs.fas.harvard .edu/pdfs/batten.pdf (accessed November 7, 2011), 49.

6   Even those few military governor lineages that survived into the Tokugawa era did so at a considerably reduced size.

7   Philip C. Brown, *Central Authority and Local Autonomy in the Formation of Early Modern Japan: The Case of Kaga Domain* (Stanford, Calif.: Stanford University Press, 1993).

8   On the complexity of paddy agriculture and maintenance of fields, see Francesca Bray, *The Rice Economies: Technology and Development in Asian Societies* (Oxford: Basil Blackwell, 1986). On Japanese social arrangements for providing and allocating water through irrigation projects, see William W. Kelly, *Irrigation Management in Japan: A Critical Review of Japanese*

Social Science Research, Cornell University East Asia Papers, No. 30 (Ithaca, N.Y.: China-Japan Program Rural Development Committee, Cornell University, 1982); and William W. Kelly, *Water Control in Tokugawa Japan: Irrigation Organization in a Japanese River Basin, 1600–1870*, Cornell University East Asia Papers, No. 31 (Ithaca, N.Y.: China-Japan Program, Cornell University, 1982). For an excellent treatment of this tendency as it applies to the American West, see Mark Fiege, *Irrigated Eden: The Making of an Agricultural Landscape in the American West*, Weyerhaeuser Environmental Books (Seattle: University of Washington Press, 1999).

9   Brown, *Central Authority and Local Autonomy in the Formation of Early Modern Japan*.

10 Hayami Akira, *Population and Family in Early-Modern Central Japan*, Nichibunken Monograph Series, 11 (Kyoto: International Research Center for Japanese Studies, 2010). See also Osamu Saito's discussion of population estimates in chapter 11 of this volume and its discussion of population estimates by William Wayne Farris and others.

11 Brown, *Central Authority and Local Autonomy in the Formation of Early Modern Japan*. William Wayne Farris, *Japan's Medieval Population* (Honolulu: University of Hawai'i Press, 2006), argues for late medieval population growth supported by widespread local agricultural improvements in addition to expansion of arable land prior to 1600.

12 *Bu, tan,* and *chō* are traditional Japanese units of area: 1 *chō* (9,917 square meters) = 10 *tan*; 1 *tan* = either 300 or 360 *bu*, as noted in the text.

13 Brown, *Central Authority and Local Autonomy in the Formation of Early Modern Japan*, especially chaps. 3 and 9, but also Philip C. Brown, "'Feudal Remnants' and Tenant Power: The Case of Niigata, Japan, in the Nineteenth and Early Twentieth Centuries," *Peasant Studies* 15, no. 1 (1987): 5–26; Philip C. Brown, "Never the Twain Shall Meet: European Land Survey Techniques in Tokugawa Japan," *Chinese Science* 9 (1989): 53–79; and Philip C. Brown, "A Case of Failed Technology Transfer—Land Survey Technology in Early Modern Japan," *Senri Ethnological Studies* 46 (1998): 83–97. Philip C. Brown, "Practical Constraints on Early Tokugawa Land Taxation: Annual Versus Fixed Assessments in Kaga Domain," *Journal of Japanese Studies* (1988): 369–401, also has implications for how well seventeenth-century administrators could estimate the agricultural wealth of their domains.

14 Fiege, *Irrigated Eden*, presents a superb analysis of the tendency toward entropy in irrigation systems in the context of Idaho agriculture.

15 Doboku Gakkai, *Meiji izen Nihon doboku shi* (Doboku Gakkai, 1936); Kozo Yamamura, "Returns on Unification Economic Growth in Japan, 1550–1650," in *Japan before Tokugawa: Political Consolidation and Economic Growth, 1550–1650*, ed. Nagahara Keiji, John W. Hall, and Kozo Yamamura (Princeton, N.J.: Princeton University Press, 1981), 327–72; Ōkurashō Shuzeikyoku and Nonaka Jun, eds., *Dai Nihon sozeishi*, 4 vols. (Chōyōkai, 1926–27); Kitajima Masamoto, *Tochi seidoshi,* vol. 2, Taikei Nihonshi sōsho 7 (Yamakawa Shuppansha, 1973).

16 The case of an island reclamation project at Yatsubunshima in modern Tōkamachi, Niigata Prefecture, is discussed in chapter 8 of Brown, *Cultivating Commons*.

17 The early nineteenth-century Echigo writer Suzuki Bokushi noted nine castle towns of varying size, but the number actually functioning as castle towns fluctuated over the course of the Tokugawa era, with the province at one time being ruled by a single daimyo for a brief interval. Suzuki Bokushi, *Snow Country Tales: Life in the Other Japan,* trans. Jeffrey Hunter and Rose Lesser (New York: Weatherhill, 1986), 159.

18 Isabella L. Bird, *Unbeaten Tracks in Japan*, Virago/Beacon Travelers (Boston: Beacon Press, 1987), 125–130.

19 The name now changes at the Nagano-Niigata prefectural border, but the dividing line was

not fixed in pre-Meiji times. The label "Chikumagawa" could be seen on maps well inside the Echigo provincial boundaries.

20 Bird, *Unbeaten Tracks in Japan*, 126.

21 R. Henry Brunton, *Building Japan 1868–1876* (Sandgate, Folkestone, Kent: Japan Library, 1991), 41–42.

22 Only the 1964 Niigata earthquake prompted massive reconfiguring of the city's water-based transportation system. Today, many parts of this area actually lie below sea level, and Niigata is known as a major region of ground subsidence in the country.

23 Niigata-ken, *Niigata-Ken no ayumi* (Niigata-shi: Niigata-ken, 1990), 179–80.

24 Niigata Shishi Hensan Iinkai, *Niigata shishi, tsūshi hen* (Niigata-shi: Niigata-shi, 1995), 1: 485–503, presents a detailed overview of this project.

25 The Ōkōtsu diversion project is better known; located upstream on the Shinano, it was the largest civil engineering project in East Asia at the time of its completion in 1922. For an overview, see Philip C. Brown, "Constructing Nature," in *Japan at Nature's Edge: The Environmental Context of a Global Power*, ed. Ian Jared Miller, Julia Adeney Thomas, and Brett L. Walker (Honolulu: University of Hawai'i Press, 2013), 90–114.

26 Kurosaki-chō, *Kurosaki-chō shi*, 8 vols. (Niigata: Kurosaki-chō, 1994–2000). *Tsūshi*, 187, presents the dike breakage incidents in convenient tabular form.

27 After the Niigata earthquake in 1964 and its accompanying tsunami, water stood in the downtown area for more than two weeks; in an 1896 flood, water was still standing more than three weeks afterward.

28 Today, much of this area is actually below sea level, and some of it may have been below sea level in the nineteenth century.

29 Philip C. Brown, "Moving Rivers: Lowland Water Management in Nineteenth Century Japan" (paper presented at the annual meeting of the Society for the History of Technology, Tacoma, Washington, September 30–October 3, 2010).

30 Kurosaki-chō, *Kurosaki-chō shi, Tsūshi*, 190, fig. 16.

31 Ibid., 185. One Kurosaki-area village vigorously opposed attempts by other villages to convert this land to paddy and dry field, contending that it would interfere with the functioning of its own agriculture; the bakufu concurred. The overflow function filled by some of the Kurosaki-area land was the product of a natural formation, not a planned, constructed *Shingen-tei*.

32 Ibid., 185.

33 Ibid., 189.

34 Suzuki, *Snow Country Tales*, 22–26.

35 Conrad Totman, *The Green Archipelago: Forestry in Preindustrial Japan* (Berkeley: University of California Press, 1989).

36 Tōkamachi-shi Hakubutsukan Tomo no Kai Komonjo Guruupu, *Tamura Tani ke shiryō (1)*, Tōkamachi kyōdo shiryō sōsho 14 (Tōkamachi, Niigata Prefecture: Tōkamachi Hakubutsukan, 2006), 20, 181–84. The whole volume documents the effects of landslides in this area.

6

High-Growth Hydrosphere

Sakuma Dam and the Socionatural Dimensions of "Comprehensive Development" Planning in Post-1945 Japan

ERIC G. DINMORE

In the years following World War II, Japanese academic economists and public policy commentators viewed "backwardness" as a primary factor behind their country's ruinous wartime pursuit of empire. They welcomed the postwar policy environment as an opportunity to cure this disorder, and their prescriptions included newfound commitments to economic plans, empirically based statistics, Keynesian full-employment policies, the promotion of domestic consumerism, and "growthism": the enrichment of Japanese society through sustained growth in the gross national product (GNP).[1] High GNP growth required a tremendous expansion of the country's manufacturing capacity, which in turn required tremendous quantities of raw material and energy resources. To overcome backwardness, Japan would need to refashion its entire relationship with the natural environment so as to maximize the amount of resources available to feed a growing economy.

Many policy commentators who endorsed growthism also advocated "comprehensive development" *(sōgō kaihatsu)* as an appropriately scientific approach to land and water use. Such advocates performed leading policy-making roles in the immediate postwar years, amid low public confidence in an economy stripped of its colonial resources and anxiety about the Cold War's effect on future global trade. They believed Japan would first need to consider how to sustain growth by maximizing the use of resources available on the four main Japanese islands before relying on overseas trade.[2] Comprehensive development combined many of the techniques of growthist economics—including long-term planning, empirical data analysis, and technocratic guidance in the name of the public good—into a unified management of the domestic landscape. As its backers saw it, the approach would serve national policy goals far better than if questions of land and water use were left up to a laissez-faire marketplace or, even worse, nature itself.

A key, if not *the* key, to early postwar comprehensive development schemes was the promotion of large, phenomenally expensive multipurpose dams. The immense scale of postwar dam projects set them apart from prior hydrological interventions. As the Kawasumi, Sano, and Brown essays in this volume indicate, Japanese had long been constructing dikes and actively managing bodies of water.[3] After premiering in the 1890s, hydroelectric dams began to proliferate in the early twentieth century.[4] Yet an estimated 55.5 percent of all "large dams" and 91.9 percent of dams over 30 meters in height appeared after 1945 and before the 1990s, when the pace of construction ebbed.[5] Japan by the 1990s was one of the most heavily dammed countries on the planet: large dams blocked a staggering 97 percent of major rivers at an operating cost of ¥200 billion per year.[6] The World Commission on Dams in 2000 estimated that Japan had constructed 2,675 of the globe's 47,655 large dams, or 5.6 percent.[7]

The backers of multipurpose dam projects in early postwar Japan justified their necessity for flood control, provision of water for irrigation and industrial applications, improvement of river transport, promotion of rural development, and hydroelectricity. This last reason factored most importantly at the outset of comprehensive development planning during the 1950s because Japan suffered from chronic electricity shortages, and the hydroelectric generation of power prevailed over then-expensive thermal power generation.[8] Moreover, dam proponents viewed river water as one of the few strategic resources Japan held in abundance, and they forcefully argued for hydroelectric development as a means of offsetting the country's dependence on imported energy. Over the past forty years, Japanese scholars and activists have certainly become aware of the lackluster economic performance and negative ecological consequences of large dams. Ecological repercussions have included collapses of riparian fisheries, erosion of river delta regions, soil salinization, and the siltation and eutrophication of reservoirs created by dams.[9] Yet few of these long-term costs were adequately understood in the 1950s and 1960s; instead, Japanese policy commentators took great interest in the evident short- and medium-term benefits of damming watersheds. Comprehensive development channeled most of Japan's hydrosphere away from historic drainage systems into reservoirs designed to power high growth in the country's gross domestic product.

Here I examine early postwar comprehensive development and dam promotion on the levels of policy discourse and actual practice. I begin by tracing the genealogy of the 1950 Law on Comprehensive National Land Development (Kokudo sōgō kaihatsu hō), the basic law that underpinned attempts to reconfigure Japanese landscapes during the 1950s and early 1960s. I then present a case study of the Sakuma Dam, which at its completion in 1956 stood as Japan's largest dam and as a concrete product of comprehensive development policy. The Sakuma Dam, I argue, illustrates the limitations of the comprehensive development approach by skewing the benefits of hydraulic exploitation toward urban

FIGURE 6.1. Sakuma Dam, Aichi, June 2013
Photograph by author.

industrial centers, by failing to encourage rural revitalization, and by upsetting
the natural environment of the Tenryū River valley in central Japan.

◼ The Context and Implementation of
Comprehensive Development

Japanese interest in large dams and comprehensive development emerged from
a mid-twentieth-century discourse on resource planning that valorized "grand
design" strategies. Such approaches sought to harness nature as a provider of
resources for human consumption, rural welfare, and high economic growth.
During the 1950s, this cognitive map of the relationship between humans and
nature saw articulation among Japanese policy commentators, first in the 1950
Law on Comprehensive National Land Development, and then in follow-up
legislation that focused on "special area development" *(tokutei chi'iki kaihatsu)*.
The end result was a policy framework that endorsed big projects centering on
Japan's river valleys and fundamentally replumbed Japan's hydrosphere to serve
human society.

The Law on Comprehensive National Land Development emerged out of dueling notions of grand design land-use planning. The first of these was national land planning *(kokudo keikaku),* which drew from German geopolitics and experiences during World War II. Students of German land planning in the Cabinet Planning Board (Naikaku Kikakuin) and, after 1943, its successor, the Home Ministry (Naimushō) National Land Bureau (Kokudokyoku), drafted a series of mobilization plans to extract raw material resources from the Japanese islands, Korea, Taiwan, and Manchuria and create an autarkic bloc economy.[10]

Dams helped energize these mobilization plans, and several large hydroelectric generation facilities operated in Korea and Manchuria by the early 1940s. Far more than their counterparts in other empires, Japanese dam builders lavished money on colonial projects that drew on American technology gleaned from 1930s observations of the Boulder Dam, Grand Coulee Dam, and the Tennessee Valley Authority (TVA). Most famous among these undertakings was the Sup'ung (Suihō) Dam, which authorities in Korea and Manchuria intended to be the first in a series of dams to subdue the unpredictable Yalu River, which separated the two colonies. Despite infighting between the colonial governments and numerous on-site mishaps, the Japan Nitrogenous Fertilizer Company (Nihon Chissō Hiryō Kabushiki Kaisha, or Nitchitsu), the construction firms Nishimatsu Gumi and Hazama Gumi, and thousands of coerced Korean and Chinese laborers managed to plan and construct what was then the world's second-largest gravity concrete dam by 1941. The comprehensive development experiences that the Japanese firms and other colonial profiteers carried away from World War II crucially informed them when they engineered high growth at home and abroad in the postwar years.[11]

In January 1945, because of declining fortunes in the war and the Allied blockade of colonial resources, the National Land Bureau shifted its focus to developing eight self-sufficient regional blocs in the four main Japanese islands. Each bloc was to plan a mobilization of local food and resources in time for an anticipated Allied invasion in October. That invasion never came, but just one month after Japan's surrender, the bureau resumed issuing comprehensive plans aimed at rebuilding infrastructure and accommodating repatriated soldiers and civilians from the wartime empire. Gone was the language of geopolitics, replaced by Potsdam-friendly phrasing like "peace industry" *(heiwa sangyō),* yet the plans retained key wartime themes, such as the intensive use of "national land" to ensure Japanese self-sufficiency.[12] The dissolution of the Home Ministry by the Supreme Commander for the Allied Powers (SCAP) in late 1947 stymied the National Land Bureau's efforts, but in 1948 the new Ministry of Construction inherited the former bureau's administrative authority over national land planning. The new ministry ran up against a powerful bureaucratic rival.

Offering a second grand design for postwar Japan's relationship to the natural environment was the Economic Stabilization Board (Keizai Antei Honbu), which backed American New Deal models of comprehensive development.[13] The board was then at the peak of its power as an economic general staff, and the 1947–48 Socialist cabinet of Katayama Tetsu and American SCAP personnel keenly promoted it as a "democratic" counterweight to the old Home Ministry.[14] During the war, America's TVA had already risen to become Japan's most famous overseas model of a public corporation steering comprehensive rural development with dams, irrigation, public facilities, afforestation, and industrial relocation schemes.[15] The Tohoku Development and Promotion Program (Tōhoku Shinkō), a wartime attempt to relieve poverty in northeastern Japan by comprehensively developing strategic raw material and electrochemical industries, even included a public electric power corporation modeled on the TVA. Between 1936 and 1941, the Tohoku Development and Power Company (Tōhoku Shinkō Denryoku Kabushiki Kaisha) managed to construct eleven modestly sized hydroelectric stations and unify the northeastern power grid despite a shrinking share of the wartime budget.[16] After 1945, the TVA model surged in popularity as a means of nurturing democratic society. Public policy discussions during the Allied Occupation (1945–1952) focused on Japan's river valleys for several reasons. Perhaps the most pressing of these was the frequency of floods, which were exacerbated by riverbed silting from excessive lumbering during and immediately after the war. Another was that hydroelectric plants were among the only reliable sources of energy left after the American bombardment of thermal generation facilities in urban centers.[17]

In response to those conditions, as well as the need to boost living standards, public intellectuals connected to the Economic Stabilization Board, men like Tsuru Shigeto, Ōkita Saburō, and Aki Kōichi, established the TVA Studies Discussion Group (TVA Kenkyū Kondankai) and investigated the possibility of scaling down the TVA to the geographic conditions of the Tadami River valley in northeastern Honshu.[18] In 1949, one of the members also produced a widely read translation of former TVA director David Lilienthal's manifesto-like *TVA: Democracy on the March*.[19] The message Tsuru, Ōkita, Aki, and others sought to convey through their published research was that state-guided, comprehensive hydrologic management could bring modern rationality to the ravaged landscapes of Japan, much as the TVA did to the depressed Tennessee Valley of the 1930s. They also believed TVA-style planning would democratize Japan by encouraging grassroots participation in the development process.[20] With members of the discussion group like Ōkita and Aki serving on the Resources Council (Shigen Chōsakai) of the Economic Stabilization Board, the TVA model shaped the comprehensive development idea after 1945.[21]

American advisers also played crucial supporting roles in selling the TVA concept to Japan. David Lilienthal briefly visited Japan and met with Prime Minister Yoshida Shigeru in March 1951, although he had little direct policy impact.[22] However, American SCAP personnel in Occupied Japan did. SCAP's Natural Resources Section maintained a particularly close relationship with the Resources Council in the Economic Stabilization Board, and for Ōkita, Aki, and others, it sponsored study missions that included visits to TVA sites.[23] The Harvard resource geographer Edward Ackerman served as special adviser to the Natural Resources Section, met regularly with the TVA Studies Discussion Group, and presented modified TVA-style development plans as Japan's only hope for postwar reconstruction.[24] The overall emphasis of American occupation policy amid an intensifying Cold War shifted away from "demilitarization" toward Japanese economic independence, with an insistence on balanced budgets and self-reliance in resources through careful, comprehensive planning.[25]

In 1949, the Economic Stabilization Board formed the Comprehensive National Land Development Deliberative Council (Kokudo Sōgō Kaihatsu Shingikai), which became a forum where national land planning and TVA models vied for preeminence in creation of a new national land-use plan.[26] Eventually, elements from both grand designs informed the June 1950 Law on Comprehensive National Land Development. The law's stated purposes were "the comprehensive utilization, development, and conservation of national land through a unified viewpoint on economic, social, and cultural policies; the appropriate planning of industrial location; and the improvement of social welfare." All of these ambitious goals were to be accomplished while "taking into account the natural conditions of national land."[27] The law would serve as the basis for subsequent comprehensive development plans focused on energy and natural resource development, disaster prevention, river management, urban-rural relations, and industrial location. These plans fell under four broad categories: nation (zenkoku), prefecture (tofuken), region (chihō), and "special area" (tokutei chi'iki).[28] By including special areas—which encompassed entire watersheds in the manner of the TVA—as a focus of development planning, the Law on Comprehensive National Land Development absorbed the Economic Stabilization Board's model. Yet as its very name indicated, it retained much of the language of the Ministry of Construction's older national land planning model.

The Law on Comprehensive National Land Development appeared to be a synthesis of Occupation-era thought on land use, but as comprehensive development materialized in the 1950s, it became clear that special area planning superseded all other levels and that it usually meant construction of large, multipurpose dams. Development of a national-level plan based on the law was delayed until 1962. Numerous factors retarded implementation. The conservative Yoshida Shigeru, who had long regarded the Economic Stabilization Board with antipathy, finally gutted it in August 1952.[29] The succeeding body, the Economic Deliberation Agency (Keizai Shingichō), was smaller and had no authority to coordinate comprehensive development planning. Even before the dissolution of the board, the special procurements boom triggered by the Korean War lessened the anxiety regarding Japan's economic future that had driven much interest in comprehensive development. Accordingly, the eagerness to integrate each of the four levels of land planning in the 1950 law diminished.[30] Perhaps most crucially, many of the Diet members who voted for the law in the first place calculated in terms of using special area plans to funnel pork to their home districts. Special area development wound up as a joint venture between the Ministry of Construction and the Construction Committee of the Diet, which collaborated in their negotiations with local authorities to devise a list of target locations.[31]

By the end of the 1950s, the scope of special area projects exceeded anything that Tsuru Shigeto and the TVA Studies Discussion Group ever planned. After almost two years of deliberations, the Japanese government designated nineteen special areas for comprehensive development in December 1951. It added three more areas in 1957, and all but one of these twenty-two locations obtained official cabinet approval by 1958. Together, the approved special areas accounted for about 30 percent of Japanese territory, as well as 29 percent of the national population. They existed in every rural part of the country except Hokkaido, which developed under separate regulations (see table 6.1).[32]

Although the Economic Stabilization Board had come up with an original list of fifty-one candidate areas across forty-two prefectures, those that already had some infrastructural work in progress rose to the top of the competition. Evaluators in the Diet and Ministry of Construction also thought in terms of anticipated cost-effectiveness, and especially in terms of the candidate areas' potential for hydraulic development. Thirteen of the special areas contained an estimated 68 percent of Japan's undeveloped hydroelectric potential.[33] As noted earlier, hydraulic development and multipurpose dams assumed central importance in the world of early postwar planning because of the immediate problems of power shortages and flooding. However, as economic growth gathered pace in the 1950s, promoters of hydraulic development marketed dam projects as a means of preserving water supplies from Japan's cascading rivers for agricultural and industrial use. They also asserted that dams would promote social development in Japan's mountainous interior because the construction process would entail infrastructural improvements, and the reservoirs behind the dams would become steady sources of tourist income.[34]

TABLE 6.1. Designated "special areas" under 1950 Law on Comprehensive National Land Development

	Prefecture(s)	Cabinet approval date	Land area (km²)	Population (1955 census)	Approved project cost (million ¥)
Kitakami	Iwate, Miyagi	Feb. 6, 1953	13,422	1,890,961	66,389
Ani-Tazawa	Akita	Oct. 16, 1953	2,038	75,537	7,669
Mogami	Yamagata	Oct. 16, 1953	2,453	218,053	7,069
Tenryū-Higashi Mikawa	Aichi, Nagano, Shizuoka	June 11, 1954	7,912	1,870,503	84,645
Daisen-Izumo	Okayama, Shimane, Tottori	June 11, 1954	4,024	840,275	18,488
Aso	Kumamoto, Oita	June 11, 1954	3,073	343,709	17,221
Minami-Kyushu	Miyazaki, Kagoshima	June 11, 1954	5,541	979,191	24,174
Kita-Kyushu	Fukuoka	Aug. 3, 1954	2,314	2,166,798	74,992
Noto	Ishikawa	Aug. 23, 1955	2,225	416,410	8,557
Shikoku-Seinan	Ehime, Kochi	Aug. 23, 1955	4,535	645,213	8,816
Geihoku	Hiroshima	Nov. 18, 1955	1,228	65,804	7,569
Nishikigawa	Yamaguchi	Nov. 18, 1955	1,767	479,665	12,608
Tadami	Fukushima, Niigata	Mar. 6, 1956	4,458	239,202	47,221
Hietsu	Gifu, Toyama	Mar. 6, 1956	5,608	885,875	85,874
Kiso	Aichi, Gifu, Mie, Nagano	Mar. 23, 1956	12,180	4,530,977	140,426
Nakagawa	Tokushima	Mar. 23, 1956	1,724	212,491	8,604
Yoshino-Kumano	Mie, Nara, Wakayama	Oct. 5, 1956	5,017	408,224	59,056
Tone	Chiba, Gunma, Ibaraki, Saitama, Tochigi, Tokyo	May 10, 1957	17,326	7,226,866	158,976
Towada-Iwakigawa	Aomori	Oct. 24, 1958	3,466	569,600	20,094
Kita-Ōu	Akita, Aomori, Iwate	Oct. 24, 1958	12,533	1,248,915	87,360
Senen	Miyagi	Oct. 24, 1958	1,631	578,828	49,386
Tsushima	Nagasaki	—	unknown	unknown	—
Total (excluding Tsushima)	36 of 47 prefectures (46 prior to 1972 Okinawa reversion)	—	114,475 (30.96% national land area)	25,893,097 (29.00% national population)	995,194

SOURCE: Satō Atsushi, Nihon no chiiki kaihatsu (Miraisha, 1965), 79–80.

121

The Law on Comprehensive National Land Development, originally conceived as a basis for a limited array of integrated land-use plans, in actuality became the legal basis for damming almost all of Japan's watersheds. "Comprehensive development" became shorthand for the multipurpose dams that began to proliferate in rural Japan as hydrologists attempted to transform the unmanageable natural flows of rivers into manageable resources for economic growth.

▦ Special Area Comprehensive Development and the Sakuma Dam

Having sketched the evolution of comprehensive development ideology, I next explore the unforeseen but damaging consequences of its implementation in the case of the Sakuma Dam project, an immense operation that stretched across the Tenryū River and the border between Aichi and Shizuoka prefectures (figures 6.2 and 6.3). Begun in 1953 and completed just three years later, to many Japanese Sakuma stood as a monument to restored self-confidence after military defeat. In 1956, it was the largest dam ever constructed in Japan, the tenth largest in the world, and the first major undertaking of the Electric Power Development Company (Dengen Kaihatsu Kabushiki Kaisha) as a public corporation charged with overseeing the comprehensive development of a major river valley. The engineering feats at Sakuma were impressive in light of the technologies of the day, and Sakuma's turbines still generate one of the largest amounts of hydroelectricity in Japan, with a maximum output of 350,000 kilowatts.[35] Yet it has become apparent almost sixty years later that the dam has functioned more effectively as a supplier of electric power to the Tokyo and Nagoya metropolitan areas than as a TVA-style engine of rural revitalization: although the Sakuma Dam project was certainly big, it was not comprehensive in the spirit of early postwar policy ideology.

Public and private sector interest in hydraulic development along the swift-flowing Tenryū River stretched back to the 1920s. However, geography and geology posed formidable challenges to engineers working for prewar Japanese companies that employed low-level technology, epitomized by straw baskets and trolleys. This was particularly true at the Sakuma Dam site, long estimated as capable of producing a full one-third of the Tenryū River valley's hydroelectric potential of approximately 1 million kilowatts. The high-volume seasonal floods of the Tenryū River severely constrained the amount of time available for crews to construct the channel around the dam site to divert river water temporarily. Sheer, brittle banks and heavy riverbed sedimentation, the result of the river's position along Japan's largest onshore fault, the Median Tectonic Line, made the prospect of dam construction all the more formidable. In addition to these

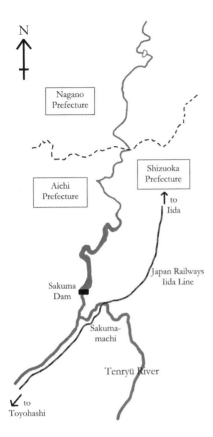

N

Nagano
Prefecture

Shizuoka
Prefecture

Aichi
Prefecture

↑ to
Iida

Japan Railways
Iida Line

Sakuma
Dam

Sakuma-
machi

Tenryū River

to
Toyohashi

FIGURE 6.2. Regional map of the Sakuma
Dam site, Aichi, shortly after completion
of the project, 1956

Map by Brian Burns

natural obstacles, the destitution of the early postwar years made it prohibitively expensive for either the state electric power monopoly, Nippon Hassōden, or any of the nine privatized electric companies that succeeded it in 1951 to undertake the Sakuma project.[36]

The situation surrounding the hydraulic development of the Tenryū River changed quickly in the early 1950s. Through the convoluted selection process detailed above, the Sakuma site in December 1951 became part of the officially designated "Tenryū-Higashi Mikawa Special Area" under the Law on Comprehensive National Land Development, which meant that it would be eligible for public development subsidies. The Ministry of Construction, the Ministry of Agriculture and Forestry, and the governments of Shizuoka, Aichi, and Nagano prefectures drew up a comprehensive ten-year plan for the special area. It entailed electric power development centered on construction of the Sakuma Dam, forestry operations, land reclamation and irrigation, flood control, and prevention of silting in the lower reaches of the Tenryū.[37] Shortly

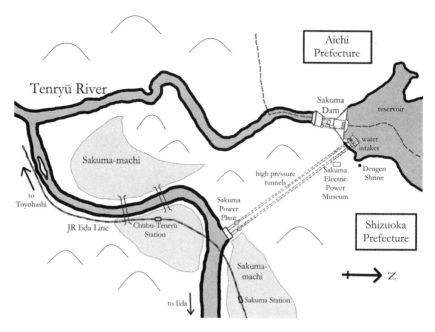

FIGURE 6.3. Local map of the Sakuma Dam site, Aichi, shortly after
completion of the project, 1956

Map by Brian Burns

after its creation in September 1952, the Electric Power Development Company
(EPDC) stepped into this framework and took on the Sakuma Dam project as
its first major operation.

Expediency factored into EPDC's decision as much as, or even more than,
long-term comprehensive development goals. Among the top reasons were the
Sakuma Dam site's proximity to sand and gravel for mixing concrete; its unusually easy access to Japan National Railways, in this case the Iida Line; the relatively
low number of homes, fields, and forestland that the Sakuma reservoir would
flood; and in a time of public exasperation over recurring electricity shortages,
the shorter time the Sakuma Dam would take compared with other large dam
projects in EPDC's plans.[38] In view of these circumstances, the Sakuma project
was ready to proceed after nearly thirty years of anticipation.

The leadership of the Sakuma Dam project shaped how the construction process
unfolded. The well-connected EPDC president, Takasaki Tatsunosuke, had lived
in the United States in the 1910s. By virtue of his American experience, Takasaki
could comfortably work in English-speaking environments. Further, by virtue

of his service as president of Manchurian Heavy Industries (Manshū Jūkōgyō) during World War II, he was well acquainted with public corporations and state-guided hydraulic development projects.[39] Based on his knowledge of Japanese construction technology in the early 1950s and the demands of the Sakuma Dam site, he concluded that damming the Tenryū would require American financial and technical assistance. Takasaki was instrumental in securing a three-year loan from the Bank of America, which covered approximately 6.5 percent of the Sakuma Dam's ¥39 billion price tag. He also brought in twenty-six technical advisers from the Guy F. Atkinson Company, an American firm with an extensive portfolio of large dam projects, including the Grand Coulee Dam.[40] Hazama Gumi took on the hazardous work of building the Sakuma Dam, and it applied engineering expertise acquired from wartime projects in Japan and the colonial empire, such as Sup'ung Dam. It also had become accustomed to working with American advisers and techniques after 1945, when it helped set up U.S. base facilities in Okinawa.[41] Thus, the firms leading the effort to redirect the Tenryū drew from extensive experiences in continental Asia and the United States.

The construction of Sakuma Dam itself marked a major turning point in the history of public works technology. Imported American machinery expedited the construction process, which the Hazama Gumi, Kumagai Gumi, and Atkinson companies spearheaded. The latest power shovels, bulldozers, cable cranes, dump trucks, and other technologically advanced machinery converged on the small mountain village of Sakuma-mura in April 1953 and tore into the landscape to prepare the dam site. Concrete pouring began in January 1955, and Hazama Gumi thrilled the Japanese construction industry by setting a new world record—5,180 cubic meters poured per day in early 1956.[42] At the time of its completion in October 1956, the concrete gravity dam stood at 155.5 meters with a reservoir capacity of about 330 million cubic meters, as well as a hydroelectric plant that could produce the equivalent of about 2.3 percent of all electric power generated in Japan at the time.[43] After the Sakuma Dam project, Japanese manufacturing firms indigenized the imported American technology and established a domestic construction machinery industry that helped build such 1960s monuments as the Kurobe Dam and the *shinkansen* "bullet train" network.[44] The Sakuma project also revolutionized dam construction and concrete mixing techniques, thus catalyzing the spread of large dams across Japan in the ensuing decades.[45]

The dam profoundly transformed life in the built environment of Sakuma-mura. It generated a short-lived "dam boom" that drew thousands of laborers, managers, vendors, and service providers from outside regions. The Sakuma district's population swelled from 4,861 at the outset of construction in 1953 to a historic peak of 10,619 in 1955, and the male-female ratio shifted from 1:1 to almost 2:1.[46] A boomtown atmosphere pervaded the area as many new arrivals as well as recipients of EPDC compensation payments for relocation away from the dam site dumped sizable amounts of cash in the village. Neon-lit shops, electronics stores, perm salons, pachinko parlors, a movie theater, and other modern spectacles proliferated along village streets. A small but bustling red-light district *(akasen)* expanded to serve the predominantly single workmen.[47]

Although Sakuma-mura saw limited industrial development prior to the 1950s, including a mine and a paper mill, the dam and the relocation of population forced a rapid reconfiguration of the village's employment structure away from agriculture and silviculture. Moreover, by blocking the flow of the Tenryū River, Sakuma and other dams destroyed long-established timber rafting *(ikada nagashi)* and commercial fishing industries, and EPDC had to pay special compensation to the structurally unemployed.[48] Census statistics indicate that although the primary sector accounted for over half of local employment, at 53.6 percent, in 1950, that percentage plummeted to 20.6 percent during the Sakuma Dam's construction and only partially recovered to 32.0 percent by 1960 before collapsing to 9.7 percent in 1990. This trend away from primary sector employment became increasingly common in rural Japan during the postwar years, but the dam project accelerated it by flushing many families out of their farms or forestland and pushing them into secondary or tertiary sector work tied to outside regions.[49]

Those dramatic changes in daily life elicited an array of responses from residents of Sakuma-mura and the surrounding area. On the positive side, the village's mayor, Kitai Mineo, embraced the coming of the dam as an opportunity for local "enlightenment." Kitai firmly believed in the power of TVA-style development as a means of bringing modernity, democracy, and industrial prosperity to his region. To further his beliefs at the community level, he circulated a hortatory newsletter that touted the progressive effects of the Sakuma project.[50] Many locals marveled at the high living standards of Tokyoite and American managers from EPDC and the Atkinson Company, and in conjunction with the New Life Movement of the 1950s, the Sakuma-mura Women's Association sponsored "Atkinson home visits" designed to modernize members' thinking about domestic life.[51] Haru Matsukata relayed this story of cultural exchanges and differential living standards to American readers of *The Saturday Evening Post*, proclaiming, "In a remote Japanese river gorge, where blue eyes and blond hair startle the indigenes, a few isolated American families run one of the happiest deals in the Far East."[52]

Among negative reactions, villagers regarded the lower-level construction workers as menacing, prone to violence, unhygienic, and lascivious. Women's associations in Sakuma-mura and across Shizuoka Prefecture mobilized against a perceived moral panic brought on by the rootless male workers, especially their patronage of brothels in Sakuma's red-light district. This campaign led to a prefectural antiprostitution ordinance, a curfew for women in Sakuma-mura, and a PTA-enforced ban for children on walking to school by way of the red-light district.[53] Another area of friction between villagers and the dam project concerned noise pollution from the construction site, and the Sakuma-mura Board of Education went so far as to petition the Ministry of Education in 1954 over the deleterious effects of the noise on children's education.[54] Clearly, not everyone saw eye-to-eye with Mayor Kitai on the "enlightening" experience of dam construction.

Despite those noteworthy local tensions, no organized resistance against the Sakuma Dam itself ever materialized, even as a bitter compensation dispute over another EPDC project, the Tagokura Dam in Fukushima Prefecture, saw wide coverage in the national media.[55] Indeed, great fanfare inside and outside the region greeted Sakuma's opening in 1956, which to many Japanese signaled a new postwar spirit where technology would bend the workings of nature to serve human society. Iwanami Productions launched a three-part film series documenting the construction process from 1954 to 1957, and audience figures totaled 3 million for the first installment, 2.5 million for the second, and 620,000 for the third. October 1956 saw the issuance of commemorative Sakuma Dam postage stamps, which depicted the dam soaring majestically over the Tenryū River valley and which the government marketed especially to elementary school children. The fanfare's climax came during a formal dedication ceremony in 1957 featuring Emperor Hirohito and Empress Nagako: the imperial couple surveyed the new facilities and composed memorial poems to the ninety-six workers who had died while building the dam.[56] Overseas readers were treated to superlative-laden English-language articles that celebrated the Sakuma Dam as the "TVA of Japan."[57]

Even after the commemorations had passed, promoters of Sakuma worked to make it into a tourist destination. Bus tours and school groups visited in the years following its completion, and a tourist boat line plied the waters of the reservoir.[58] EPDC built the Sakuma Electric Power Museum, which still perches on a hill overlooking the dam, provides an observation deck, and narrates an affirmative technological history of the project. It added the Sakuma Dengen Jinja, a small Shinto shrine dedicated to hydroelectric generation, along the road leading up to the museum. Hikers and campers have visited Sakuma Dam while exploring the surrounding Tenryū-Okumikawa Quasi-National Park, established in 1969. Recreational fishing has drawn a steady stream of weekend anglers who populate the banks of the Tenryū below the Sakuma reservoir in search of *ayu* (sweetfish) and nonnative rainbow trout. The promotion of recreational fishing has required routine fish stocking because the dam has blocked the natural flow of the Tenryū River.[59] Thus, since the late 1950s, the dam has attracted moderate numbers of visitors who have beheld the transformation of a turbulent river into a predictable power source.

◼ Rural Development

Though highly successful as a generator of electricity and modestly successful as a tourist draw, the Sakuma Dam failed to revitalize the local area over the long term: the "dam boom" of the 1950s proved ephemeral. This shortcoming is important to note because promoters of comprehensive development highlighted rural revitalization and self-reliance when marketing the land-use approach to the public. At the time of the dam's construction, EPDC took pride in its TVA-inspired model of relations with local communities affected by the project. A company history from the 1980s continued to relate how EPDC personnel broke with earlier heavy-handed practices when negotiating compensation for local communities. Among other policies, EPDC held open discussions with each village in the construction zone, and it consulted displaced persons on moving and finding new employment. It furthermore spent ¥1.7 billion on the construction of public works in the dam's vicinity and elsewhere in the Tenryū-Higashi Mikawa Special Area.[60]

Although such policies appeared to be the very essence of the comprehensive development idea, other studies of the Sakuma Dam region during and after the construction have indicated less favorable economic outcomes. Foremost among issues affecting the Sakuma region, not to mention most of rural Japan after the 1950s, has been its depopulation. The *Asahi shinbun* reported in 1975 that the population of Sakuma-mura since the mid-1950s had dropped 60 percent and was rapidly graying.[61] Indeed, after peaking above 10,000 during the dam boom of 1953–1956, the population of Sakuma-mura fell to 3,194 in 1975, and to 1,835 in 2000.[62] Those leaving the village were disproportionately young people who took up industrial or service sector employment in nearby cities like Hamamatsu and Toyohashi, leaving locals hard-pressed to maintain the area's cypress and cedar forests. The improved transportation infrastructure that accompanied comprehensive development in the Tenryū Valley only accelerated this outmigration.[63] Partly as a consequence, the formerly autonomous villages of the Sakuma region were amalgamated into much larger urban entities. Shortly after completion of the dam in 1956, Sakuma-mura joined with neighboring villages to become the town of Sakuma-machi. Subsequently, Sakuma-machi became Sakuma-chō, part of the larger Tenryū-ku ward of Hamamatsu City.

What is more, by the 1960s and 1970s, the Sakuma region surrendered its economic self-reliance, as well as its natural relationship with the Tenryū Valley. The Sakuma Dam, its hydroelectric generation plant, its frequency conversion station, and related public works provided important sources of local tax revenue, but the benefits they generated accrued primarily to large property owners and depreciated over time. Owners of forestland, for instance, steered a significant percentage of EPDC's compensation funding toward the construction of little-used access roads for logging.[64] For most others, the dam's construction and the submergence of farmland and forests by the reservoir meant a restructuring of socioeconomic life, away from locally oriented agriculture and other primary sector jobs. In their place came jobs linked to the national construction industry, or to small-scale manufacturers in textiles, light electronic goods, and automobile parts. Labor shortages during the 1960s pushed these manufacturers into the Sakuma region, and they employed women on a part-time basis, as opposed to more secure full-time employment for either men or women. Young men saw few new employment opportunities, and they sought work elsewhere, accelerating outmigration.[65]

Comprehensive development altered the human communities around Sakuma Dam, but it failed to revitalize them as boosters had envisioned. Moreover, as the sociologist Machimura Takashi noted through polls and interviews conducted in 2002, residents no longer celebrated the dam unambiguously; rather, most paid it only passing attention and regarded it as one of many dated concrete features in their landscape. Enthusiasm for comprehensive development, alongside other fixtures of the early postwar, faded.[66]

▨ Ecological Costs

Less immediately apparent than regional depopulation but no less relevant have been the ecological consequences of Sakuma Dam's construction. The 2002 poll from Machimura indicated that many residents of the Sakuma region lamented the loss of the free-flowing Tenryū River, particularly those who remembered life before comprehensive development. Whereas younger residents tended to view continued, albeit attenuated, downstream flows as evidence of the river's good health, many elderly residents regarded Sakuma Dam as the "death" of the Tenryū.[67] Evidence over the nearly sixty years since the dam's completion has shown that the elder residents were right to be concerned. The short-term gains of multipurpose dam construction under the Law on Comprehensive National Land Development had significant long-term ecological costs.

One of the knottiest problems to confront dam operators and Tenryū Valley residents has been the accumulation of driftwood and refuse behind the dam wall (figure 6.4). This problem is common to many hydroelectric dams in forested areas, but the immensity of the dam wall, 34-kilometer length of the reservoir, and geographic conditions along the river have made it acute at Sakuma. From its headwaters at Lake Suwa in Nagano Prefecture, the Tenryū travels 138 kilometers before reaching Sakuma Dam. Illegally disposed industrial waste and household refuse from human settlements along the route have routinely

FIGURE 6.4. Accumulated driftwood on the Sakuma reservoir, June 2013
Photograph by author.

wound up in the Sakuma reservoir. After storms, snowmelts, and high-water season, downed trees and fallen branches have rolled down the sheer banks of the Tenryū Valley, mixed with refuse in the river, and accumulated in large floats behind the dam wall. Whereas timber was once rafted down the free-flowing Tenryū, humans since the 1950s have struggled to manage these masses of driftwood, which can obstruct the discharge of water through sluice gates. Sakuma operators once let driftwood flow through the gates, but understandable protests from downstream fisheries in the early 1980s compelled them to keep the logs behind the dam wall. By the 1990s, 25,000 cubic meters of driftwood had piled up.[68] The driftwood has begun to alter the ecology of the reservoir. A power station chief in 2006 wryly noted in an interview that the accumulations appeared as distinct "biotopes" because some were old enough to sprout grasses.[69]

Sakuma Dam's operators in J-Power, the privatized successor to EPDC, have developed methods to manage the driftwood buildup. In the 1990s, dam workers installed equipment to make salable wood chips for plywood manufacturers and mushroom cultivators, as well as composting facilities for saplings, leaves, and other organic matter unsuited to chip production. Although chip production reportedly was successful and reasonably efficient by the 2000s, workers at Sakuma had to spend long hours separating usable logs from saturated sinker logs and refuse, which typically included discarded tires, PET bottles, cans, Styrofoam, and plastic packaging and needed to be sent to an outside contractor for proper incineration.[70] As a consequence, the chipping operations have not been commercially sustainable, even though they have featured prominently in "greenwashing" exhibits online and at the Sakuma Electric Power Museum.

Reservoir siltation and coastal erosion near the mouth of the Tenryū River provided two further, interlinked predicaments generated by construction of Sakuma and other nearby dams. Before the dam, the Tenryū was fast-flowing and its current carried a considerable amount of alluvium. From the start planners sought to end sedimentation in the lower reaches of the river by constructing Sakuma and other large dams to filter out river muck.

They succeeded too well: by the 1990s, the Sakuma Dam was one-third silted up, and other Tenryū River dams had reached even higher levels of siltation. Measurements from 2005 recorded 117.2 million cubic meters of sediment in the Sakuma reservoir.[71] Such siltation ultimately will impair effective production of hydroelectricity, but more immediately, the siltation of the Sakuma and its associated dams contributed to the disappearance of the famed Enshūnada Coast near Hamamatsu. Tenryū sediments once replenished the Enshūnada beaches, but the Sakuma Dam has all but halted the river's alluvial flows by trapping nearly 2 million cubic meters of earth each year since 1956.[72] In the mid-2000s, local activists scrambled to rescue the coastline, especially the Nakatajima Dunes, an important breeding ground for endangered loggerhead turtles. Meanwhile, hydraulic engineers attached to J-Power instituted expensive methods for transporting sediment from the Sakuma and other reservoirs to the lower reaches of the Tenryū River. Such methods included flushing upstream sediments through the reservoir and sluice gates, as well as dredging the reservoir bottom with boats and hauling the silt downstream by truck.[73] The effectiveness of these measures remains to be seen.

NOTES

1   Scott O'Bryan, *The Growth Idea: Purpose and Prosperity in Postwar Japan* (Honolulu: University of Hawai'i Press, 2009), 172–77.

2   For an expanded analysis of this important postwar discussion, see Laura Hein, *Fueling Growth: Energy and Economic Policy in Postwar Japan, 1945–1960* (Cambridge, Mass.: Harvard University Council on East Asian Studies, 1990), 159–60; Eric G. Dinmore, "A Small Island Nation Poor in Resources: Natural and Human Resource Anxieties in Trans-World War II Japan" (Ph.D. diss., Princeton University, 2006), chap. 3; and Satō Jin, *"Motazaru kuni" no shigenron—jizoku kanō na kokudo o meguru mō hitotsu no chi* (Tōkyō Daigaku Shuppankai, 2011), chaps. 2–4.

3   See chapters 2, 4, and 5 of the present volume. Flood control is also an important topic in Philip C. Brown, *Cultivating Commons: Joint Ownership of Arable Land in Early Modern Japan* (Honolulu: University of Hawai'i Press, 2011).

4   Indeed, dam construction was already a contentious topic in debates over land and water use. On pre-1945 protests over dam projects, see Michael Lewis, *Becoming Apart: National Power and Local Politics in Toyama, 1868–1945* (Cambridge, Mass.: Harvard University Asia Center, 2000), and Murakushi Nisaburō, *Kokuritsu kōen seiritsushi no kenkyū: Kaihatsu to shizen hogo no kakushitsu o chūshin ni* (Hōsei Daigaku Shuppankyoku, 2005).

5   Japan Commission on Large Dams, *Dams in Japan: Past, Present, and Future* (London: CRC Press, 2009), 182. The International Commission on Large Dams defines "large dams" as those with walls 15 or more meters in height from the foundation, or with walls between 5 and 15 meters in height and a reservoir capacity of more than 3 million cubic meters.

6   Alex Kerr, *Dogs and Demons: Tales from the Dark Side of Japan* (New York: Hill and Wang, 2002), 26. Also see Alex Kerr, *Inu to oni—shirarezaru Nihon no shōzō* (Kōdansha, 2002).

7   World Commission on Dams, *Dams and Development: A New Framework for Decision-Making* (London and Sterling, Virginia: Earthscan Publications, 2000), 370.

8   Gotō Kunio, "The National Land Comprehensive Development Act," in *High Economic Growth Period 1960–1969*, vol. 3, *A Social History of Science and Technology in Contemporary Japan*, ed. Nakayama Shigeru, Gotō Kunio, and Yoshioka Hitoshi (Melbourne: Trans Pacific Press, 2006), 339.

9   John Robert McNeill, *Something New Under the Sun: An Environmental History of the Twentieth-Century World* (New York: W.W. Norton & Company, 2000), 182.

10   Gotō, "National Land," 334–35. For an English-language account of the intellectual context of wartime national land planning, see Janis Mimura, *Planning for Empire: Reform Bureaucrats and the Japanese Wartime State* (Ithaca, N.Y.: Cornell University Press, 2011), especially 191–94.

11   Aaron Stephen Moore, "'The Yalu River Era of Developing Asia': Japanese Expertise, Colonial Power, and the Construction of Sup'ung Dam," *Journal of Asian Studies* 72, no. 1 (February 2013): 115–39. Also see Aaron Stephen Moore, *Constructing East Asia: Technology, Ideology, and Imperialism in Japan's Wartime Era, 1931–1945* (Stanford, Calif.: Stanford University Press, 2013).

12   Satō Atsushi, *Nihon no chi'iki kaihatsu* (Miraisha, 1965), 42–43; Mikuriya Takashi, *Seisaku no sōgō to kenryoku—Nihon seiji no senzen to sengo* (Tōkyō Daigaku Shuppansha, 1996), 229–30.

13   Shimokōbe Atsushi, *Sengo kokudo keikaku e no shōgen* (Nihon Keizai Hyōronsha, 1994), 44–45.

14   Miwa Ryōichi, "Reorganization of the Japanese Economy," in *A History of Japanese Trade and Industry Policy*, ed. Sumiya Mikio (Oxford: Oxford University Press, 2000), 196–97.

15   For a detailed history of the TVA model in wartime and early postwar Japan, see Eric G. Dinmore, "Concrete Results? The TVA and the Appeal of Large Dams in Occupation-Era Japan," *Journal of Japanese Studies* 39, no. 1 (Winter 2013): 1–38. For briefer treatments in Japanese, see Machimura Takashi, *Kaihatsushugi no kōzō to shinshō—sengo Nihon ga damu de mita yume to genjitsu* (Ochanomizu Shobō, 2011), 42–52; Miyata Ichirō, "'Risō tsuikyū e no hi'—TVA shisō, minshuka, so shite jiritsu," in *Kaihatsu no jikan, kaihatsu no kūkan—Sakuma damu to chi'iki shakai no hanseiki*, ed. Machimura Takashi (Tōkyō Daigaku Shuppankai, 2006), 29–50; and Satō Jin, "Motazaru kuni" no shigenron, 83–93.

16   Iwamoto Yoshiteru, *Tōhoku kaihatsu 120-nen* (Tōsui Shobō, 1994), 95–96, 105–19.

17   Gotō, "National Land," 339.

18   See Tsuru Shigeto, *Ikutsu mo no kiro o kaiko shite: Tsuru Shigeto jiden* (Iwanami Shoten, 2001), esp. 217, 255–56.

19 See David E. Lilienthal, *TVA: Minshushugi wa shinten suru*, trans. Wada Koroku (Iwanami Shoten, 1949).

20 Mikuriya, *Seisaku no sōgō to kenryoku*, 232. The TVA, of course, has been criticized for its *lack* of meaningful grassroots participation. See Michael J. McDonald and John Muldowny, *TVA and the Dispossessed: The Resettlement of Population in the Norris Dam Area* (Knoxville: University of Tennessee Press, 1982). Japanese proponents of the TVA model tended to take its claims of grassroots participation at face value.

21 Mikuriya, *Seisaku no sōgō to kenryoku*, 231.

22 See David Lilienthal, *The Venturesome Years, 1950–1955*, vol. 3, *The Journals of David Lilienthal* (New York: Harper & Row, 1966), 117–29.

23 Memorandum for Record: Study Abroad of Japanese Technicians Connected with the Resources Committee and Students Interested in Resources Utilization and Planning, October 7, 1948, National Archives and Records Administration, College Park, Md., RG 331 (Supreme Commander for the Allied Powers), Box 9132, Folder #7.

24 Edward A. Ackerman, *Japan's Natural Resources and Their Relation to Japan's Economic Future* (Chicago: University of Chicago Press, 1953), 559–65, 574.

25 Shimazaki Minoru, "Dengen kaihatsu sokushin hō: Sakuma damu no baai," *Jurisuto* 533 (1973): 61.

26 Okada Tomohiro, *Nihon shihonshugi to nōson kaihatsu* (Kyoto: Hōritsu Bunkasha, 1989), 270.

27 Ministry of Land, Infrastructure, Transport, and Tourism, "Kokudo Sōgō Kaihatsu Hō (Law on Comprehensive National Land Development)," Article 1, Clause 1, http://www.kokudokeikaku .go.jp/document_archives/ayumi/11.pdf (accessed February 13, 2011).

28 Okada, *Nihon shihonshugi to nōson kaihatsu*, 270–71.

29 Chalmers Johnson, *MITI and the Japanese Miracle* (Stanford, Calif.: Stanford University Press, 1982), 220.

30 Gotō, "National Land," 337.

31 Mikuriya, *Seisaku no sōgō to kenryoku*, 233.

32 Satō, *Nihon no chi'iki kaihatsu*, 79–80.

33 Okada, *Nihon shihonshugi to nōson kaihatsu*, 271–72.

34 Gotō, "National Land," 339–40.

35 Okamoto Takuji, "The Reconstruction of the Electric Power Industry," in *Road to Self-Reliance 1952–1959*, vol. 2, *A Social History of Science and Technology in Contemporary Japan*, ed. Nakayama Shigeru, Gotō Kunio, and Yoshioka Hitoshi (Melbourne: Trans Pacific Press, 2005), 432; Japan Commission on Large Dams, *Dams in Japan*, 64–65.

36 Dengen Kaihatsu Kabushiki Kaisha, *Denpatsu 30-nenshi* (Dengen Kaihatsu, 1984), 80; Kikuchi Koichirō, Muranaga Mineo, and Itagusu Katsukuni, "Sakuma damu no taisha jōkyō to taisaku," *Denryoku doboku* 291 (2001): 41–42.

37 Nihon Jinbun Kagakukai, *Sakuma damu: Kindai gijutsu no shakaiteki eikyō* (Tōkyō Daigaku Shuppankai, 1958), 24.

38 Dengen Kaihatsu, *Denpatsu 30-nenshi*, 80–82.

39 Hazama Gumi Hyakunenshi Hensan Iinkai, *Hazama Gumi hyakunenshi*, vol. 2 (Hazama Gumi, 1989), 205.

40 Okamoto, "Electric Power Industry," 428–29.

41 Kawamura Masami, "Damu kensetsu to iu 'kaihatsu pakkēji,'" in *Kaihatsu no jikan, kaihatsu no kūkan*, ed. Machimura Takashi, 80.

42 *Hazama Gumi hyakunenshi*, vol. 2, 218.

43 Japan Commission on Large Dams, *Dams in Japan*, 64–65.

44　"Sakuma damu no kibo," *Asahi shinbun*, August 31, 1975.

45　*Hazama Gumi hyakunenshi*, vol. 2, 201–203, 218.

46　Kamiyama Ikumi, "Sakuma damu kaihatsu to chi'iki fujinkai katsudō—sengo Nihon ni okeru minshuka to josei," in *Kaihatsu no jikan, kaihatsu no kūkan*, ed. Machimura Takashi, 135.

47　Kamiyama, "Sakuma damu kaihatsu to chi'iki fujinkai katsudō," 136–40.

48　Dengen Kaihatsu Kabushiki Kaisha, *Denpatsu 30-nenshi*, 87; Shizuoka-ken, *Shizuoka kenshi tsūshihen*, vol. 6 (Shizuoka: Shizuoka-ken, 1997), 535–36.

49　Machimura Takashi, "Posuto-damu kaihatsu no hanseiki—chi'iki shakai ni kizamareru Sakuma damu kensetsu no inpakuto" in *Kaihatsu no jikan, kaihatsu no kūkan*, ed. Machimura Takashi, 184–85. As of 2000, the primary sector accounted for only 9.1 percent of employment.

50　Machimura Takashi, "Chi'iki shakai ni okeru 'kaihatsu' no juyō—dōin to shutaika no jūsōteki katei," in *Kaihatsu no jikan, kaihatsu no kūkan*, ed. Machimura Takashi, 97–101, 114–17.

51　Kamiyama, "Sakuma damu kaihatsu to chi'iki fujinkai katsudō," 140–44.

52　Matsukata, Haru, "They're Taming the Heavenly Dragon," *Saturday Evening Post* (November 19, 1955): 34. Matsukata was working as a U.S.-based reporter before her famous marriage to Edwin O. Reischauer in 1956.

53　Kamiyama, "Sakuma damu kaihatsu to chi'iki fujinkai katsudō," 144–48.

54　Shizuoka-ken, *Shizuoka kenshi tsūshihen*, vol. 6, 536.

55　For example, see "Kojireta Tagokura no hoshō mondai—damu kōji no miokuri mo," *Asahi shinbun*, May 19, 1954, 4.

56　Dengen Kaihatsu, *Denpatsu 30-nenshi*, 90–92. For an extended study of the Iwanami films and "the politics of development imagery," see Machimura Takashi, *Kaihatsushugi no kōzō to shinshō—sengo Nihon ga damu de mita yume to genjitsu* (Ochanomizu Shobō, 2011), chaps. 5–9.

57　"T.V.A. of Japan: Gigantic Sakuma Dam Nearing Completion," *New Japan* 8 (1955): 54–55.

58　See "'Kankōchi' Sakuma damu," *Asahi shinbun*, March 3, 1957; "Sakuma damu ni kankōsen," *Asahi shinbun*, August 7, 1959. No bus route serves Sakuma Dam today.

59　Shizuoka-ken, *Shizuoka kenshi tsūshihen*, vol. 6, 535.

60　Dengen Kaihatsu Kabushiki Kaisha, *Denpatsu 30-nenshi*, 87.

61　"Sakuma damu no kibo," *Asahi shinbun*, August 31, 1975. See also the feature-length "Hitokage no taeta mura—Dengen Kaihatsu būmu no ato ni," *Asahi jānaru*, March 3, 1972: 31–35.

62　Machimura, "Posuto-damu kaihatsu no hanseiki," 176. Sakuma-mura had legally existed as an independent village only since 1889.

63　Machimura, "Chi'iki shakai ni okeru 'kaihatsu' no juyō," 127.

64　Nihon Jinbun Kagakukai, *Sakuma damu*, 481–82; Shimazaki, 65–66. Sakuma is one of three locations in the national power grid where a station converts the frequency from 50 Hz (eastern Japan) to 60 Hz (western Japan).

65　Machimura, "Posuto-damu kaihatsu no hanseiki," 186.

66　Machimura Takashi, "Kioku no naka no Sakuma damu—'bure' to 'nigori' no sōhatsuryoku" in *Kaihatsu no jikan, kaihatsu no kūkan*, ed. Machimura Takashi, 233–58.

67　Machimura, "Kioku no naka no Sakuma damu," 251–52.

68　Tashiro Kiichi, Nakagawa Takeshi, and Hoshino Hitoshi, "Sakuma damu ni okeru ryūboku shori," *Denryoku doboku* 263 (1996): 33–35.

69　"Suiryoku hatsuden no kinjitō no 'ima' o tazuneru—Sakuma damu unten kaishi gojū shūnen," *Kōken* 519 (2006): 120. In one extraordinary episode from 1961, heavy rainfalls washed houses, livestock, crops, and a large quantity of driftwood into the Sakuma reservoir. The accumulations covered the entire surface, and provided breeding locations for a tremendous swarm of mosquitoes that plagued the area until the sluice gates were opened a week later. See

Sakakibara Masazumi, "Shōwa 36-nen 6-gatsu Ina dani no shūchū gōu ni yoru kōzui ni ki'in shita Sakuma damu-ko ni okeru ka no daihassei," *Nagano daigaku fūdobyō kiyō* 7:2 (June 23, 1965): 130–41.

70 Tashiro et al., "Sakuma damu ni okeru ryūboku shori," 34–35, 39; "Suiryoku hatsuden no kinjitō no 'ima' o tazuneru," 120.

71 Gavan McCormack, *The Emptiness of Japanese Affluence* (Armonk, N.Y.: M.E. Sharpe, 2001), 46; "Suiryoku hatsuden no kinjitō no 'ima' o tazuneru," 121.

72 Masato Wada and Shuzo Shikano, "Famed Shizuoka Dune Doomed by Rising Ocean Levels, Waves," *Japan Times*, August 4, 2006.

73 Motoyuki Inoue, "Promotion of Field-Verified Studies on Sediment Transport Systems Covering Mountains, Rivers, and Coasts," *Science and Technology Trends Quarterly Review* 33 (October 2009); Kikuchi et al., 44–45; "Suiryoku hatsuden no kinjitō no 'ima' o tazuneru," 121. Trucks also have transported dredged sediments to asphalt producers and golf courses.

PART III

Life:
Flora, Fauna, Fertilizer

7

Japan as an Organic Empire

Commercial Fertilizers, Nitrogen Supply, and Japan's Core-Peripheral Relationship

TOSHIHIRO HIGUCHI

All life forms need various chemical elements for existence. The overall volume, distribution, and flow of these nutrients, especially nitrogen (N), phosphorus (P), and potassium (K), determine the basic characteristic of biomass in the terrestrial and aquatic environments in a given location. Because humans have long used plants and animals for food, feed, fiber, and other products, biogeochemical cycles have always shaped the course of human civilization. Human activities, however, are part of the cycles, adding, subtracting, and transferring nutrients. Harvesting removes them from soil and water, and consumed food and other organic matter is ultimately returned to the environment in the form of raw waste or as processed fertilizer. In the modern era, humans also have been in a position to increase the overall nutrient budget through industrial processes. This artificial intervention, in turn, is reshaping the rhythm of the natural environment. In short, biogeochemical cycles constitute a feedback loop in which human activities and natural forces influence one another.

When agricultural development in Japan progressed apace during the early modern era (ca. 1600–1868), the nutrient cycle feedback relationship became increasingly deliberate and structured. Studies have revealed the "visible hand" of human intervention in nutrient cycles. The rise of the "night soil" trade in greater Edo, for example, functioned as a human-driven feedback loop of soil nutrients between consuming cities and farming villages.[1] A similar mechanism also mediated the terrestrial and aquatic zones of Japan's freshwater regions. As Shizuyo Sano explains in chapter 4 of this book, communities surrounding Lake Biwa periodically dredged sediment and plants from the lake, thereby returning otherwise lost nutrients to the soil while maintaining the quality of water.[2] These examples represent a microcosm of the nationwide nutrient recycling that buttressed Japan's agricultural development. In fact, noncommercial manures were a pillar of nutrient recycling and consistently supplied more nitrogen to the soil than commercial sources until as late as the end of World War II.[3]

Its long tradition of compost making might create the impression that Japan had once enjoyed a sustainable and harmonious human-nature relationship.

An exclusive focus on a closed biogeochemical loop, however, has a danger of overlooking the obvious: the overall pool of soil nutrients in Japan constantly expanded with the inflow of nutrients from abroad. Until the mid-twentieth century, much of the exogenous nitrogen input in Japan took the form of commercial fertilizers: fish meal from Hokkaido, then soybean cake from Manchuria, and later ammonium sulfate from Europe and Korea. Japan's long record of importing soil nitrogen from abroad raises a serious question about the concept of resilience, which presupposes a closed space within which humans negotiate with nature. Japan's nitrogen cycle rather describes an open system in which its agricultural core managed to avoid a biogeochemical crisis by sucking up nitrogen and other soil nutrients from outside.

To be sure, Japan was not unique among the advanced countries in its importation of nitrogen from abroad to enlarge the overall budget at home, but it was distinctive. From the mid-nineteenth century, European countries turned to South America for the excrement of seabirds (guano) collected in Peru, and later for sodium nitrate mined in Chile.[4] Japan was unique, however, in its distinctly organic orientation. By the end of the nineteenth century, the Western countries had made the transition from nitrogen fertilizers of animal origin (guano) to those mined or manufactured (Chilean nitrate and ammonium sulfate). In contrast, Japan continued to import fish meal and soybean cake from its northern outlying territories and sphere of influence. As a result, Japan's commercial nitrogen fertilizers remained predominantly plant- and animal-based until the late 1920s (figure 7.1). This organic practice in Japan stands in sharp contrast to that of the West. In 1913, approximately 70 percent of commercial nitrogen fertilizers used in Japan came from animal and vegetable sources, compared with only 2 percent in major Western countries.[5]

Japan's "organic" solution to soil nutrient depletion, extracting nitrogen from animals and plants instead of mining and industrial processes, indicates something unique about Imperial Japan—something that leads me to propose an ecological definition of Japanese colonialism: Japan as an organic empire. The massive transfer to mainland Japan of fish meal from Hokkaido and soybean cake from Manchuria suggests that the imperial core aggressively exploited its peripheral regions not only for minerals and fossil fuel but also for plant- and animal-based products such as food, fiber, and fertilizer. As this essay will show, the interregional movement of fish meal and soybean cake changed the metabolism of Imperial Japan. The agricultural core sucked up the organic material while transferring the environmental costs of its production to the colonial frontier.

Although the intensive extraction of organics in Japan's colonial frontiers partly reflected the inability of nature at home to meet the growing demand, this neo-Malthusian narrative overlooks a socioeconomic aspect of resource choice.

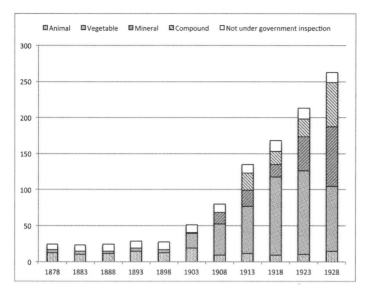

Animal   Vegetable   Mineral   Compound   Not under government inspection

FIGURE 7.1. Japan's nitrogen input from commercial fertilizers, 1878–1928, in thousands of tons

SOURCE: Kazushi Ohkawa, Nobukiyo Takamaysu, and Yuzo Yamamoto, eds., *Estimates of Long-Term Economic Statistics of Japan since 1868*, vol. 9 (Tōyō Keizai Shinpōsha, 1966), 196–97.

I will argue that the continuous use of organic nitrogen fertilizer underlay Japan's strength as a commercial empire. This essay will show that merchant ships and colonial railroads successfully marketed fish meal and soybean cake and also dominated their interregional shipment. The production of these soil nutrients, however, still relied on countless harvesters and middlemen as well as the ecological dynamics of fish and soybeans at the edge of the empire. This decentralized mode of nitrogen extraction at Japan's colonial frontier in Northeast Asia differed from that in South America, where indentured labor was deployed in a concentrated form to transfer nitrogen to North America and Western Europe. In short, the rise of Japan as an organic empire reflected its notable success in interregional commerce.

Herring Meal and Hokkaido

The earliest commercial sources of soil nutrients used in Japan during the early modern period were rice bran, cottonseed, and oilseed. These residues were closely related to the local consumption of rice, cotton, and vegetable oil. Their production was thus locally based and small in scale.[6] The shortage of

FIGURE 7.2. Drying of herring meal in Hokkaido

SOURCE: Postcard entitled "Otaru kinkai nishin ryō" (Kawasaki Shōten, n.d.) in collection of Historical Museum of Hokkaido, Sapporo.

soil nutrients, however, occurred at a faster pace in some regions than in others. During the Tokugawa period, the Kinki (Osaka-Kyoto) region and its surrounding areas expanded cultivation of cotton, hemp, indigo, tea, and other commercial crops that required much more labor and demanded increased soil nutrients. Here, as elsewhere, farmers managed to solve the labor part of the bottleneck through a so-called industrious revolution—to put it simply, by working harder.[7] The question remained, however, whether local ecological systems could meet increased demands for nutrients. This dimension was particularly problematic because the local stock of materials used to make compost and green manure grew scarce as villagers reclaimed more and more grasslands and open forestland for agricultural use.[8]

Japan's agricultural core initially managed to overcome the soil nutrient crisis by using fish as fertilizer. The earliest type of fish manure was made of sardines *(iwashi)*, in particular the Japanese sardine, *Sardinops melanostictus (maiwashi)*. This small, silvery fish, at most 25 centimeters (10 inches) in length, was processed into fertilizers in one of two forms, either dried *(hoshika)* or as cakes of scrap left after the extraction of oil *(iwashi-shimekasu)*. The making of cakes began with boiling sardines, after which they were crushed, the water and oil were pressed out, and the solids were dried in cake form. The use of sardines as fertilizer proved to be an ideal solution to the inherent fluctuation of the sardine catches, which in the absence of modern food preservation technologies often far exceeded what people could consume before spoilage. As late as the Taisho era, about 40 percent of the sardine catch was processed into fertilizer.[9]

The Kanto region initially served as a major supplier of sardine manure for the Kinki region. In the early eighteenth century, fishermen from Kishū (south of Osaka) migrated or organized seasonal expeditions to the Kanto region. Merchants from the Kinki also handled purchasing and shipping fish fertilizer to their home market. By the middle of the century, however, the Kanto region no longer provided an inexpensive and reliable source of soil nutrients. The harvest of sardines collapsed along the Pacific, and even when it recovered to some degree, an increasing portion of sardine fertilizer was consumed directly in the Kanto region.[10]

When the Kinki region was forced to look beyond the Kanto for an inexpensive source of soil nutrients, fertilizers made of Pacific herring, *Clupea pallasii (nishin)*, began to arrive from Hokkaido. This silver, foot-long fish returned to Hokkaido's shorelines each spring to spawn on beds of a kind of sea grass called *sugamo*. Until the mid-seventeenth century the Ainu people captured this fish for food with a hand net, called *tamoami*, made of the bark of a lime tree.[11] When Japanese traders began to export dried herring fillet *(migaki-nishin)*, the wastes from fish processing were also shipped as fertilizer *(do-nishin, or ha-nishin)*. Small herring unfit for food were processed into meal *(nishin-shimekasu)*, whose nutrient content was lower in phosphorus than its sardine counterpart (5 percent versus 7 percent) but richer in nitrogen (10 percent versus 8 percent).

The driving force behind the import of herring meal from Hokkaido was the thriving trade along the Sea of Japan. By the late seventeenth century, Ōmi merchants from the Kinki region had established their presence in the Matsumae Domain, located in the southwest part of Hokkaido, which controlled the island's small Japanese zone. From the latter half of the eighteenth century, the seamen from the Hokuriku region (northeast of Kyoto), originally employed by Ōmi merchants, entered trading business with their own vessels, called Kitamae ships. Acting as entrepreneurial traders, these Kitamae shipowners aggressively marketed herring fertilizer along the navigation routes.[12] Their business thrived on the robust demand for nitrogen. A good indicator is the price of herring heads, which increased by nearly 150 percent in central Japan from the middle to the end of the eighteenth century.[13] In the meantime, production costs in Hokkaido remained low because the Ainu people were a source of cheap labor. Herring catches also increased as Japanese fishermen began to penetrate into the Ainu zone when the harvest of herring collapsed in the Japanese area during the mid-eighteenth century.[14]

Initially, herring fertilizer was a byproduct from food processing; however, it became a dominant form of herring use once multiple obstacles to large-scale fishing were removed. The herring harvest in the Japanese zone recovered by the 1820s and continued to grow throughout the island until it reached about 1 million tons in 1898.[15] The full exploitation of the upward trend in the herring population was made possible by an increase in the labor force in Hokkaido. After 1854, the Edo shogunate extended its control to Hakodate and part of the Matsumae Domain and encouraged the immigration of Japanese fishermen and settlers as a counterweight to a growing threat of Russian encroachment from the north. After the Meiji Restoration (1868) and the establishment of a modern central government, the Hokkaido Development Commission reported 3,324 fishers and 33,630 hired hands exploiting the island's ocean bounty.[16]

The influx of workers from Japan proper coincided with a technological breakthrough for large-scale herring fishing: the introduction of the pound trap *(tate-ami)* in the beginning of the nineteenth century and its spread during the Tenpō years (1830–1843).[17] Whereas fishermen could mount the relatively small traditional gill net *(sashi-ami)* on one boat and chase the school of herring, three or more boats were required to operate the pound trap, and it needed to be installed near shore. The operation of a single pound trap required twenty to thirty workers, most of them seasonal migrants hailing from Japan's northeastern region. Each trap could catch up to 300 *koku*, or about 83.6 cubic meters, a year.[18] Although the use of the pound trap was initially banned in favor of gill nets, the prohibition was gradually lifted after 1857. In 1881, the Hokkaido Development Commission reported 1,303 pound traps in total, along with 33,382 gill nets. By 1898, the total number of nets to catch herring rapidly increased to 8,585 pound traps and 314,282 gill nets.[19]

When fish-catching capacities increased along with the population of herring around Hokkaido, the production of herring fertilizer multiplied. According to historian Nakanishi Satoshi, Hokkaido produced between 200,000 and 300,000 *koku* (55,710 to 83,565 cubic meters) of fish fertilizer each year from 1854 to 1870. The annual production continued to rise up to its peak yields of 700,000 to 900,000 *koku* (194,986 to 250,696 cubic meters) in 1888–1898.[20] This enterprise drove Hokkaido's economic development during a crucial period in its colonization. In 1893, the productive value of the fishing industry totaled 8.75 million yen, 5.8 times greater than the value of agricultural output. Some 70 percent of the fish sales in Hokkaido came from herring, about 90 percent of which was processed into fertilizer.[21] Supplying soil nutrients in the form of herring meal brought Hokkaido fully into Japan's economic orbit.

Until the early Meiji era, the use of herring meal had remained limited to central Japan and the island of Shikoku and was restricted to the cultivation of cash crops; by the late 1880s, however, its use gradually spread to the Kanto, Tōsan (Yamanashi, Nagano, and Gifu), and Tokai (Nagoya) regions.[22] Despite the aggressive expansion of herring fishery, the production of herring meal could not keep up with consumption, leading to price hikes for herring meal.

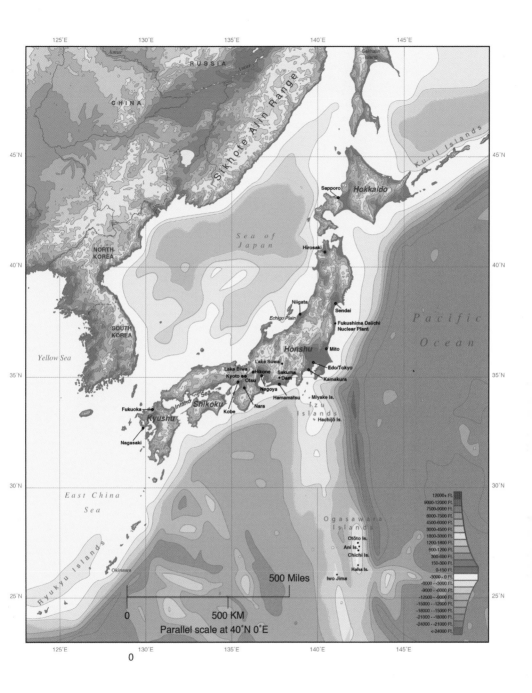

PLATE 1. Map of Japan

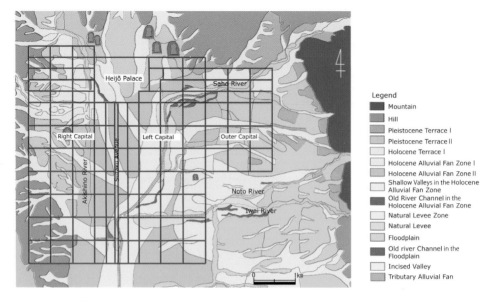

PLATE 2. Landform classification of Heijō-kyō (Nara, modern)
Note the narrow floodplains in particular.

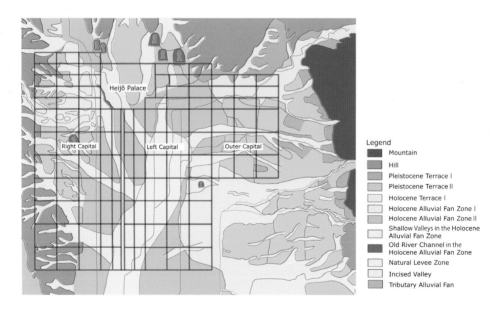

PLATE 3. Landform classification of Heijō-kyō (Nara, Nara period)
Note that the floodplains are wider than in present-day Nara.

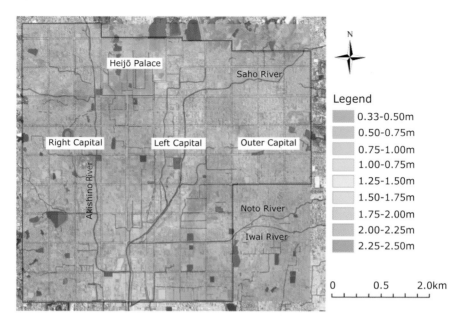

PLATE 4. Depth distribution of Heijō-kyō (Nara, Nara period)
Depths shown are relative to modern land surfaces: lower depth indicates higher Nara-period elevation, and greater depth indicates higher levels of sedimentation and/or landfill.

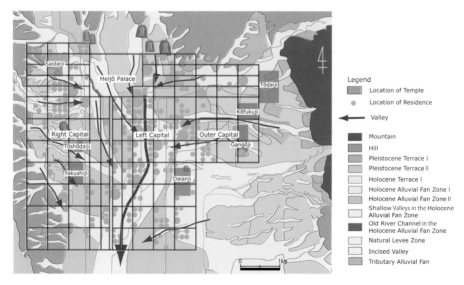

PLATE 5. Relationship between residential area and landforms in Heijō-kyō (Nara)
Note that residences are clustered on land close to water, represented by green and blue, and larger buildings are on higher, more solid ground, represented by orange and red.

PLATE 6. The *eri* fishing method

Photograph by Masanari Matsuda

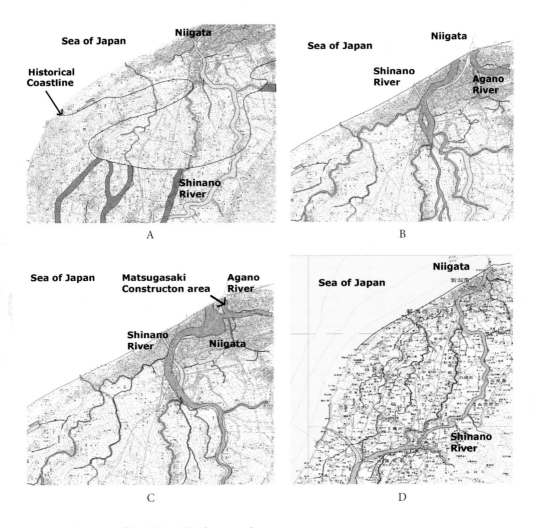

PLATE 7. Evolution of the Echigo Plain's geography

Historical shoreline, river courses, and lakes overlaid on a modern topographical map. Map scales differ. In (A), note the omega-shaped intrusion of the Sea of Japan on present-day Niigata City, a situation not evident in (B), which also shows the Agano (north) and Shinano (south) rivers entering the Sea of Japan separately. In (C) only the Shinano empties into the sea. All show historical streams wider than their modern counterparts, a situation less evident in (D) than in (C).

SOURCES: Maps by P. C. Brown. Data underlying changes in shorelines and waterbodies derived from standard geological surveys and from core samplings, excavations, and other techniques employed in civil engineering projects. Kokudo Kōtsūshō Kokudo Chiriin, Kokudo Kōtsūshō Hokuriku Chihō Seibikyoku, *Kochiri ni kansuru chōsa: Kochiri de saguru Echigo no hensen—Arakawa, Aganogawa, Shinanogawa, Himekawa* (Niigata: Kokudo Kōtsūshō Hokuriku Chihō Seibikyoku, 2004), and its accompanying CD of geological and geographic survey data and map images (edited with Arc Explorer and Photoshop).

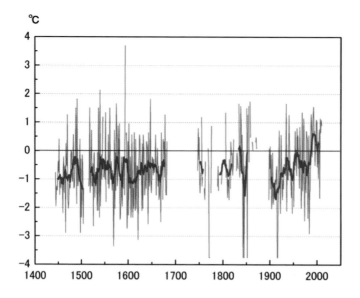

PLATE 8. Year-to-year variations in December–January temperatures at Lake Suwa, 1444–1870 (reconstructed) and 1891–2010 (observed)
Thick blue lines indicate eleven-year running means.

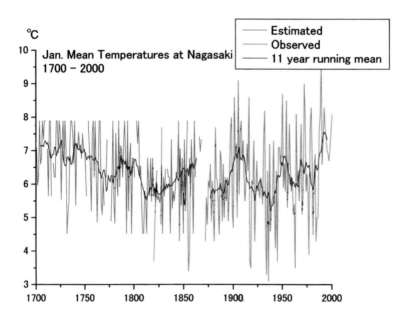

PLATE 9. January mean temperature variations at Nagasaki, 1700–2000

SOURCE: T. Mikami, M. Zaiki, G. P. Können, and P. D. Jones, "Winter Temperature Reconstruction at Dejima, Nagasaki Based on Historical Meteorological Documents during the Last 300 Years," in *Proceedings of the International Conference on Climate Change and Variability*, edited by Takehiko Mikami (Hachiōji: International Geographical Union Commission on Climatology and Tokyo Metropolitan University, 2000), 105, fig. 3.

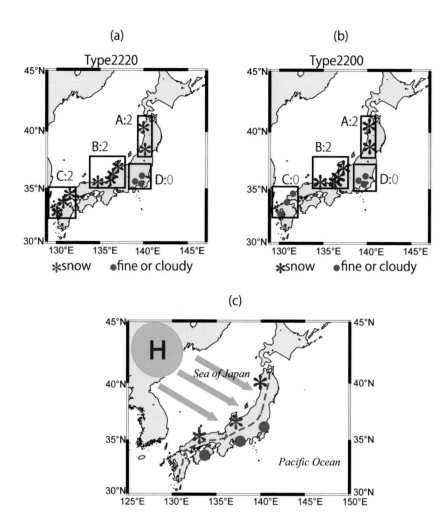

PLATE 10. Major weather distribution types for winter monsoon patterns
Type 2220 and Type 2200 on the upper charts (snowy weather on the Japan Sea Coast
and fine weather on the Pacific Coast) correspond to the typical strong winter monsoon
pressure patterns (lower chart).

SOURCE: Junpei Hirano and Takehiko Mikami, "Reconstruction of Winter Climate Variations
during the 19th Century in Japan," *International Journal of Climatology* 28 (2008): 1421, fig. 4.

PLATE 11. A portion of the Shiosaito complex at Shiodome, Tokyo, viewed from across Hamarikyū Park.

This image was taken with the harbor at the back of the photographer.

Photograph by Scott O'Bryan, October 2010.

PLATE 12. The Tokyo Shiodome Building, a tower in the Shiosaito complex.

The dominating wall seen here at the base of the building complex faces the street along Kaigan Road between the main cluster of Shiosaito buildings and Hamarikyū Park.

Photograph by Scott O'Bryan, October 2010.

From 1879 to 1894, the price of herring meal consistently remained above its 1879 value, except for a few years during the severe Matsukata deflation period of the 1880s.[23] In 1888 a poor catch and a strong demand for herring meal pushed its price so high that it sold for about the same price as rice.[24] By the end of the nineteenth century, the fishing industry in Hokkaido, well aware of the coming of soybean cake from Manchuria as a cheap substitute for fish meal, decided to market herring as human food and animal feed.[25]

The transfer of nitrogen from Japan's coastal frontiers to mainland farms in the form of fish meal fertilizer had clear environmental consequences. The most immediate result was the clearing of woods along the coast. One estimate in the late 1880s indicated that in Hokkaido about 14,250,000 *kan* (53,437.5 tons) of firewood was annually consumed to boil herring.[26] Little woodland remained in areas where herring fisheries flourished. In 1895 fertilizer makers had to go as far as 3 to 4 *ri* (12 to 16 kilometers) to procure firewood.[27] The excessive use of firewood was considered not only wasteful but also detrimental to fish catches. Fishermen knew from experience that coastal woodlands, providing calm, shallow, and nutritionally rich waters, served as a protective shelter for fish spawning. Indeed, contemporaries during the Meiji era blamed destructive lumbering for poor herring catches along the coast of Ajigasawa in Aomori Prefecture. Fearing the same fate, the fishing industry in Hokkaido campaigned for the use of coal-powered boilers to help protect herring populations.[28]

A more devastating environmental development may also be related to overfishing. Following the peak year of 1897, a long-term, gradual decline in herring catches began. This trend eventually led to the complete collapse of the harvest in 1937 and again by the end of the 1950s. It is tempting to blame overfishing, driven by an insatiable demand for soil nutrients in Japan's core region, for this outcome; however, the underlying reason has remained in dispute. A recent investigation, for example, found that historical fluctuations in herring catches along the western coast of Hokkaido were related to changes in water temperatures. The more frequent influx of warm currents into Hokkaido's waters over the past century may have pushed herring's migratory route north of Hokkaido.[29] Another major hypothesis is advanced by forest scientist Miura Masayuki, who in the 1950s proposed that the destruction of forests in Hokkaido diminished the flow of minerals from the terrestrial to the aquatic ecosystems. This change might have caused the widespread demise of kelp along the coast, which in turn deprived herring of their spawning beds.[30]

Hokkaido fishermen have long tried to make sense of fluctuations in harvests. As early as 1892, fishermen in Hokkaido noted that the size of the herring catch was diminishing, prompting an investigation of possible overharvesting.[31] In 1907, when the harvest collapsed along the southwestern coast of Hokkaido, another probe was conducted. Its conclusion, however, ruled out overfishing as a cause and instead focused on change in water temperatures.[32] Not until thirty years later, with the complete collapse of the herring catch throughout Hokkaido, were the variations in harvests and their possible link to overfishing finally addressed in a systematic manner.[33]

Soybean Cake and Manchuria

By the time of the Russo-Japanese War of 1904–1905, fish fertilizer had gradually yielded to soybean cake as a leading commercial source of nitrogen. The soybean, *Glycine max (daizu)*, is a crop rich in protein, which constitutes about 36 percent of the total weight of its mature raw seed. To synthesize this much protein, the soybean takes up nitrogen from the soil, a process assisted by its symbiosis with nitrogen-fixing bacteria in the nodules on its roots. The protein in its seeds serves as human food and animal feed but also can work as a fertilizer. For this purpose, soy is usually processed into cakes, after the extraction of oil and water. This cake product, easily decomposable, contains 6.7 percent nitrogen by weight.

FIGURE 7.3. Soybean cake awaiting shipment at Dairen (Dalian) port

SOURCE: Minami Manshū Tetsudō Kabushiki Kaisha, *Minami Manshū tetsudō ensen shashin chō* (Dairen [Dalian]: Nanman Tetsudō, 1940).

The world's leading soybean producer in the early twentieth century was Manchuria. A 1919 estimate indicated that this northeastern region of China had about 70 percent of China's 4.4 million hectares of soybean fields, dwarfing the second- and third-largest producers, Korea (0.46 million hectares) and Japan (0.43 million hectares).[34] Manchuria had once been a vast land of forests and grasslands, but a growing number of Han Chinese farmers gradually moved into this sparsely populated region, at times by invitation and at times defiantly. In a striking parallel with the Japanese colonization of Hokkaido, the late-nineteenth-century Qing administration encouraged farming settlements in Manchuria to forestall Russia's influence in the region. By the end of the century Han Chinese constituted about eleven million of the twelve million people in Manchuria.[35]

Migrants from northern China introduced their traditional three-crop rotation system to this newly developed region. The main crops for household consumption, sorghum (kaoliang) and millet, tended to exhaust soil nutrients, particularly nitrogen. To restore soil fertility, Chinese farmers planted soybeans every third or fourth year.[36] Planting soybeans had the additional benefit of suppressing the weeds that often grew alongside kaoliang and millet.[37] The soybeans were then sold. This commercial characteristic turned soybeans into the locomotive of Manchuria's modern economic development.[38]

A substantial portion of the soybeans raised in China was used as a source of vegetable oil. The oil-making process produced soybean cake as a byproduct that was sold to other parts of China. The trade in soybean cake initially functioned as one of the growing links binding Manchuria with the rest of China. From the late eighteenth century, the increasingly widespread cultivation of cotton in central China was largely made possible by soybean cake from Manchuria. In the mid-nineteenth century, however, the Taiping Rebellion ruined this agriculturally advanced region and destroyed the major source of domestic demand for the fertilizer. This economic shift redirected the flow of soybean cake toward southern China, via Xiamen and Shantou, for the production of sweet potato.[39] By the late 1880s, however, the nitrogen flow was reoriented toward Japan as Chinese merchants, aware of the high price of herring meal, began to market soybean cake as an inexpensive substitute.[40]

Japan's imports of soybean cake from Manchuria initially increased slowly, to 25,000 tons in 1893. The number then soared thanks to favorable trading conditions in the aftermath of the Sino-Japanese War (1894–1895). As Japan extended its influence into Manchuria, interregional shipping thrived by taking advantage of soybean cake's tariff-free status.[41] Favorable foreign exchange rates also promoted the transfer of nitrogen across the sea. When Japan switched to the gold standard in 1898, the strong value of gold vis-à-vis silver, still used in China, meant that Japanese consumers could buy soybean cake at much cheaper prices.[42] As a result, soybean cake imports rapidly expanded, hitting a record of

200,000 tons in 1903, with the nitrogen content (13,400 tons) exceeding that of herring meal used in Japan (6,370 tons).[43]

The switch in Japan's nitrogen source from fish to soybean by the end of the nineteenth century was well timed because Japan's imperial core experienced an agricultural boom following the Russo-Japanese War (1904–1905). As industrialization accelerated, more people moved into cities to work, in turn stimulating the market demand for rice. In response, farmers turned to commercial fertilizers to squeeze as much rice as possible from their small paddies. Although data showing fertilizer input for rice are lacking, it is possible to infer its rapid growth because overall expenditures for commercial fertilizer grew 2.5 times between 1905 and 1913.[44] The heavy fertilization can be also inferred from a rise in productivity. From 1905 to 1920, even though rice-planted areas expanded by only a small extent—from 2.8 million to 3.1 million hectares—total yields jumped from 5.7 million to 9.48 million tons.[45] It is estimated that fertilizer accounted for as much as 80 percent of the increase in rice production between 1916 and 1922.[46]

Another crop that critically hinged on the nitrogen budget in Japan was silk. The silkworm, *Bombyx mori,* is the larva of a domesticated moth that feeds only on mulberry leaves. As a nutrient that is essential for plant growth, nitrogen affects leaf supplies for the silkworm—and ultimately silk production. This biogeochemical characteristic had serious economic implications because raw silk had been Japan's top export product since the dawn of the Meiji period. In fact, silk exports served as a main engine for major segments of Japan's industrialization. The earned dollars from silk exports to the United States funded the import of cotton, which textile mills in Japan processed into thread, yarn, and cloth for export to China and Southeast Asia. Profits from silk and textile sales, in turn, increased the overall capital for investment in the heavy industry sector.[47]

The elaborate interlocking of silk, cotton, and capital for Japan's industrialization worked well while the silk trade was booming after the Russo-Japanese War. From 1905 to 1913, silk exports to the United States nearly tripled, from 3,377 to 9,160 tons.[48] To keep up with the robust demand abroad, more farmers in Japan entered the silkworm business. Traditionally confined to mountainous hinterlands in central Japan, sericulture spread to other parts of the Kanto region, the Tokai (Nagoya) region, and western Japan. These areas also witnessed a generous use of commercial fertilizer. In 1920, Nagasaki consumed 61 *kan*, or 228.75 kilograms, of soybean cake per *tan* (approximately 1,000 square meters) of mulberry field, followed by Kagawa (37 *kan*), Nara (36 *kan*), and Aichi (34 *kan*). Traditional regions applied much less: 16 *kan* of soybean cake was applied to each *tan* in Gunma, 10 *kan* in Nagano, and only 8 *kan* in Fukushima.[49] In short, fresh nitrogen supplies helped newcomers in sericulture and buttressed the boom in silk exports.

As the agricultural boom continued on Japan's mainland, Manchuria's terrestrial biomass was called upon to meet the exponential increase in demand for soil nitrogen. Soybean cake imports from Manchuria quadrupled from 184,000 tons in 1905 to 727,000 tons in 1913. The volume almost doubled again during World War I, hitting an all-time high of 1.34 million tons in 1919.[50] In 1921, the nitrogen contributions of soybean cake reached 84,550 tons, accounting for 57 percent of the total nitrogen inputs from all commercial fertilizers and nearly matching those of compost (90,000 tons) at the 1918 level.[51]

The infrastructure that made possible the massive transfer of nitrogen from northeastern China to Japan was the modern railroad network in Manchuria. Russia first laid the Eastern Chinese Railway in 1902, but three years later Japan took over its southern part as one of the spoils of the Russo-Japanese War. The Russian rail line and Japan's Southern Manchuria Railway (SMR) competed to promote soybean production along their trunks and ship the product to the ports of Dalian (Dairen) and Vladivostok. As one observer noted, the soybean became "almost the only source of life" for the Japanese imperial enterprise in Manchuria.[52] Indeed, between 1907 and 1919, cargoes of soybeans and soybean by-products accounted for 37 to 40 percent of SMR's freight sales.[53] As historian Kaneko Fumio has explained, soybeans were "the most important cargo item, one whose growing freight volume supported SMR's business and thus Japan's control of Manchuria."[54]

Although soybean cake at once reflected and consolidated Japan's commercial empire in Manchuria, its production was largely left in Chinese hands. Soybeans were raised by Chinese farmers and collected by local agents, called *liangzhan*. Even Mitsui and other powerful Japanese trading companies could not displace these middlemen.[55] Chinese enterprises also claimed a significant share in soybean processing. Although their factories tended to be small and poorly equipped, the total output from Chinese manufacturers continually surpassed that of their better-capitalized Japanese rivals.[56] Overall, countless agents competed or collaborated in Japan's outer sphere of influence, together underwriting the imperial core's insatiable appetite for nitrogen.

The decentralized mode of nitrogen extraction in Manchuria differed from the concentrated form it took in Chile. Unlike soybeans, which required arable land and skillful farmers, sodium nitrate was a nonrenewable mineral buried in the forbidding Atacama Desert. Historian Edward Melillo has explained how nitrate mining gave rise to a new labor regime based on the migration of unskilled workers. Miners were recruited through the so-called *enganche* system, in which poor peasants would sign labor contracts for wages and supplies, only to realize that their passage to the mines imposed a huge debt to repay. Families who moved with contract workers also had to toil under the unbearable conditions of the desert for coupons to procure supplies from the company.[57] This coercive labor regime built around Chilean nitrate was absent from Manchurian soybean production. As discussed at the beginning of this section, it was small farmers, not plantation workers, who tended the soybean fields. Although soybean producers must have suffered from unequal exchanges, debt service, and other abuses, the nature of the relationship was likely to have been more commercial than coercive.

Chilean nitrate, however, also opened up a new type of workers association that the Manchurian soybean cake production failed to catalyze. Timothy Mitchell has noted that in the energy sector, the mining of coal was uniqiuely conducive to labor activism and democratic politics because of its idiosyncratic extraction system. Coal moved through limited paths from mines to market, each depending on large numbers of workers, in a rigid structure that had an empowering effect for workers: they could organize themselves and seize control of choking points during labor unrest.[58] Nitrate mines similarly became a cradle of the labor movement in Chile. In 1907, twenty thousand nitrate miners went on strike for higher wages, and thousands of them died under fire from government troops.[59] Soybean cake in Manchuria, in contrast, seemingly led to neither concerted action nor a sense of solidarity on the part of farmers. In this sense, Japan's choice of soybean cake, not sodium nitrate, might have hampered the rise of labor activism and mass politics at the outer edge of its empire.

As the socioeconomic outcomes of soybean cake and Chilean nitrate clearly diverged, a question arises: what held Japan back from using Chilean nitrate? A major reason was climatic. Chilean nitrate was a highly reactive form of nitrogen, best suited for upland fields in a cool and dry climate, such as in northern Europe. Japan's warm and humid environment, however, tended to dissolve Chilean nitrate too quickly, favoring instead organic fertilizers, whose slower rate of decomposition allowed crops to absorb nutrients over time. The extreme solubility of Chilean nitrate also made it ill suited for rice paddies. For this reason, its use in Japan was largely confined to Hokkaido, where in a climate similar to northern Europe, cash crops such as sugar beet were raised on upland fields.[60]

Unlike Chilean nitrate, ammonium sulfate, another popular nitrogen fertilizer in the West, was suited for the climate in Japan. Its price was also competitive, being roughly equal to that of soybean cake from 1903 to 1912.[61] Indeed, the import of ammonium sulfate, chiefly from Europe, rose alongside that of soybean cake. The nitrogen input of imports in 1913 (23,800 tons) was almost half that of Manchurian soybean cake.[62] World War I, however, checked further increases. With the trade with Europe disrupted, the domestic production of ammonium sulfate soared in Japan but still failed to meet the prewar import levels, much less the strong wartime demand for nitrogen fertilizers. The manufacturing process was still a challenge for Japan's fledging chemical industry. Moreover, wartime inflation and electric power shortages in Japan pushed production costs upward.[63] If the cheap and abundant supplies of soybean cake from Manchuria underscored Japan's commercial strength, the stagnant Japanese production of ammonium sulfate during World War I exposed the country's industrial backwardness at home.

Manchurian soybean cake, however, gradually lost its leading position as a nitrogen supplier. As Figure 7.4 shows, the growth of nitrogen inputs from soybean cake halted shortly after the war. Following a period of waxing and waning, its contributions plummeted. In the meantime, ammonium sulfate uses expanded rapidly, and by the end of the 1920s, the synthetic product finally surpassed soybean cake as the leading commercial nitrogen fertilizer.

The underlying reason for the transition from the organic to the inorganic era in nitrogen supply was a growing price gap between soybean cake and ammonium sulfate. The price of ammonium sulfate declined far below Japan's general

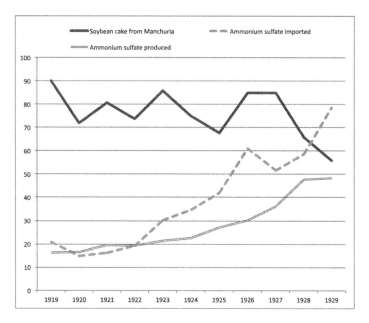

FIGURE 7.4. Japan's nitrogen input from commercial fertilizers, 1919–1929, in thousands of tons
Estimates based on nitrogen content of soybean cake (6.7 percent) and that of ammonium sulfate (20.6 percent).
SOURCE: Nōrinshō Daijin Kanbō Sōmuka, *Nōrin gyōsei shi*, vol. 1 (Nōrin Kyōkai, 1957), 843.

price index. After World War I, European manufacturers converted their wartime capacities of ammonia synthesis from gunpowder to fertilizer. As a result, the production of ammonium sulfate quickly saturated the world market, triggering a fierce price war among the chemical giants.[64] In the meantime, the global demand for soybeans surged. In the late 1920s, Germany began to import raw soybeans from Manchuria to extract vegetable oil for various industrial uses. This new development pushed the price of soybeans upward, which in turn raised the cost of soybean cake production.[65] In 1923, soybean cake was approximately 20 percent more expensive, by nitrogen weight, than ammonium sulfate, but the margin exceeded 60 percent in 1924–1926 and widened to as much as 80 percent in 1928–1930.[66] As a result, SMR and other soybean-related enterprises in Manchuria switched their marketing approach and began to sell soybeans as animal feed—just as the Hokkaido fishermen had done with herring when it became too expensive as fertilizer.[67]

By the time the era of soybean cake ended, the environmental consequences of soybean cultivation in Manchuria were extensive. The demand for soybeans as a cheap source of vegetable oil and soil nutrients motivated Chinese peasants to aggressively reclaim forests and grasslands. This horizontal expansion resulted from various limits to intensive soybean cultivation in Manchuria. Part of the problem was biological. Continuous cropping of soybeans often led to significant reductions in yield.[68] Socioeconomic practices also worked against intensive cultivation. Few local farmers completely specialized in soybeans because sorghum, another major crop raised in Manchuria, was useful for various purposes—not only food but also animal feed, fuel, liquor, and even building materials. As one observer aptly summarized, the sorghum fields served as "rice paddies, woodlands, and bamboo forests, all combined in one single concentrated form."[69] Soybeans thus remained a rotation crop for sorghum and millet instead of displacing them. According to one estimate, soybeans' share of cultivation acreage in Manchuria grew only slowly, from approximately 20 percent in 1910 to 27 percent in the south and 34.5 percent in the north in 1927–1928.[70]

Given the limits to intensive cropping in the existing fields, land reclamation was the chief means to produce more soybeans. From 1908 to 1930, cultivated areas in Manchuria expanded from approximately 8 million to 13 million hectares.[71] This colossal reclamation drastically changed Manchuria's landscape from a mosaic of forests and grasslands into a monotonous expanse of agricultural fields. Railroads accelerated this trend by promoting lumbering along the line, with the cleared sites converted into farmlands by Chinese settlers coming from the south. This combination of lumbering and farming stripped Manchuria of about half its foliage cover in the first half of the twentieth century.[72] Land reclamation also affected animal populations. For example, by the late nineteenth century, the expansion of agricultural fields in Jilin Province had diminished the habitat of deer and disrupted their migratory trails. The deer population plummeted, so much that the local hunting declined and hunting grounds eventually were turned into farmland.[73] As Manchuria became a manufacturer of vegetable oil and soil nutrients, its biodiversity was lost, creating its now familiar image as a vast agricultural frontier.

Conclusion

When Japan faced a biogeochemical soil nutrient bottleneck in its agricultural development, local communities employed various techniques to maintain soil fertility; this microregional conservation was then supplemented by the interregional importation of soil nutrients. The marine and terrestrial biomass in the nation's imperial frontiers was exploited and commoditized to replenish the nitrogen in the core regions. The surplus of biologically available nitrogen in Hokkaido and Manchuria was unlocked, and the geographical circulation of this soil nutrient expanded through a geopolitically driven network of labor, capital, transportation, and technologies.

The rise of Japan as an organic empire reflected the nation's socioeconomic and environmental conditions. The warm and humid climate in most parts of Japan made the use of organic matter for fertilizers more effective than in Europe. But the continuous use of herring meal and soybean cake also mirrored Japan's strength as a commercial empire: the peripheral areas provided the core regions with cheap, abundant biomass resources. Moreover, Japan's organic choice to replenish its soil nutrients resulted from the country's industrial weakness, which postponed the transition to ammonium sulfate until the late 1920s.

The exploitation of the untapped nitrogen pool in the northern frontiers helped Japan overcome not only the domestic biogeochemical bottleneck to agriculture but also obstacles to industrialization by boosting rice and silk production. However, the ecological outsourcing of the imperial core pushed socioeconomic and environmental costs onto the peripheral regions. Making herring meal consumed much firewood and overtaxed Hokkaido's coastal forests. It also imposed enormous pressure on the fishery, far beyond that imposed by the amount of fish consumed as human food, although its ultimate effects remain uncertain. The import of soybean cake from Manchuria as a substitute for herring meal led to its own ecological consequences at the place of origin. Chinese farmers cultivated soybeans in a crop rotation to keep their existing farmland productive, but massive reclamation to increase soybean production diminished the size of forests and grasslands and thereby reduced the region's biodiversity.

When we speak of ecological scarcity, collapse, and resilience, we tend to approach a human community as if it were confined to a fixed territorial space. The expansive production and distribution of organic nitrogen fertilizers consumed in Japan since the dawn of the modern era, however, show that the boundaries of the community are flexible and dynamic. Political conquest and commercial networks in Japan's frontier, combined with the introduction of technologies, enlarged the geographic circulation of soil nutrients, such that Japan's core agricultural regions removed ecological bottlenecks to their further development. This solution, however, only postponed the problem through externalization; it was not a sustainable solution. Today, the nature of the problem has shifted from scarcity to excess, and the overuse of chemical fertilizers has reshaped Japan's ecological landscape. Perhaps a solution to the oversupply of soil nutrients lies not in a return to Japan's mythical "ecofriendly" past but rather in a creative approach that takes into full account the nation's mixed historical record on the problem and its broad geographical ramifications.

NOTES

1   David L. Howell, "Fecal Matters: Prolegomenon to a History of Shit in Japan," in *Japan at Nature's Edge: The Environmental Context of a Global Power*, ed. Ian Miller, Julia Adeney Thomas, and Brett L. Walker (Honolulu: University of Hawai'i Press, 2013), 137–51; Anne Walthall, "Village Networks: *Sōdai* and the Sale of Edo Nightsoil," *Monumenta Nipponica* 43, no. 3 (1988): 279–303.

2   Also see Sano Shizuyo, *Chū-kinsei no sonraku to mizube no kankyōshi: Keikan, nariwai, shigenkanri* (Yoshikawa Kōbunkan, 2008), 257–58, 271–74, 305–307, 309–10, 316–18. Also see Hiratsuka Jun'ichi, "1960 nen izen no nakaumi ni okeru hiryōmo saishūgyō no jittai: Satoko to shite no katako no yakuwari," *EkoSophia* (*Ekosofia*) 13 (2004): 97–112.

3   Kurokawa Kazue, *Nihon ni okeru Meiji ikō no dojō hiryō kō* (Nihon ni okeru Meiji Ikō no Dojō Hiryō Kō Kankōkai, 1978–1982) 2: 4–5; 3: 182–83.

4   Gregory T. Cushman, *Guano and the Opening of the Pacific World: A Global Ecological History* (Cambridge: Cambridge University Press, 2013); Brett Clark and John Bellamy Foster, "Ecological Imperialism and the Global Metabolic Rift: Unequal Exchange and the Guano/ Nitrates Trade," *International Journal of Comparative Sociology* 50, no. 3–4 (2009): 311–34.

5   Kurokawa, *Nihon ni okeru meiji ikō no dojō hiryō kō*, 3: 65.

6   Nōrinshō Daijin Kanbō Sōmuka, *Nōrin gyōsei shi*, vol. 1 (Nōrin Kyōkai, 1957), 840.

7   For the concept of an "industrious revolution" in Japan, see Hayami Akira, "Kinsei Nihon no keizai hatten to Industrious Revolution," in *Tokugawa shakai kara no tenbō: Hatten, kōzō, kokusai kankei*, ed. Hayami Akira, Saitō Osamu, and Sugiyama Shin'ya (Dōbunkan, 1989).

8   Furushima Toshio, *Nihon nōgyō gijutsu shi*, vol. 6, *Furushima Toshio chosakushū* (Tōkyō Daigaku Shuppankai, 1975), 347, 349.

9   Kurokawa, *Nihon ni okeru Meiji ikō no dojō hiryō kō*, 1: 83.

10   Furuta Etsuzō, *Kinsei gyohi ryūtsū no chiiki teki tenkai* (Kokon Shoin, 1996), 247–49; Nakanishi Satoru, *Kinsei-kindai Nihon no shijō kōzō: "Matsumae nishin" hiryō torihiki no kenkyū* (Tōkyō Daigaku Shuppankai, 1998), 60–61, 134–35.

11   Kaitakushi, *Kaitakushi jigyō hōkoku*, vol. 3 (Ōkurashō, 1885), 297; Hokkaidō, *Hokkaidō gyogyō shi*, vol. 1 (Sapporo: Hokkaidō Suisanbu, 1957), 22.

12 Nakanishi, *Kinsei-kindai Nihon*, 42, 60–61, 72–73, 95.

13 Ibid., 98.

14 David L. Howell, *Capitalism from Within: Economy, Society, and the State in a Japanese Fishery* (Berkeley: University of California Press, 1995), 24–49; Nakanishi, *Kinsei-kindai Nihon*, 33, 35, 61, 83.

15 Hokkaidō Suisanbu, *Hokkaidō gyogyō shi*, 1: 117–18.

16 Kaitakushi, *Kaitakushi jigyō hōkoku*, 3: 296; Hokkaidō Suisanbu, *Hokkaidō gyogyō shi*, 1: 29.

17 Howell, *Capitalism*, 68–77, 106–18.

18 Hokkaidō Suisanbu, *Hokkaidō gyogyō shi*, 1: 336, 347, 353–56.

19 Ibid., 229–30.

20 Nakanishi Satoshi, "Bakumatsu-Meiji ki kinai hiryō shijō no tenkai," *Keizaigaku kenkyū* (Hokkaidō Daigaku) 47, no. 2 (1997): 283–84, footnote 26; Nakanishi, *Kinsei-kindai Nihon*, 39–40.

21 Murao Motonaga, *Nishin hiryō gaiyō* (Murao Motonaga, 1895), 1–3.

22 Nakanishi, *Kinsei-kindai Nihon*, 42–43, 46–51; Murao, *Nishin*, 6–7, 35, 45–46; Sakō Tsuneaki, *Nihon hiryō zensho*, rev. 2nd ed. (Yūrindō, 1894), 168–77, 198.

23 Kurokawa, *Nihon ni okeru meiji ikō no dojō hiryō kō*, 1: 85, 97.

24 *Hokusui kyōkai hōkokusho*, no. 32 (March 1888): 29–30.

25 Hokusui Kyōkai, *Hokusui kyōkai hyakunenshi* (Sapporo: Hokusui Kyōkai, 1984), 324.

26 The original text shows the number as 1,425 *kan*, but the subsequent discussion strongly indicates that the print mistakenly omitted *man* (10,000). I would like to thank one of our anonymous reviewers for pointing out this potential error.

27 Murao, *Nishin*, 12.

28 Hokusui Kyōkai, *Hokusui kyōkai hyakunenshi*, 24.

29 Tanaka Iori, "Hokkaidō seigan ni okeru 20 seiki no engan suion oyobi nishin gyokakuryō no hensen," *Hokkaidō suisan shikenjō hōkoku* 62 (2002): 41–55.

30 Wakana Hiroshi, "Gendai uotsuki-rin to 'Nishin yama ni noboru': Miura Masayuki, Ōtaki Shigenao ra no "mori to umi" ni kansuru hukusōryū," *Muroran kōgyō daigaku kiyō* 51 (2001): 147–58; Hoshikawa Yutaka, Tajima Ken'ichirō, and Kawai Tadashi, "Nishin sanransho no keisei ni oyobosu shokusei to chikei no eikyō," *Hokkaidō suisan shikenjō hōkoku* 62 (2002): 105–11.

31 Hokusui Kyōkai, *Hokusui kyōkai hyakunenshi*, 236–37.

32 Ibid., 381.

33 Yokohama Shōkin Ginkō Chōsabu, *Hokkaidō nishin gyogyō ni tsuite* (Yokohama Shōkin Ginkō Tōkyō Shiten Fuzoku Insatsubu, 1943), 10–11.

34 Minami Manshū Tetsudō Shomubu Chōsaka, *Manshū ni okeru yubōgyō* (Dairen [Dalian]: Minami Manshū Tetsudō Shomubu Chōsaka, 1924), 31–32.

35 Christopher M. Isett, *State, Peasant, and Merchant in Qing Manchuria, 1644–1862* (Stanford, Calif.: Stanford University Press, 2007); Komine Kazuo, *Manshū: Kigen, shokumin, haken* (Ochanomizu Shobō, 1991), 148–56.

36 Komine, *Manshū*, 189–90; Minami Manshū Tetsudō Nōji Shikenjo, *Manshū daizu narabini mamekasu* (Gongzhuling: Minami Manshū Tetsudō Nōji Shikenjo, 1921), 15, 17, 19–23; Minami Manshū Tetsudō Shomubu Chōsaka, *Manshū ni okeru yubōgyō*, 2, 17–21.

37 Sanka Isao, *Daizu no saibai* (Dairen [Dalian]: Minami Manshū Tetsudō Kōgyōbu Nōmuka, 1924), 293.

38 Kaneko Fumio, *Kindai Nihon ni okeru tai Manshū tōshi no kenkyū* (Kondō Shuppansha, 1991), 21.

39 Adachi Keiji, "Daizu kasu ryūtsū to Shin dai no shōgyō teki nōgyō," *Tōyōshi kenkyū* 37, no. 3 (1978): 35–63; Komine, *Manshū*, 137–38; Yasutomi Ayumu, "Kokusai shōhin to shiteno Manshū

daizu," in *"Manshū" no seiritsu: Shinrin no shōjin to kindai kūkan no keisei*, ed. Yasutomi Ayumu and Fukao Yōko (Nagoya: Nagoya Daigaku Shuppankai, 2009), 291–95.

40 Komine, *Manshū*, 137, 203–204; Yasutomi, "Kokusai shōhin," 296.

41 Komine, *Manshū*, 268–69.

42 Kaneko, *Kindai Nihon*, 45–46.

43 Kurokawa, *Nihon ni okeru Meiji ikō no dojō hiryō kō, chū kan*, vol. 2, 4–5.

44 Kazushi Ohkawa, Nobukiyo Takamatsu, and Yūzō Yamamoto, eds., *Estimates of Long-Term Economic Statistics of Japan since 1868*, vol. 9 (Tōyō Keizai, 1966), 186–87.

45 "Nōsakumotsu sakutsuke menseki oyobi seisanryō," Table 7.14 of Sōmushō Tōkeikyoku, *Nihon no chōki tōkei keiretsu*, http://www.stat.go.jp/data/chouki/ (accessed October 31, 2013).

46 Kurokawa, *Nihon ni okeru Meiji ikō no dojō hiryō kō*, 2: 93, 188; 3: 66, 116–18.

47 This classic analysis of the "silk-cotton" exchange was originally put forward by Nawa Tōichi, *Nihon bōseki gyō to genmen mondai kenkyū* (Osaka: Daidō Shoin, 1938).

48 Ueyama Kazuo, "Dai-ichiji taisen mae ni okeru Nihon kiito no tai-Bei shinshutsu," *Jōsai keizai gakkai shi* 19, no. 1 (1983): 48.

49 Ikawa Katsuhiko, *Kindai Nihon seishigyō to mayu seisan* (Tōkyō Keizai Jōhō Shuppan, 1998), 69–72.

50 Yasutomi, "Kokusai shōhin," 298.

51 Kurokawa, *Nihon ni okeru Meiji ikō no dojō hiryō kō*, 2: 4–5; 3: 182–83.

52 Komai Tokuzō, *Manshū daizu ron* (Sendai: Tōhoku Teikoku Daigaku Nōka Daigaku Nai Kamera Kai, 1912), 50, 112, 119, 127–30.

53 Kaneko, *Kindai Nihon*, 105–106, 231, 234.

54 Ibid., 21.

55 Ibid., 328–30.

56 Ibid., 78, 189.

57 Edward D. Melillo, "The First Green Revolution: Debt Peonage and the Making of the Nitrogen Fertilizer Trade, 1840–1930," *American Historical Review* 117, no. 4 (2012): 1048.

58 Timothy Mitchell, "Carbon Democracy," *Economy and Society* 38, no. 3 (2009): 399–432.

59 Melillo, "The First Green Revolution," 1048–49.

60 Kurokawa, *Nihon ni okeru Meiji ikō no dojō hiryō kō*, 2: 4–5, 67.

61 Calculated from Kurokawa, *Nihon ni okeru Meiji ikō no dojō hiryō kō*, 1: 106.

62 Kurokawa, *Nihon ni okeru Meiji ikō no dojō hiryō kō*, 2: 3–4, 66–67.

63 Hashimoto Jurō, *Jūkōgyōka to dokusen*, vol. 2, *Senkanki no sangyō hatten to sangyō soshiki* (Tōkyō Daigaku Shuppankai, 2004), 137–38. For the expansion of ammonium sulfate production by Nihon Chisso Hiryo during World War I, see Barbara Molony, *Technology and Investment: The Prewar Japanese Chemical Industry* (Cambridge, Mass.: Harvard University Press, 1990), 85–146.

64 Sakaguchi Makoto, "Senkanki Nihon no ryūan shijō to ryūtsū rūto: Mitsui Bussan, Mitsubishi Shōji, Zenkōren o chūshin ni," *Rikkyō keizaigaku kenkyū* 59, no. 2 (2005): 156–60.

65 Sakaguchi Makoto, "Kindai Nihon no daizu kasu shijō: Yunyū hiryō no jidai," *Rikkyō keizai kenkyū* 57, no. 2 (2003): 62–66.

66 Hashimoto, *Senkanki no sangyō hatten to sangyō soshiki*, 2: 131–32, 141–42.

67 For example, see Minami Manshū Tetsudō Kabushiki Kaisha Chihōbu Shōkōka, *Manshū daizu kasu to sono shiryōka ni tsuite* (Dairen [Dalian]: Minami Manshū Tetsudō, 1932).

68 Sanka, *Daizu*, 293.

69　Kantō Totokufu Minseibu Shomuka, *Manshū daizu ni kansuru chōsa* (Dairen [Dalian]: Kantō Totokufu Minseibu Shomuka, 1912), 16.

70　Kaneko, *Kindai Nihon*, 47.

71　Ibid., 46.

72　Fukao Yōko, "Baikofu ni sasagu," in *"Manshu" no seiritsu*, 6–7; Nagai Risa, "Taiga no sōshitsu," in Ibid., 22.

73　Irie Hisao, *Manshū kanjin shokumin chiiki* (Dairen [Dalian]: Minami Manshū Tetsudō Kabushiki Kaisha, 1937), 28–29.

8

Struggling with Complex Natures in the Ogasawara Islands

COLIN TYNER

After almost a decade of campaigning, in June 2011 the Ogasawara Islands became one of the newest members of Japan's growing list of United Nations Educational, Scientific and Cultural Organization (UNESCO) World Heritage Sites. Situated along the Izu-Ogasawara Arc Trench, the islands span 400 kilometers from the Mukojima Group in the north to the Volcano Islands in the south. Because of the high proportion of species endemic to the islands and evidence of marine species' evolution into terrestrial species (such as the endangered land snails found on the island of Anijima), the Ogasawara Islands have been branded the "Galapagos of the Orient" (Tōyō no Garapagosu) since the mid-1990s. One issue that had previously undermined the islands' chances for designation as a World Heritage Site was that endemic species of flora and fauna on the islands had a lot of unwanted company, in the form of a substantial population of invasive species. Despite the successes of the campaign to remove many problematic species from islands, the UNESCO World Heritage Committee noted in its decision that the future status of the islands could still be threatened by nonindigenous organisms, whose "[f]uture invasions have the potential to compromise the very values the Ogasawara Islands have been recognized for and therefore need careful and continuous attention."[1] The encouragement of many of these unwanted introduced species occurred during a sustained period of large-scale agriculture in the nineteenth and twentieth centuries. From the mid-1880s, numerous species were brought to the islands by successive waves of Japanese settlement and government-initiated projects to render the islands profitable.

This chapter examines the history of these introductions during the period of industrial agriculture in the Ogasawara Islands, which by the early 1920s had transformed the islands into "Tokyo's largest natural greenhouse" (Tōkyō ichidai shizen no onshitsu).[2] From the late 1880s to the early 1920s, the Japanese central government used the structure of a plantation complex to change the demographic and ecological landscape of the Ogasawara Islands. This agricultural system, which lasted from 1883 to the colonization of the islands by the Japanese Imperial Military in the early 1930s, appealed to important stakeholders in the agricultural development of the Ogasawara Islands because it helped bind the terrestrial ecologies of the islands into a disciplined "instrument of production."[3] As with most "organic machines," however, the latent power to produce capital was difficult to maintain because of the challenges of enrolling heterogeneous human practices and the nonhuman logics and agencies of things that could not be easily reasoned with, browbeaten, or terrified.[4] The Japanese state and its human collaborators powered the complex not only by human work but also through a series of what Timothy Mitchell calls nonhuman "partnerships," which act through a logic and energy that human expertise does not fully command.[5] In the Ogasawara Islands, these nonhuman partners enrolled in the making and maintenance of the plantation complex included such things as potentially invasive plants and animals, altered soil conditions, and climate. The chapter ends with an examination of the months before the return of the islands to Japanese sovereignty in June 1968. It was then that former stakeholders in the maintenance of industrial agriculture had to come to terms with how these nonhuman partners had managed to transgress borders, thereby revealing an uneasy tension between national anxieties and local realities within which the landscape of the Ogasawara Islands was built—a borderland created from the promotion of a model of development that was meant to settle the landscape and yet unintentionally opened possibilities for unsettling the land.

"Settling" the Islands in the Extractive Space of the "Japan Ground"

In the first years of human settlement, botanical exchanges on the Ogasawara Islands were often haphazard and directed less by human intention than by the consumptive and world-altering appetites of animals that were introduced to the islands in the early nineteenth century. From the early 1820s to the late 1850s, the islands were situated in the "Japan" whaling ground. Pelagic whalers frequently visited the islands for shelter and supplies during their hunts for oil and fat, fur, and feathers for the markets of London, Paris, and New Bedford, Massachusetts. This routing of marine biota to metropolitan markets of Europe and North America was enabled by the openness of island and marine ecologies. Before being incorporated into the Empire of Japan in 1875, the Ogasawara Islands remained outside the explicit territorial boundaries of nation-states. John Gillis has suggested that the power of islands in the nineteenth century was not in their bounded natures but in their openness. Located within the spaces of maritime empires and extractive open-access resource frontiers, islands in the North Pacific—like those of the Atlantic world—were less important for the maintenance of security than for their strategic importance in the projection of mercantilism.[6]

The islands' first settlement was established by a group of thirty Euro-American former whalers and Pacific Islanders, who landed on the largest of the then-uninhabited islands in June 1830. Settling the islands was not easy work, despite reports that there was "water in almost every valley."[7] Recognizing that more capital could be drawn from the extraction of marine resources than from cultivating the land, the settlers made up their calorie deficit in the early years by continuing the hunting practices they had learned on whaling ships in the waters around the islands. The former whalers primarily hunted green sea turtles (*Chelonia mydas*), which were "so numerous [on the islands] that they hide the colour of the shore."[8] They also hunted their "menagerie on deck"[9] of pigs and goats from the Hawaiian Islands, as well as the already flourishing populations

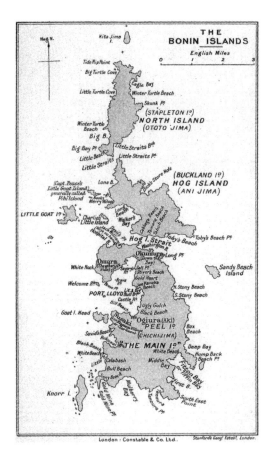

FIGURE 8.1. Map of the Ogasawara Islands

SOURCE: Lionel Berners Cholmondeley, *The History of the Bonin Islands from the Year 1827 to the Year 1876, and of Nathaniel Savory, One of the Original Settlers; To Which Is Added a Short Supplement Dealing with the Islands after Their Occupation by the Japanese* (London: Constable, 1915), http://mysite.du.edu/~ttyler /ploughboy/boninchol.htm (accessed November 1, 2013).

of feral pigs that had been ravaging the roots of trees and sea turtle eggs since their introduction by whalers in 1826.[10]

The first settlers and their descendants kept the environmentally detrimental behavior of goats and pigs in check through hunting, or through the exiling of animals to the smaller islands of Buckland (Anijima) and Stapleton (Otōtojima). These two islands were so defined by these ungulates that the settlers renamed them "Hog Island" and "Goat Island," respectively.[11] Surveyors from Commodore Matthew Perry's landing on the Ogasawara Islands in 1854 marveled at the fecundity of the goats and their ability to overwhelm "indigenous productions": "The goats which had been introduced [to Stapleton Island] had increased marvelously, to the extent, it was supposed, of several thousands, and had become very wild in the course of their undisturbed wanderings through the secluded ravines."[12] Perry's crew also took note of "sheep, deer, hogs, and goats with an infinite number of cats and dogs. The cats and hogs, having lost some of their quiet domestic virtues, had strayed into the jungle, and were dignified by the inhabitants with the title of wild animals."[13]

▓ The Construction of the Plantation Complex

With the beginning of the Japanese colonization of the Ogasawara Islands in 1875, the central government of Japan approached the islands as an environmentally precarious place where failure to control invasive humans, nonhumans, and pathogens could have catastrophic consequences for their development and security. Officials responsible for the management of island environments at the Ministry of the Interior (Naimushō) intervened as much as possible to hold back a plethora of anthropogenic, biophysical, and climatic threats that came with commercial agriculture and mass settlement of the islands. Established human practices that promoted neither the economic development of the islands nor their stabilization posed a special problem for the central Japanese state. The administrators of the islands had to deal with social practices that were less oriented to the land than to the open, ill-defined marine ecosystems surrounding the islands. To begin the material reclamation of the islands in 1875, more than a decade after the Tokugawa shogunate's failed attempt to colonize the islands between 1862 and 1863, the central government regulated land-use practices to prevent the clearing of woodlands, overgrazing, and soil erosion. The Ministry of the Interior reestablished regulations for islands (shima kisoku), regulations for the harbor (minato kisoku), and duties and tariffs for the import and export of goods (yushutsunyūhin zeisei), regulations that had been first put in place by the Tokugawa government in 1862.[14] These regulations were further strengthened in April 1883, when officials from the Ministry of Finance noted that upland soil in

Chichijima, the largest, most populated island in the Ogasawaras, was suffering from overgrazing and overuse.

In the first decade of open settlement, new settlers, most of them from the island of Hachijōjima, managed to exhaust much of the woodlands in search of green fertilizer and building materials.[15] In response, the government restricted the reclamation *(kaikon)* of groundwater. It also protected up to 70 percent of the forested area in Chichijima, limiting foraging by settlers to the small strips of "commoner lands" *(min'yūchi)* situated near the water's edge.[16] The boundaries worked to discourage foraging in the restricted forests *(seigenrin)* inland.[17] Most of the wood, of both use and market value, was located in the interior of the island.[18]

Despite the land-use restrictions state officials placed on the islands, hunting and foraging in the woodlands persisted well into the 1890s. As late as 1888, one of the stakeholders in the management of natural and human resources of the islands complained in *Ogasawara yōran (Perspectives on the Ogasawara Islands)* that after nearly a decade of Japanese development, the descendants of the first islanders remained "dull-witted" *(orokamono)* layabouts who "don't

FIGURE 8.2.
"Naturalized" Japanese on the beach, Ogasawara.

SOURCE: Shinkōsha, *Nihon chiri fuzoku taikei*, vol. 1 (Shinkōsha, 1931), 286, 287.

farm," "fish little," and were often "up to mischief."[19] Without government "guidance" to encourage them to cultivate *(kaitaku)* the land, the islanders would continue to "live freely with little concern for their own education or material progress."[20] The writers of the book did not hold out much hope for these naturalized Japanese *(kikajin)*:

> Stated simply, you could say that their way of living in the simplest terms was old fashioned, similar to the ways people lived in ancient times … [and] for a person of the nineteenth century, it would seem that they are stubbornly protecting their indigenous ways of living. For those who grew up on the islands there are a number who walk dirty and shoeless among the goats of the islands.[21]

These are examples of some of the "destructive side effects" that came with the inscription of national development policies onto the social and ecological terrain of the Ogasawara Islands.[22] One of the most destructive, and purposeful, reactions to the imposition of the Meiji state's "culture of control" over the islands was the persistence of arson, which government officials blamed on the settlers' "fears of fierce poisonous beasts that lived in the forest."[23]

State officials continued to be frustrated with human practices and logics that pulled at the edges of their authority until the introduction of mass sugarcane cultivation and the complex of labor and machinery that accompanied it. The plantation complex modified the land of the islands to make it agriculturally productive, transforming the terrestrial ecosystem of the islands into what Marx might call the "monopolizable force of Nature."[24] The process began with the introduction of hundreds of laborers between the late 1880s and the early 1920s. Sugarcane cultivation and the refinement of cane into processed sugar are extremely labor intensive, requiring approximately one worker for each acre of cultivated land.[25] The islands had soil and climate that were suitable for sugarcane cultivation but not enough people to clear land and process the cane, which must be done within about twenty-four hours of cutting.[26]

The government used the plantation system to give direction, organization, and control to migration, which hitherto had been quite modest. With only 531 people living on the islands in 1885, the Ogasawara Islands as a settler colony can only be classified as a demographic failure. Things began to change with the introduction of industrialized sugar farming in that year. Within a decade (1895), the civilian population had leaped to 4,018 people. Five years later (1900), the population had risen to 5,550. It stabilized thereafter and was still around 5,500 in 1921, at the time of the global sugar collapse.[27] Most of the people who came to the islands from the early 1890s to the mid-1920s were enlisted to work in the sugar plantations as either managers or laborers. The mass cultivation of sugarcane began in 1889, using the species of sugarcane that the first settlers had brought from the Hawaiian Islands in 1830.[28] However, the structure

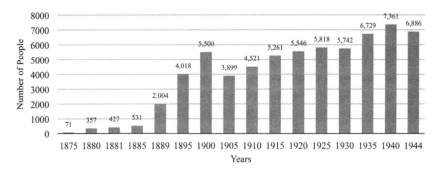

FIGURE 8.3. Population change in the Ogasawara Islands, 1875–1940

SOURCES: Compiled from Ishihara Shun, *Kindai Nihon to Ogasawara shotō: Idōmin no shimajima to teikoku* (Heibonsha, 2007), 268, unnumbered table; Dangi Kazuyuki, "Ogasawara shotō iminshi," in *Umi to rettō bunka: Kuroshio no michi,* edited by Amino Yoshihiko, Ōbayashi Taryō, Hasegawa Kenichi, Miyata Noboru, and Mori Kazuo (Shōgakukan, 1991), 256, unnumbered table; and Shinkōsha, *Nihon chiri fuzoku taikei,* vol. 1 (Shinkōsha, 1931), 272, unnumbered table.

of cultivation and the scale of sugar processing in the Japanese plantation system differed greatly from the case in Hawaii. Instead of being turned into rum, the cane was processed in refineries into brown and white sugar for metropolitan markets. Cane harvesting was done in the dry season, normally starting in February and ending at the beginning of May.[29] Workers were brought to the islands through a system of consignment. Initially, laborers came from Hachijōjima and poorer agricultural areas in Japan, such as Tokushima and Shizuoka prefectures.[30] The cheapest labor was provided by boys between the ages of eight to nineteen years of age; they came from sugarcane plantations in Okinawa or the Tokyo juvenile reformatory system in Tokyo.

The most dramatic transformation of the demographic and physical landscape of the Ogasawara Islands took place on the second-largest island, Hahajima. Since the islands' settlement in 1830, Hahajima remained underpopulated and undercultivated until the introduction of commercial sugar cultivation in the mid-1880s. Steamer service to the islands, initiated in 1885, allowed for the immigration of settlers and laborers from Hachijōjima and Okinawa Prefecture to work in the sugar fields owned by Yasui Mankichi, a farmer from Hachijōjima, who began to test-plant sugarcane in 1887.[31] By 1900, three years after the start of sugarcane cultivation on the island, the population of Hahajima (3,000) was larger than that of Chichijima (2,250).[32]

Early successes eventually led to the transformation of the island into an enormous plantation complex, which exhausted the nitrogen in both the forests and the soil, changing the microclimate of the islands. But this change did not proceed without resistance. One of the consequences of altering the ecological and social landscape of the islands wholesale was the increased appearance of "invasiveness," or what foresters and Ministry of the Interior officials marked down as resistance to "naturalization" *(kika)*. Introduced species of plants marked as invasive were threatening because of their demands on the growing space of the islands. The concern was that their growth could compete for the available nitrogen. In 1930, *Aspects of Life on Ogasawara Islands (Ogasawara shotō seibutsu)*, published by the Japanese Biogeography Association (Nihon Seibutsu Chiri Gakkai), stated that "on the Ogasawara Islands, there are more kinds of plants that flourish on their own than the kinds of plant life that are purposely grown on the islands" and that "there is a need to differentiate between plants that are naturalized *[kika shokubutsu]* and those that are contaminative *[konkō]*."[33]

Opportunities for further border breaching by introduced plants increased over time with the deterioration of growing conditions on the islands. By 1890, harvestable growth on the islands was decreasing. In 1893, farmers reported that sugarcane fields on the island of Hahajima yielded less sugarcane than in the past. By 1897, the fields were described as being barren. Each year, hillsides collapsed and washed away, making large-scale farming impossible.[34] The rapid denuding of the island's forests set in motion further ecological changes to the islands.

FIGURE 8.4. Sugarcane fields on the island of Hahajima, Ogasawara

SOURCE: Shinkōsha, *Nihon chiri fuzoku taikei*, vol. 1 (Shinkōsha, 1931), 297.

Faced with a critical shortage of fuel and ecological collapse, the Tokyo Forestry Bureau (Tōkyō Eirinkyoku) began what could be called the first ecological restoration project on the islands. Foresters concentrated on the introduction of plants that could be used for further agricultural development. The system employed for these introductions at the beginning of the twentieth century was more sophisticated than that used by the Ministry of the Interior in the initial stages of settlement. It was developed as a varied system of management to maximize the potential of the islands' landscape. Okinawan pine *(Pinus luchuensis)* was brought in from Okinawa for construction purposes, rosewood *(Dalbergia latifolia)* was introduced from Southeast Asia for its high commercial value, and crape myrtle *(Lagerstroemia indica)* from the Korean peninsula and ironwood *(Eusideroxylon zwageri)* from Borneo were brought in for the building of bridges.[35] Through agricultural science and controlled botanical transfers, managers were able to repair some of the damage done to a number of the keystone species of flora through the introduction of surrogates.

But the movement of these surrogate plants often followed a logic that worked against the intentions of the foresters who planned their introduction. Many of the introduced plants had high dispersal rates and filled the ecological niches vacated by plants that had been cut to levels too low to harvest for sugar production.[36] By following the movement of these species along the edges of the plantation complex, just on the edge of human control, we can see the limits of the plantation complex in making the Ogasawara Islands productive. One species of plant that quickly filled the ecological niches left empty by the appetites of sugar production was the lead tree *(Leucaea glauca)*. Native to South America, it was first brought to Chichijima in 1879 from colonial India to prevent erosion and provide fuel for sugar refinement, feed for livestock, and green fertilizer. A 1914 book, *The General Conditions of the Ogasawara Islands and Their Forests (Ogasawarajima no gaikyō oyobi shinrin)*, noted that the plant was particularly hardy and that "there was no reason to transplant the plant by seedlings as the plant naturally flourished."[37] Forestry officials were able to keep the lead tree in check, for the most part, by confining it to 837 locations on Chichijima.[38] Another flourishing tree species introduced purposefully through the front door, but from an intercolonial route, is today on Japan's "most-wanted" list of noxious invasive species.[39] Bishopwood *(Bischofia javanica)*, a tree endemic to Southeast Asia, was cultivated on Dutch East Indies plantation complexes for fuelwood production. Its cultivation in Chichijima and Hahajima was limited to five plantation areas across the islands, but it was dispersed quickly by wind and by birds that consumed its seeds.[40]

However, no one was able to anticipate or control the effects that the collapse of the sugar industry at the end of the 1920s and the increased militarization of the islands would have on their ecology. In 1927, the sugar industry on the islands brought in ¥305,000 of taxable revenue to the islands. In 1933, it brought in only ¥81,905.[41] By the early 1930s, the laborers on the sugar plantations were following commodities to other plantation complexes like those on the South Sea Islands (Nanyō Guntō) because of the displacement caused by the sugar collapse.[42]

▓ From Island Garden to Island Fortress

By the early 1930s, the capacity of foresters and farmers to repair the built environment of the plantation complex was being challenged by the appropriation of labor and nonhuman resources by the Japanese imperial military. The militarization of the Ogasawara Islands began modestly, with the Japanese Imperial Navy's construction on Chichijima of a communications center in 1914 and a base in 1916.[43] Militarization was not limited to the state's improvement of its capacity to defend or attack in the case of war. It was also a process by which the island chain as a system of production became "controlled by, dependent on, [and] derived its value from the military as an institution."[44] Farmers were still able to continue their work, but by July 1927, with the coordination of village leadership, the forest bureau, the fortification command bureau (yōsai shireibu), the naval communication station, the military police (kenpei bunchūjo), the police, the post office, the regional judiciary, the local militia, and primary schools, the fortification and appropriation of labor to fit the Ogasawara Islands within national defense plans had moved into high gear.[45]

Because plantation work is organized on the basis of labor gangs that carry out repetitive and physically demanding tasks under the supervision of managers who sequence and synchronize their movements, it was readily appropriated by the military. The Japanese Imperial Army effectively took over the role of enlisting the labor and mobilized the entire population to fortify the Ogasawara Islands. By the early 1940s, it had transformed the islands into the "Gibraltar of the Pacific."[46]

The Japanese Imperial Army had an insatiable appetite for labor. By 1940, the islands hosted a civilian population of 7,361 and a military population of more than 20,000 soldiers. Able-bodied men and women who made their homes on the islands were enlisted to perform backbreaking work for the military. The army began by directing people to dredge Futami Harbor in Chichijima to allow warships to anchor and to help dig hundreds of tunnels for defensive and supply purposes.[47] In 1939, the Imperial Navy constructed an airstrip, a navy defense base, and an air base on the islands.[48] The military appropriation of the islands was completed in 1944, when the army attaché in consultation with the chief administrator (shichochō) of the Ogasawara Islands and the police chief decided to evacuate most—but not all—of the islanders to the main island of Honshu to conserve resources (both human and nonhuman) needed for the defense of the islands.[49] In February 1944, the remaining civilians, many of whom were suffering from outbreaks of epidemic, "eruptive" typhus and tuberculosis, were forcibly evacuated from the islands.[50] Many of them would not return home until June 1968.

Without farmers to work the fields, the agricultural system began to unravel, and the soldiers began to ravage and deplete the landscape. In addition to lacking adequate drinking water, the islands had difficulty accommodating the caloric and nutritional needs of twenty thousand soldiers. Most of their protein was obtained through eating Ogasawara flying fox *(Pteropus pselaphon)*, which the soldiers drove out of caves, and feral goats that remained on some of the outer islands.[51]

Although the insatiable appetites of the imperial military depleted the environments of the islands of caloric content, there is some evidence that military presence and activity increased the opportunities of some species to flourish. *Leucaea glauca*, for example, quickly filled in the spaces vacated by the clearing of trees for fuel. The military encouraged the growth of lead tree because of its ability to provide cover over concrete structures like radio stations, bunkers, and artillery mounts.[52]

One of the stranger purposeful introductions that took advantage of the coming of the military, and one that was likely linked to the spread of the lead tree, was the giant African snail *(Achatina fulica)*, introduced to Chichijima sometime between 1935 and 1937 by an entrepreneur, Tazawa Shingo. Tazawa asserted that "this light-purple snail can be used for food." He went on to state that "[i]ts fleshy, pink meat has earned it the name land abalone *[riku no awabi]* and within one year a person can produce to up to five thousand snails. It is a farmer's dream come true."[53] His optimism proved shortsighted. The snails turned out to be a commercial flop—and an agricultural and hygienic hazard. Giant African snails are known to carry rat lungworm *(Angiostrongylus cantonensis)*, a vector for eosinophilic meningitis.[54] On the mainland, government sanitation officers quickly became concerned with the health risks of the snail because of reports that people were eating the "fleshy, pink meat" of the snails raw. In 1936, after hearing word overseas that the snail caused damage to agricultural products, the government prohibited the importation and use of the snails in Japan. Yet an article about the snails in the first issue of the flagship journal for research on the Ogasawara Islands, *Ogasawara Islands Annual Research Review*, carried the provocative subtitle "The Ecology of an Aggressor," suggesting that the Imperial Japanese soldier's taste for escargot was perhaps not such a bad thing.[55] During the Pacific War, the breeding of snails by soldiers was encouraged as a food source.[56]

Conclusion: An Imagined Future of the "Galapagos of the Orient"

From the end of World War II until 1968, the Ogasawara Islands were under the administrative control of the U.S. Navy as a Pacific Trust Territory. Each family of the 137 descendants of the original settlers of 1830, who were allowed to return to the islands because they were not considered a "security threat," was provided with a 60-square-meter house, education, and medicine. Besides hauling freight for the military, the islanders worked as fishermen, divers, carpenters, or construction crews for the building of roads.

One scholar of the islands aptly summed up the effects of the U.S. Naval administration of the Ogasawara on the environment as follows: "They couldn't have cared less about the environment."[57] When I asked a person who grew up on the islands during the American interlude how the environment fared under the U.S. Navy, he joked that other than using the goats for target practice, throwing grenades in abandoned tunnels, and storing nukes in the tunnels with copper doors, the land was something that the military took for granted.[58]

Taken as a whole, the U.S. military's attitude toward the environment on the Ogasawara Islands can be characterized as indifference or neglect. Of course, a lot can result from doing nothing, and a lot can be imagined by those whose access is thwarted. When the first Japanese government surveyors returned to

FIGURE 8.5. World War II plane wreck covered with foliage
on the island of Chichijima, Ogasawara

Photograph by author, November 2010.

the islands in May 1968, in preparation for the eventual handover of the islands on June 6, 1968, they encountered an alien landscape, markedly different from the one that some of them remembered from their trips as students in the prewar period.

Opinions on these changes were mixed. Some people with memories of the prewar landscape of Ōmura, the main settlement in the Ogasawara Islands, complained that the line of handsome Alexandrian laurels *(Calophyllum inophyllum),* cultivated in the 1910s to provide a windbreak, had been cut down and replaced with a grassy field and picnic tables.[59] Yamazaki Satoshi, a geographer, wrote that "walking under the thick leaves of the trees gave you the feeling that you were in a completely different country *[mattaku chigatta ikoku]* from Tokyo." He continued: "In place of the forest where Ōmura used to be was a large field, and around houses bloomed tropical flowers like hibiscus, bougainvillea, and buttercups, making it really feel like we had come to the tropics."[60] On the other hand, the fact that reestablishment of human communities had been limited to Chichijima meant that most of the islands went through a flourishing process of regrowth that the natural scientists working in the islands through the 1970s and 1980s called "jungle-ification" *(janguruka).* Hahajima went through this change in spectacular fashion, with thick foliage covering the former settlement. *The Ogasawara Islands Have Returned!—Ah, Iwo Jima (Kaette kita Ogasawara—Chichijima, Hahajima—Ā Iōtō),* published by the *Tōkyō shinbun* company one month before the handover, suggested that Hahajima's reversion to an uninhabited island state *(mujintōka sareta jōtai)* provided an opportunity for "a new start" for the metropolitan government's management of the islands. In 1977 Kazaki Yukio, the first chair of the Ogasawara Islands Research Group, explained that because of their "unique position," the Ogasawara Islands represented to scientific researchers an example of a "pristine" and "unpolluted environment."[61]

Government officials, tour operators, and academics have attempted to promote the islands as one of the last places of "wild Japan" by citing its high rates of endemicity and distance from the metropolitan center.[62] But since the mid-nineteenth century, one could certainly make the claim that most of what could be described as "wild" has been produced not by endemic species in isolation but by introduced species of plants and animals that have worked their way into presettlement ecological niches and gaps created by the destruction of the environment through more than a century of economic development and resource extraction. Nigel Clark has suggested that species of plants and animals introduced to islands since the first Cook voyages have shown a remarkable capacity to "naturalize" not only in the spaces to which they were introduced but also outside spaces planned or managed by human beings. These species show that "nature" is not always something that seeks to be rooted, but plenty of "natural,"

weedy things that are unwanted by human beings can move in.[63] These movements may have begun with human beings; however, the human colonizers set
into motion machinery beyond their ultimate control, and "no legislation can
regulate the dissemination of seeds."[64]

NOTES

1   World Heritage Committee, "Convention Concerning the Protection of the World Cultural
    and Natural Heritage, 35th Session" (Paris: UNESCO, 2011), 180, http://whc.unesco.org
    /archive/2011/whc11-35com-20e.pdf (accessed July 20, 2011).
2   Tōkyō-fu Eirin-kyoku, *Ogasawara-jima kokuyūrin shokubutsu gaikan* (Tōkyō-fu, 1929), 14.
3   David Harvey, *The Limits to Capital*, new ed. (London: Verso, 2006), 334–36.
4   Christopher R. Henke, "Situation Normal? Repairing a Risky Ecology," *Social Studies of Science*
    37, no. 1 (2007): 135–42; Richard White, *The Organic Machine: The Remaking of the Columbia
    River* (New York: Hill and Wang, 1995). Michel Callon used the concept of "enrollment" to
    describe the devices and the processes of negotiation used by key stakeholders in the St.
    Brieue Bay, France, scallop fishery to coax investors, fishermen, scallops, and other key actors
    to make the fishery work. Michel Callon, "Some Elements of a Sociology of Translation:
    Domestication of the Scallops and the Fishermen of St. Brieue Bay," in *Power, Action, and
    Belief: A New Sociology of Knowledge*, ed. John Law (London: Routledge & Kegan Paul, 1986),
    216–17.
5   Timothy Mitchell, *Rule of Experts: Egypt, Techno-Politics, Modernity* (Berkeley: University of
    California Press, 2002), 299.
6   John R. Gillis, "Islands in the Making of an Atlantic Oceania, 1400–1800" (paper presented
    at "Seascapes, Littoral Cultures, and Trans-Oceanic Exchanges," Library of Congress,
    Washington, D.C., February 12–15, 2003, http://www.historycooperative.org/proceedings/
    seascapes/gillis.html, accessed September 9, 2013); John R. Gillis, *Islands of the Mind: How the
    Human Imagination Created the Atlantic World* (New York: Palgrave Macmillan, 2004), 91–92.
7   Frederick William Beechey, John Richardson, Nicholas Aylward Vigors, George Tradescant
    Lay, Edward Turner Bennett, Richard Owen, John Edward Gray, George Brettingham Sowerby,
    William Buckland, Edward Belcher, and Alexander Collie, *The Zoology of Captain Beechey's
    Voyage* (London: H. G. Bohn, 1839), 209.
8   Beechey et al., *The Zoology of Captain Beechey's Voyage*, 210.
9   Wade Graham, "Traffick According to Their Own Caprice: Trade and Biological Exchange in
    the Making of the Pacific World, 1766–1825" (paper presented at "Seascapes, Littoral Cultures,
    and Trans-Oceanic Exchanges," Library of Congress, Washington, D.C., February 12–15, 2003,
    http://www.historycooperative.org/proceedings/seascapes/graham.html, accessed September
    9, 2013).
10 Tanaka Hiroyuki, *Bakumatsu no Ogasawara: Ōbei no hogeisen de sakaeta midori no shima*
    (Chūō kōronsha, 1997), 33–35.
11 The most widely recognized names of the Ogasawara Islands were designated by the captain
    of the H.M.S. *Blossom*, Frederick William Beechey, who visited the islands in 1828. Frederick
    William Beechey, *Narrative of a Voyage to the Pacific and Beering's Strait: To Co-operate with
    the Polar Expeditions: Performed in His Majesty's Ship Blossom, under the Command of Captain
    F. W. Beechey, R.N. in the Years 1825, 26, 27, 28* (Philadelphia: Carey & Lea, 1832); Beechey et al.,
    The Zoology of Captain Beechey's Voyage.

12   Francis L. Hawks, *Narrative of the Expedition of an American Squadron to the China Seas and Japan, Performed in the Years 1852, 1853, and 1854 under the Command of Commodore M.C. Perry, United States Navy* (New York: D. Appleton and Company, 1856), 242.

13   Ibid., 246.

14   Tsuji Tomoe, *Ogasawara shotō rekishi nikki*, vol. 2 (Kindai bungeisha, 1995), 131.

15   Tanaka, *Bakumatsu no Ogasawara*, 186–87.

16   Yamagata Ishinosuke, *Ogasawarajima shi* (Tōkyōdō, 1906), 318–22.

17   Tōkyō-fu Ogasawaratō-chō, *Ogasawara-jima no gaikyō oyobi shinrin* (Ogasawara-mura: Tōkyō-fu Ogasawaratō-chō, 1914), 20; Kimura Jun, "Ogasawara no ginnemu hayashi," *Ogasawara kenkyū nenpō* 2 (1978): 19–28.

18   Tōkyō-fu Dobokubu, *Ogasawara no shokubutsu* (Tōkyō-fu, 1935), 37.

19   Isomura Teikichi and Tsuda Sen, *Ogasawara-jima yōran* (Ben'ekisha, 1888), 156.

20   Ibid., 157.

21   Ibid., 157.

22   Nigel Clark, "Wild Life: Ferality and the Frontier with Chaos," in *Quicksands: Foundational Histories in Australia and Aotearoa New Zealand*, ed. Klaus Neumann, Nicholas Thomas, and Hilary Ericksen (Sydney, New South Wales: NSW Press, 1999), 152.

23   Ogasawaratō-chō, *Ogasawara-jima no gaikyō oyobi shinrin*, 51.

24   Cited in Harvey, *The Limits to Capital*, 336.

25   Philip D. Curtin, *The Rise and Fall of the Plantation Complex: Essays in Atlantic History*, 2nd ed. (Cambridge: Cambridge University Press, 1998), 4.

26   Jason W. Moore, "*The Modern World-System* as Environmental History? Ecology and the Rise of Capitalism," *Theory and Society* 32 (2003): 347–48.

27   Ishihara Shun, *Kindai Nihon to Ogasawara shotō: Idōmin no shimajima to teikoku* (Heibonsha, 2007), 268; Dangi Kazuyuki, "Ogasawara shotō iminshi," in *Umi to rettō bunka: Kuroshio no michi*, ed. Amino Yoshihiko, Ōbayashi Taryō, Hasegawa Kenichi, Miyata Noboru, and Mori Kazuo (Shōgakukan, 1991), 256; Shinkōsha, *Nihon chiri fuzoku taikei*, vol. 1 (Shinkōsha, 1931), 272.

28   Ishihara, *Kindai Nihon to Ogasawara shotō*, 248.

29   Midori Arima, "An Ethnographic and Historical Study of Ogasawara/the Bonin Islands, Japan" (Ph.D. diss., Stanford University, 1990), 48.

30   Ibid., 48.

31   Ibid., 85–90.

32   Funakoshi Masaki, "Ogasawara ni okeru ginnemu hayashi no seiritsu—I'nyū to bunpu no kakudai o meguru oboegaki—sono 2," *Ogasawara kenkyū nenpō* 11 (1987): 43.

33   Kimura, "Ogasawara no ginnemu hayashi," 20.

34   Tōkyō-fu Eirinkyoku, *Ogasawara-jima kokuyūrin shokubutsu gaikan*, 53, 136; Tōkyō-fu, *Ogasawarajima sōran* (Tōkyō-fu, 1929), 154–55.

35   Ogasawaratō-chō, *Ogasawara-jima no gaikyō oyobi shinrin*, 178.

36   John Robert McNeill, *Mosquito Empires: Ecology and War in the Greater Caribbean* (Cambridge: Cambridge University Press, 2010), 26–32.

37   Cited in Kimura, "Ogasawara no ginnemu hayashi," 22.

38   Tōkyō-fu Eirinkyoku, *Ogasawara-jima kokuyūrin shokubutsu gaikan*, 28.

39   Nihon Seitai Gakkai, *Gairaishu handobukku* (Chijin Shokan, 2002), 362–63.

40   Nobuyuki Tanaka, Keita Fukasawa, Kayo Otsu, Emi Noguchi, and Kumito Koike, "Eradication of the Invasive Tree Species *Bischofia javanica* and Restoration of Native Forests on the Ogasawara Islands," in *Restoring the Oceanic Island Ecosystem: Impact and Management of*

Invasive Alien Species in the Bonin Islands, ed. Isamu Okochi and Kazuto Kawakami (London: Springer, 2010), 162.

41 Tōkyō-fu Dobokubu, *Ogasawara no shokubutsu*, 13.

42 Ishihara, *Kindai Nihon to Ogasawara shotō*, 61.

43 Arima, "An Ethnographic and Historical Study of Ogasawara Islands," 51.

44 Cited in Michael Szonyi, *Cold War Island: Quemoy on the Front Line* (Cambridge: Cambridge University Press, 2008), 3.

45 Tsuji, *Ogasawara shotō rekishi nikki,* 287.

46 Gavan McCormack and Nanyan Guo, "Coming to Terms with Nature: Development Dilemmas on the Ogasawara Islands," *Japan Forum* 13, no. 12 (2001): 183.

47 Arima, "An Ethnographic and Historical Study of Ogasawara Islands," 52.

48 Ibid., 53.

49 Ogasawara Kyōkai, *Ogasawara shotō gaishi: Nichi-Bei kōshō o chūshin to shite* (Ogasawara Kyōkai, 1967), 27.

50 Ibid., 24–26

51 Tsuji, *Ogasawara shotō rekishi nikki,* 367.

52 Kimura, "Ogasawara no ginnemu hayashi," 23.

53 Toyama Kiyonori, "Ogasawara no Afurika maimai," *Ogasawara kenkyū nenpō* 11 (1985): 8–16.

54 Albert R. Mead, *The Giant African Snail: A Problem in Economic Malacology* (Chicago: University of Chicago Press, 1961), 151.

55 Toyama, "Ogasawara no Afurika maimai," 1–6.

56 Yoshikawa Kenji, "Ogasawara no Afurika maimai—Shinrakusha no seitaigaku," *Ogasawara kenkyū nenpō* 1 (1977): 49–56.

57 Personal correspondence with Gavan McCormack in Mitaka, Tokyo, October 2005.

58 Interview conducted by Colin Tyner in Chichijima, Tokyo, November 2009.

59 Tōkyō shinbun, *Kaette Kita kita Ogasawara—Chichijima, Hahajima—Ā Iōtō* (Tōkyō Shinbunsha, 1968), 86.

60 Yamazaki Satoshi, "Henkan 15 nengo no Ogasawara no shokubutsu," *Ogasawara kenkyū nenpō* 8 (1984): 5–6.

61 Kazaki Hideo, "Ogasawara kenkyū no igi," *Ogasawara kenkyū nenpō* 1 (1977): 1–4.

62 Gavan McCormack and Nanyan Guo, "Coming to Terms with Nature: Development Dilemmas on the Ogasawara Islands": 177.

63 Nigel Clark, "The Demon-Seed: Bioinvasion as the Unsettling of Environmental Cosmopolitanism," *Theory Culture Society* 19, no. 1–2 (2002): 104.

64 H. Guthrie-Smith, *Tutira: The Story of a New Zealand Sheep Station* (Seattle: University of Washington Press, 1999), 294, cited in Nigel Clark, "The Demon-Seed," 101.

9

When the Green Archipelago
Encountered Formosa

The Making of Modern Forestry in Taiwan under
Japan's Colonial Rule (1895–1945)

KUANG-CHI HUNG

In 1895, Taiwan became the first colony of the Empire of Japan. Long known as Formosa, "beautiful island," this enchanting land of high mountains, dense forests, dramatic beaches, and diverse ethnic groups thereafter experienced a multitude of changes. Among them was a shift in the relationship of Taiwanese society to its forests. Over the preceding two centuries or so, Taiwan's previous overlord, the Chinese Qing Empire, thought of most of Taiwan's forests as a "land of savages." Prior to 1875, the Qing Empire had neither the ability to manage this vast and rugged wilderness nor an interest in doing so. Yet, after this time, chiefly because of the huge profits generated by Taiwan's camphor and tea industries, the Qing Empire drastically altered its attitude. Under the motto *kaishan fufan*, or "open up mountains and pacify savages," the Qing Empire encouraged Chinese settlers to enter Taiwan's mountains, to wrest the forests away from the indigenous peoples and turn them into arable land and, above all, to make good use of the island's abundant forest resources. It was against this backdrop that the Japanese Empire took over Taiwan. Coming from a civilization with a remarkable history of reforestation and forest protection, the colonial government had no doubt that Japan's governance would revolutionize the Qing's feudal, exploitative, and shortsighted forestry regime. The effort to integrate Formosa into Japan's "Green Archipelago" now began. As the first governor-general of Taiwan, Kabayama Sukenori, famously said, "To subjugate Taiwan, we must conquer its forests."[1]

This essay offers an overview of how the Japanese Empire "conquered" Taiwan's forests and established so-called scientific forestry in the mid-1930s. I focus on actors who can be conveniently grouped under the colonizing class: the colonial government, forest scientists, technical bureaucrats, and forestry entrepreneurs. I aim to reveal the conflicting interests and contested views within this class regarding Taiwan's forest management.[2] As we shall see, Japan's eventual conquest of Taiwan's forests was by no means a linear process dictated by the colonial state in collaboration with entrepreneurs and forest scientists aiming for better exploitation of the colony's natural resources. Rather, the conquest experienced multiple "paradigm shifts," and the process endured remarkable discontinuities and disjunctures, with the forestry scheme at each stage being pushed ahead by significant challenges and conflicts. Only as we appreciate such drastic shifts can we effectively grasp the environmental and social changes that forestry induced in the context of colonialism and in the age of empire.

▦ Contested Views, Conflicting Policies

The first year of colonial rule witnessed Japan's attempt to regulate Taiwan's forest-society relationship. In 1895, despite having only a remote idea of what constituted a Taiwanese forest, the colonial government issued a law from its highest court protecting Taiwan's forests from the destructive forces of "capitalist entrepreneurs" from Japan. In the law the colonial government described this particular class of people as "profit seekers" and expressed serious doubt about their ability to manage the forests properly. Driven by potential profits from Taiwan's forests, the colonial government maintained, entrepreneurs were bound to disturb Taiwan's forest-society relationship, causing floods and other unpredictable disasters. The colonial government concluded that Taiwan's forests should be government-owned, and its forestry should be undertaken with government supervision. The colonial government should retain control over forest resources at all times and by any means necessary. The state—and the state alone—should serve as a gatekeeper or overseer, with control over access to Taiwan's forests.[3]

But when the colonial government's financial problems later became critical, it was forced to revise its rather idealistic forestry policy. Around the turn of the twentieth century, the colonial government began regarding most natural forests located in Taiwan's hills (nowadays characterized as the Machilus-Castanopsis Zone), as well as secondary or artificial forests close to Chinese settlements, as "wastelands" or "degraded forests." It then proposed that such forests be reclaimed and reforested with camphor, cypress, and other economically valuable trees.[4] The government had to bolster its argument as Japanese governance in Taiwan confronted extensive resistance from local society. In fact, no sooner had the colonial government stabilized itself than the forest become a theater for "bandits" and other forest dwellers to unite in opposition to government forces. Worse, the indigenous peoples of Taiwan had a notorious history of employing headhunting and other violent methods to protect the forests against incursion. Were the Japanese to attempt to exploit Taiwan's forests, these peoples would hunt the intruders down with little reluctance. To be sure, as

Governor-General Kabayama declared, "To subjugate Taiwan, we must conquer its forests," and in his view, this conquest could not be achieved without taking down the indigenous peoples—the "raw savages," to use his words.[5]

Colonial foresters played a pivotal role in shaping the colonial government's attitudes toward Taiwan's forests. Trained at professional forestry schools in Europe or Japan in an era when Western sciences were hailed as the dominant way of studying nature, they came to Taiwan with the ambitious goal of bringing the forces of modern science and commerce to bear on Taiwanese forestry.[6] Upon their arrival, they endeavored to carry out a survey to reveal the value of Taiwan's forests. Although its scope was confined to the plains and hills, this survey gathered rich firsthand information about Taiwan's forest-society relationship. Among the survey's striking conclusions was that Taiwan's forests were degraded and unregulated. The foresters, after intense study of major Chinese settlements, became convinced that Taiwan's plains and hills had been too aggressively developed, resulting in desolation. Beyond this, they also found that even forests that had escaped ruthless Chinese land-use methods were in a worrisome situation. Such natural forests, they observed, were so diverse in terms of tree species that no forestry operations could be effectively undertaken. In 1898 and 1899, when the survey concluded, the foresters brought the above observations to the colonial government's attention. It had become imperative, they argued, for the colonial government to adopt certain steps to solve Taiwan's forest problems. Of importance was to divide the forests into compartments and assign each of them "a solid, immovable working plan." With these plans established, they then pointed out, the colonial government should be able to put the "normal (hōsei) forest model" into practice. The foresters called the normal forest model a cornerstone of modern forestry. Its essence and beauty were that by systematically logging and reforesting, a forest manager could improve a forest's composition, growth, and shape, administering the trees as if they were crops and eventually achieving a sustained yield of timber and other forest products. Foresters argued that it was the colonial government that should assume responsibility for improving and normalizing Taiwan's forests. Of course, they added, as the colonial government gradually brought Taiwanese forestry toward a sustained yield, they should stand at the government's side and ensure every step was taken to meet the standards of modern forestry.[7]

Taking all the above factors into consideration, the colonial government decided to bankroll entrepreneurs from Japan, enlisting them to manage Taiwan's forests. Besides replacing the degraded and unregulated forests with artificial forests based on foresters' instructions, the colonial government stipulated that entrepreneurs should assist the colonial government in eliminating uprisings and pacifying indigenous rebels. The colonial government promised that if entrepreneurs could fulfill these objectives, they would be entitled to receive a portion of the government-owned forests. In other words, the colonial government expected entrepreneurs to become Taiwan's first generation of private forest managers. Entrepreneurs would own their own forestlands, and the colonial government would ensure their plots were large enough for them to run efficient forestry businesses with prospects for long-term success.[8]

The effort to integrate the entrepreneurs into Taiwan's forestry worked well at the start. During the first decade of the twentieth century, the colonial government stoked enthusiasm among Japanese entrepreneurs toward Taiwanese forests. Indeed, under the colonial government's patronage, Taiwan offered ample opportunities for entrepreneurs to diversify their businesses, expand their markets, enlarge their production bases, and secure a niche in Japan's rapidly industrializing economy. Taiwan's entrepreneur-based forestry system blossomed. Particularly following the market push and production requirements caused by World War I, increasing numbers of entrepreneurs plunged zealously into timber production in Taiwan. Sizable stretches of the forest were cleared, and multiple reforestation projects followed. The ideal of "normal" forests and sustainable yields seemed to be on the horizon.[9]

In the meantime, the colonial government waged war against the indigenous peoples. Under the resolution known as *gonen riban keikaku,* or "the five-year project for controlling the savages," the colonial government deployed tremendous military power to force the indigenous peoples to retreat to the mountainous areas of central Taiwan. What was once the "land of the savages" was then turned into government-owned forests, which meant that they were now subject to the entrepreneur-based forestry scheme. The colonial government then demarcated a boundary that separated the area now occupied by indigenous peoples from the rest of Taiwan. Adopting the name popular during Qing rule, the colonial government called this line the *bankai,* or "Savages' Boundary." It should be noted that this line did not exist just on paper or in policy; it was a boundary guarded by armed police stations stocked with guns, cannons, and high-voltage defense installations. With such infrastructure installed, the colonial government expected, this boundary would once and for all bring about the savages' compliance with government ordinances and supply the power to suppress any potential rebellions. Industrious loggers, settlers, and planters therefore no longer needed to worry about headhunting savages.[10]

From 1915 to 1925, a total of 258,380.04 hectares of government-owned forest was turned over to private managers. In the meantime, when pushing back the indigenous peoples to make room for logging and reforestation, the colonial government completed a survey that covered 696,228.33 hectares of forestland.[11] Figure 9.1 shows a modern reconstruction of the survey's results: three decades after Japan took over Taiwan, the colonial government had mostly conquered western Taiwan, with the area surveyed and documented on the map. Although

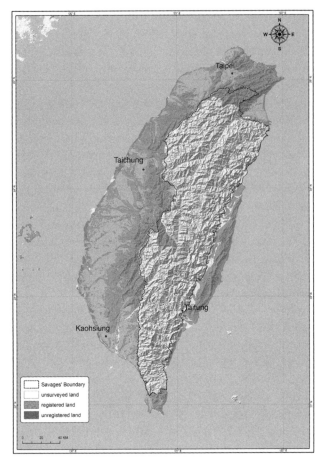

FIGURE 9.1.
Reconstruction
of the map
showing the region
designated for the
1915–1925 forest
survey, Taiwan

the same undertaking remained to be carried out inside the Savages' Boundary, the completion of the survey of 1915–1925 marked the first time in Taiwan's history that its forests had been clearly mapped. Takekoshi Yosaburō, a renowned policy analyst of the day, was impressed by how much the colonial government had achieved in rendering Japan's first colony understandable and governable. As such land-surveying projects continued, he speculated, the colonial government would look upon Taiwan's lands as being in the "palm of their hand," literally and figuratively.[12] Taiwan's forests, it seemed, would never again be a danger zone filled with bandits and savages. The colonial state, bolstered by scientists and entrepreneurs, would gain eventual prominence.

▓ Such God-Gifted Nature!

The colonial government clearly intended to conquer the unsurveyed parts of the colony (those left blank in figure 9.1). In part because the forests there were home to a variety of valuable trees, and in part because of the region's remoteness and isolation, the colonial government regarded the forests inside the boundary as a place where it could experiment with its most ambitious forestry undertakings. An analysis of these undertakings offers a complementary view to those enacted outside the boundary.

In 1896, a Japanese police officer serving in central Taiwan happened to hear some people chatting about a tree called *songluo*. According to what he understood, *songluo* meant a tree similar to the Japanese cypress, one of the most highly valued trees in Japan. If this perceived similarity were indeed the case, the officer surmised, then Japanese forestry certainly would have an auspicious future on this island. Excited by the great economic implications of this gossip, the officer reported his surmise to the colonial government. A team was then dispatched to look into the matter. Led by local people, the team of foresters, engineers, and governmental bureaucrats eagerly sought to determine whether the cypress in question actually grew in Taiwan. After days of trekking through thick forests, the team came upon a great number of trees of incredible quality. The cypress was indeed bountiful in Taiwan, and in its stature and the mass of its growth, this woodland was certainly comparable to the best in Japan.[13]

The discovery of this region, later called *Arizan,* or Ari Mountain, by foresters, was surely good news to the colonial government. In the years to come, the colonial government not only drafted "solid, immovable working plans" for the management of the region but also established specific institutions to ensure the plans were put into practice. Not all the people who got involved in the effort hailed from the colonial government. Japanese entrepreneurs were excited as well. Upon realizing that the flora on Ari Mountain included such valuable trees as fir, cypress, and hemlock, entrepreneurs began calculating possible profit yields from this divine natural gift (figure 9.2). Aware of entrepreneurs' rising enthusiasm, the colonial government became anxious. Could the government reasonably sell a place like Ari Mountain to entrepreneurs? Could entrepreneurs manage such valuable forest resources appropriately—that is to say, without detracting from the mountain's public benefits? Regardless, the colonial government maintained that management of a place like Ari Mountain could not be treated as an isolated matter but would have to be connected with the development of Taiwan as a whole.

In 1903, after consulting with Kawai Shitarō, a Tokyo Imperial University professor specializing in forest use, the colonial government made a bold decision.

FIGURE 9.2. *Chamaecyparis formosensis* on Ari Mountain, Taiwan, photographed by Japanese forestry officer Sasaki Shun'ichi, 1921

SOURCE: Eastern Asian Historical Photograph Collections, Arnold Arboretum Horticultural Library Archives, Harvard University.

The right to manage Ari Mountain could not be left to anyone else, nor could development of the mountain be "contaminated" by private interests, selfish considerations, and commercial exploitation. To be sure, this approach completely contradicted the policy that the colonial government was now implementing in the forests outside the Savages' Boundary. Unsurprisingly, the entrepreneurs complained of being excluded from Ari Mountain. They argued that the colonial government had made the wrong decision. What it should do, they argued, was open this gift from the gods to the public rather than lock it up.[14]

In the face of vigorous challenges, Kawai published his manifesto in the *Tōyō jihō*, the *East Asia Times*. He began his analysis by comparing the forestry of Taiwan to that of Bosnia. Kawai argued that Bosnia, located in southern Europe, had suffered from unregulated deforestation since the early nineteenth century. As the situation worsened and as the government found itself unable to control the decline, Bosnia was incorporated into Austria in the late 1870s. Austria then initiated a revolution in forestry. The most essential changes were the Austrian state's regulation of forestry and its attempt to eradicate local forestry companies. Austrian foresters argued that deforestation in Bosnia was due largely to competition among forestry companies. Dozens or even hundreds of tiny forestry companies crowded into forests, competed with each other, rendered forests a mess, and disturbed the market equilibrium. The state's seizure of valuable forests represented the first major step toward a kind of "rational forestry." A state-directed company was set up thereafter. With its huge scope of operations, the Austrian government successfully monopolized the timber market. Kawai then argued that the situation in Taiwanese forestry bore a striking resemblance to what had taken place in Bosnia. The Empire of Japan should learn from history and play the role that Austria had played: Japan could either nationalize the forests and monopolize the related markets, or it could forfeit Taiwan's future value to the empire. Further, the nationalization of Ari Mountain was just the first step, not the last. The colonial government would harness the most advanced science and technology to make nature submit to scientific and economic governance.[15]

But what kind of technology should the empire deploy to realize this marvelous dream? Interestingly, although the examples that Kawai had drawn were from the European continent, the technology that eventually shaped forestry in Taiwan was imported from the other side of the Atlantic, the United States. The reasons behind Japan's preference for this U.S. import were manifold. First, on the global timber market in the 1910s, timber exported from the United States had replaced timber from the British Empire: a young country had defeated an old empire, thereby seizing its place as a leader in world forestry. The timber that helped the United States gain prominence on the market was Douglas-fir *(Pseudotsuga menziesii)*, western redcedar *(Thuja plicata)*, and Lawson's cypress *(Chamaecyparis lawsoniana)*, each of which boasted a high quality and large yield. Especially attractive to European consumers was Douglas-fir, given its incomparable length and low price. If anyone in the 1910s wanted to use wood as industrial material, U.S. timber was definitely the principal—if not the only—choice.[16]

In addition to the United States' unusually rich natural endowment, U.S forestry companies' technology also contributed critically to U.S. success in the world timber market. Logging railroads and the "steam donkey," or locomotive, played the decisive role. Rail transport for timber was originally adopted in the Great Lakes area but ultimately found its best applications on the West Coast. The first logging railroad in this region was built in the Douglas-fir forests around Puget Sound and the Olympic Mountains during the last decades of the nineteenth century. The logging railroad replaced other means of transportation, solving the problem that had annoyed local loggers for decades: how to connect the timberlands to the marketplace.[17]

Impressed and inspired by the success of U.S. forestry, Japanese foresters gave particular strategic consideration to the development of Ari Mountain. They saw three competitive advantages to Taiwanese wood. First, from Japanese foresters' point of view, U.S. timber, Douglas-fir included, was not very resistant to damage by termites and other harmful insects. This problem severely curtailed the use of U.S. timber in humid environments. By contrast, several species distributed on Ari Mountain, the Taiwanese cypress (*Chamaecyparis obtusa* var. *formosana*) in particular, were largely immune to termite attacks and other humidity-related problems. Second, given that the forests on Ari Mountain had not been harvested for centuries, the length of the timber produced there was comparable if not superior to that of its American counterparts. Finally, in botanical terms, the genus *Chamaecyparis* could be found only in Japan, Taiwan, and North America. This remarkable biogeographical pattern convinced the colonial government that Japan could compete in the cutthroat timber market.[18]

Realizing the economic and political importance of Ari Mountain, the colonial government immediately ordered sets of cableway skidders from the Lidgerwood Company, a giant in U.S. timbering technology. Steam powered and equipped with steel cables, each skidder could collect timber dispersed in a circle whose radius reached 1 kilometer. Once the timber was piled up, it was transported by a logging railroad stretching from the cutting areas to the plains, where the timber was collected again and moved in greater quantities. At the end of this long, winding road was a factory with saws, kilns, planers, and mills, all of them either electricity or steam powered. As a Japanese forester at the time asserted, this vast factory was definitely the "greatest in East Asia."[19]

Using U.S. technology, the project on Ari Mountain began to shape the course of Taiwanese forestry as a whole. Yet Ari Mountain was far from the only reserve from which the Japanese Empire could appropriate precious coniferous forests. From the 1910s onward, at least two places in Taiwan were reported to have forests as abundant as those on Ari Mountain. In the years to come, backed with support from the motherland, the colonial government began complementary projects in other locations. Miles of logging railroad were laid and more advanced technology was put into place—and in fact, many of these devices helped maintain Taiwanese forestry's reputation as "the greatest in East Asia."[20]

▪ "The Shame of Taiwan"

Let us shift focus back to the forests outside the Savages' Boundary, where to manage the degraded and unregulated forests, the colonial government had recruited entrepreneurs from Japan, demanded they undertake various projects, and rewarded them with parcels of the government-owned forests. Afterward, in part because of the economic prosperity and rising demand for timber in Japan stimulated by World War I, the colonial government's scheme drew sustained attention from entrepreneurs. Foresters were delighted at the entrepreneurial opportunities that arose. For the administration's part, the cooperation of entrepreneurs was surely a precondition for the realization of the government's goal of "normal" forests.

Surprising as it might seem, however, entrepreneurs' enthusiasm for Taiwanese forestry soon chilled. They found that a variety of seemingly insurmountable problems impeded their undertakings, despite the government's patronage. Of particular importance were technical difficulties peculiar to Taiwan's broadleaf forests. Because hardness is characteristic of wood from broadleaf forests (which is why broadleaf trees are often called hardwoods), entrepreneurs were forced to spend more money than planned procuring the technology needed to harvest this tougher timber. For most entrepreneurs, the added investment and necessary changes in operations were prohibitively expensive.[21] Hence, instead of setting up advanced logging and transportation systems to systematically harvest and reforest, as the colonial government and foresters had expected, entrepreneurs ended up having to hire local Taiwanese to retrieve valuable timber from the forests. The Taiwanese for the most part had no capital, no modern forestry equipment, and no knowledge of scientific forestry, but they could be hired at low wages.[22]

Foresters became worried that entrepreneurs would overstep their bounds. It appeared to them that the involvement of entrepreneurs would eventually lead to "the tragedy of deforestation."[23] A forester reported that to make one utility pole, loggers would rather fell a two- or three-hundred-year-old oak, taking what they wanted from it and leaving the rest on the floor of the forest, a wasteful process. More importantly, it now appeared to foresters that entrepreneurs' reforestation projects had not produced the expected results. Entrepreneurs seemed to care more about extracting revenue from the land than cultivating a well-ordered and robust artificial forest. In many cases, foresters observed, the so-called reforested lands remained desolate, scattered with dying seedlings and scrawny saplings. Confronted with this sort of flagrant misuse of the forests outside the Savages' Boundary, foresters became convinced that entrepreneurs were the main cause of wasted resources. Commercial forestry in Taiwan had sounded a warning bell.[24]

A more disturbing warning was sounded in the forests inside the Savages' Boundary. At the turn of the century, after overcoming tremendous technical obstacles, the colonial government began transporting timber harvested in forestlands such as Ari Mountain to Japan. For the first time, Japanese society fully recognized the abundance of Taiwan's forest resources. It seemed that Taiwan's forestry would now play its role: to supply timber for the mother country's benefit.

However, in the years after the Russo-Japanese War (1904–1905), the dramatic industrialization and urbanization of Japan forced the Japanese government to realize that current levels of timber production could not meet rising domestic demand. To solve the timber shortage, the Japanese government adjusted the tariff duties in the early 1920s and imported timber from the United States. This seemingly slight adjustment generated astonishing consequences. In the years to come, "endless waves" of U.S. timber poured into Japan. From the Japanese people's point of view, the U.S. timber was inferior to that harvested in Japan, Korea, and Taiwan, but its cheap price and large quantity made it extremely suitable for Japan's many industrial uses. In a market where cheap, mass-produced timber was the main product, the fate of Taiwan's timber seemed easy to predict. Indeed, an entrepreneur once complained that, compared with prices for U.S. timber delivered from across the vast Pacific, the price of Taiwan's timber was ridiculously high. No serious industrialist would use such expensive timber as material, nor could the common people, since they could not afford it.[25]

Timber pricing put pressure on Taiwan's timber market in two ways. First, the colonial government, perceiving that the domestic timber market in Japan had been restructured by U.S. timber imports, decided to retreat from the Japanese market, making Taiwan into an alternative marketplace. Second, forest managers in Japan, having suffered from the competition with U.S. timber, adopted the same strategy as the colonial government. From their perspective, Taiwan should serve as a buffer zone where Japan proper could find a salve for its financial crisis. Nevertheless, for the colonial government and foresters in Taiwan, it was evident that the flood of Japanese timber into Taiwan would further weaken the economic viability of Taiwan's timber and perhaps Taiwan's forestry as a whole. As a forester observed, Taiwan's timber could hardly compete with its Japanese counterparts because of its high price. Sooner or later, he warned, Japan's timber would take control of the timber market in Taiwan. The outcome for Taiwan's timber industry, overwhelmed by imports from Japan and the United States, would be certain death.[26]

Critics of the Ari Mountain project could not resist the chance to ridicule the colonial government. They asserted that the whole project had gone bankrupt, and that the colonial government had monopolized all valuable forest resources in ways that hardly benefited the public. They derided the operations undertaken on Ari Mountain, arguing that even a modestly capable manager should have been able to generate profits under such a monopoly. Instead of becoming a model of successful colonial governance, the project turned into a real-life demonstration of the absurdity of forestry under the colonial government.[27]

Criticism extended beyond the bankruptcy of forestry projects inside the Savages' Boundary to include complaints about the tragedy of deforestation. The Japanese government, which had invested so much money to help the colonial government launch forest surveys, procure forestry equipment, and establish governing regimes, began to question the effectiveness of these undertakings, asking, "What on earth is the governor-general of Taiwan doing?" The Japanese government further asked itself, "If Taiwan is indeed the 'Forest Kingdom' of the empire, why should it not supply the mother country with her needed timber?"[28] A forestry scholar took an even stronger position. Forests, he told a nationwide conference of forestry scholars, occupy more than 70 percent of Taiwan, its forest resources are much more abundant than Japan's, and forest management has been in place in Taiwan for decades, but this colony isn't meeting the mother country's urgent needs for timber. Revealing a harsh and uncompromising attitude, he said forestry had become "the shame of Taiwan."[29]

By the 1920s, the state of Taiwan's forests had become so deplorable that some sort of reform had to be enacted, and it was against this backdrop that "scientific forestry" acquired its eventual prominence. The director of the Office of Agriculture and Industry, Kita Takaharu, publicly stated that the government would manage the forests guided by scientific principles and thereby procure a sustainable timber yield.[30] In 1924, the colonial government allocated a budget of 3 million Japanese yen to create a feasible plan of action for the management of Taiwan's forests.[31] The director of the Japan Forest Service, Kamiyama Mitsunoshin, was appointed governor-general of Taiwan and announced his intention to draft and implement government policies of forest conservation (hoiku).[32]

This new emphasis on the conservation, not just development (kaihatsu), of Taiwan's forests elevated the position of the island's forestry scholars. Azebu Sadakuma, in an effort to raise the morale of his colleagues, stated, "We, the forestry scholars, must consistently strive for sincerity and enthusiasm; with a fiery passion and the determination of a patriotic heart, we must shoulder the burden of this great responsibility."[33]

With support from both the colonial government and the mother country, foresters threw themselves into the field to carry out various surveying and mapping projects, not the least of which focused on lands within the Savages' Boundary. Given that the colonial government believed it had already conquered the indigenous peoples, foresters aspired to place forests inside the boundary under a comprehensive management scheme. From 1925 to 1935, a total of 877,629.46 hectares of forest was allocated for these working plans (figure 9.3). This area was then further divided into thirty-two working units

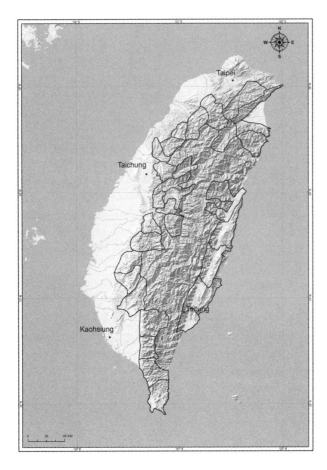

FIGURE 9.3.
Reconstruction of
the division and
distribution of the
working units in
Taiwan, 1931

(jigyōku) (some of which were later combined, so the final number of the working units was twenty-nine). With this fundamental scheme set out, foresters then compiled and assigned to each working unit specific plans that stipulated levels of intensity for logging and reforestation, identified the locations of intended operations, and documented the relationships between the forests and local society, along with voluminous maps and statistics.

They then divided the rest of the forests into two categories: "nonreserved forests" and "semireserved forests" (figure 9.4). The nonreserved forests, because they were in ecologically stable areas and relatively immune to floods, landslides, or other disasters, were considered suitable for more diversified undertakings like farming, fruit tree orchards, and animal husbandry. Following the forestland policy in the early years of colonial rule, foresters suggested that the colonial government recruit entrepreneurs to take charge of these nonreserved forests.

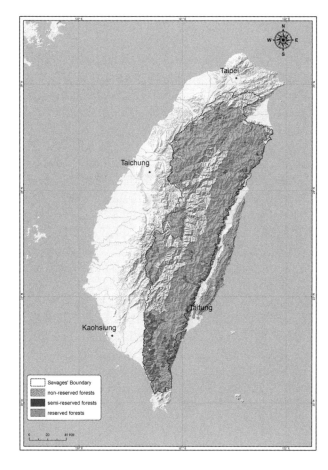

FIGURE 9.4.
Reconstruction of the division and distribution of reserved forests, semireserved forests, and nonreserved forests in Taiwan, 1935

The colonial government, foresters maintained, should grant land rights to those qualified entrepreneurs. By doing so, entrepreneurs could be spurred to improve and develop Taiwan's forestlands.

The semireserved forests, also located in ecologically stable regions, were designated for the use of indigenous peoples. However, these semireserved forests would never become indigenous peoples' private property. The dominant view at the time was that indigenous peoples had not been "civilized" enough to play by the rules of a market economy. Hence, it was necessary to maintain these forests under the colonial government's surveillance so that indigenous peoples would not lose their means of production.

By the time foresters completed these three undertakings, the whole project had required a total financial investment of 2,681,778.17 yen. In addition, a 335,087-person workforce and a police force numbering 12,083 had been mobilized.[34] Once and for all, foresters expected to be able to resolve the problem of managing Taiwan's forests.

Comprehensive as it was, the whole project's focus was on the forests included in the twenty-nine working units. In accordance with the results of an investigation into Taiwan's forest growth, forestry scholars at the time believed that these particular forests were "overripe," and that if they were not harvested in short order, their natural resources would undoubtedly go to waste.[35] Forestry scholars further advised the colonial government that it was absolutely vital that they implement comprehensive changes to improve the situation. To this end, they consulted the population statistics of England, the United States, and various other countries with "advanced forestry policies." They then determined that the population of Taiwan was increasing at a rate of 1.5 percent per year, and on these grounds they estimated Taiwan's future timber consumption. The results showed that within ten years, Taiwan would be faced with a serious shortage of timber. They advised the government to confront this crisis directly and resolve it as quickly as possible.[36]

The solution was large-scale clear-cutting of the forests. Foresters believed that because Taiwan's forests had generally overmatured, a large-scale felling of trees would not damage them. To the contrary, clear-cutting could help advance forest regeneration, removing much of the forest's overgrowth, making room for *Cryptomeria* or teak to be planted along with other economically valuable species. Foresters optimistically predicted that within ten years, the current diverse state of Taiwan's natural forests would be remedied, producing well-ordered and regulated man-made forests where "normal conditions" prevailed.[37]

Once again, the question remained: how would the colonial government operationalize the foresters' scheme, and at what pace? Here the empire's growing tensions with Western powers and subsequent militarization during the 1930s provided the ultimate stimulus. When Japan mobilized its economy for war, Taiwan, as the empire's only tropical colony, fell under close scrutiny. Korea and Manchuria most decidedly could not provide both hardwood and tropical forest products. In the 1940s, the colonial government went to great lengths to nurture entrepreneurial activity in the forests, wholeheartedly encouraging timber production. Forestry scholars continued to draft even more wide-reaching plans for the clearing of the forests, intensifying the production of timber to advance their grand plans for forest renewal. What were once three separate entities—the warring state, entrepreneurs, and scientific forestry scholars—had begun to form an interdependent system. Their intertwining sped the transformation of Taiwan's forests to an unprecedented degree.[38]

Conclusion

Environmental historians have long recognized the important role of scientific forestry in colonizing the environment. Chiefly drawing examples from regions under European colonial rule, increasing numbers of studies have argued that to understand how scientific forestry emerged and functioned in a colonial context, historians must first understand the complex interactions between colonial governance and capitalist expansion. Taking Japan's colonization of Taiwan as an example, I have focused on three actors in the colonizing class: the colonial government, entrepreneurs, and forestry officials and scholars. As we have seen, associations between the colonial government and entrepreneurs were at times mutually beneficial, but at other times they were at odds, since the demands and logical underpinnings of colonial governance and capitalist expansion frequently contradicted one another. The foresters were alternately hailed and slighted as the forests became increasingly contested. Not until the 1930s did scientific forestry acquire its ultimate authority over Taiwan's forest environment. From this perspective, although there is no denying that scientific forestry served as a "tool of empire," historians still have to pay attention specifically to how actors who contributed to empire building (for example, entrepreneurs and colonial bureaucrats) competed to control this tool. On the other hand, although it is clear that scientific forestry represents a framework by which social actors can "see like a state," we still need to answer how this particular way of seeing affected the physical environment.[39]

In retrospect, there is no doubt that Japan's colonial forestry left complex legacies for Taiwan's environment. On the one hand, it was through Japan's intensive and extensive forest surveys that Taiwan's biodiversity and floral abundance became clear to the world of science. On the other hand, the establishment of colonial forestry gave rise to unprecedented deforestation and reforestation that dramatically reshaped Taiwan's landscape. Despite appearances, these two forces did not always conflict with each other. Taken together, they represent humanity's common desire to promote the resilience and sustainability of the environment by knowing it and intervening in it. There is no doubt that the colonial government embraced sustained yield as the governing principle for Taiwan's forest environment from the very start. As far as Japan was concerned, in Taiwan there was no scenario in which a ruthless colonial government would greedily exploit the colony's natural resources for the benefit of the mother country. As my analysis shows, it is remarkable how the colonial government's effort to achieve sustained-yield forestry was compromised by both broad international conditions and those in the metropolis.[40]

Finally, we should not forget that "sustainability" is itself a concept that merits further historical scrutiny. Japan's colonial forestry in Taiwan was similar to other (not necessarily colonial) forestry regimes of the day in that it aspired to apply a universal definition of sustained yield to tackle a diverse local environment. Consequently, the colonial government and forestry scholars discarded traditional forestry of every kind, including Japan's own traditions. Interestingly enough, as Margaret A. McKean, Elinor Ostrom, and other scholars have shown, Japan's forestry traditions achieved sustainability levels perhaps unprecedented in human history. Contrary to so-called modern, scientific forestry, which considers the government and private entities as the ultimate facilitators and gatekeepers of sustainability, Japan's decentralized, village-based, and context-dependent forestry traditions helped generate the distinctive society-environment relationship that gave Japan its name "the Green Archipelago." I am not advocating a nostalgic approach to contemporary environmental problems. Rather, what I am arguing for is the importance of maintaining a historically sensitive view toward such a seemingly objective, value-free concept as sustainability. Nowadays, when forestry scholars discuss new possibilities for sustainable forestry, Japan's forestry traditions remain an important source to which scholars frequently refer, but Japan's experiences embracing and deploying modern forestry provide a remarkable window through which we can reconsider the varied meanings of sustainability.[41]

NOTES

*   In the course of preparing the essay, I received useful comments and suggestions from scholars in Taiwan, Japan, and the United States, especially from Liu Ts'ui-jung, Zheng Chinlong, Fan I-chun, Ka Chih-ming, Li Wen-liang, Lin Chih-cheng, Adelheid Voskuhl, and Yang Daqing. I also would like to thank Li Yu-ting and Liao Hsiung-ming at the Center for Geographic Information Science, Research Center for Humanities and Social Sciences, Academia Sinica, Taiwan, for providing essential assistance in creating the maps.

1   For relevant discussions on Taiwan's forest-society relationship under the rule of the Qing Empire, see John Shepherd, *Statecraft and Political Economy on the Taiwan Frontier, 1600–1800* (Stanford, Calif.: Stanford University Press, 1993); Ka Chih-ming, *Fan toujia: Qingdai Taiwan zuqun zhengzhi yu shufan diquan* (Taipei: Zhongyang Yanjiuyuan Shehuixue Yanjiusuo, 2001); Emma Teng, *Taiwan's Imagined Geography: Chinese Colonial Travel Writing and Pictures, 1683–1895* (Cambridge, Mass.: Harvard University Asia Center, 2004). For Japan's forestry history prior to the Meiji era, the most remarkable study in English still belongs to Conrad Totman, *The Green Archipelago: Forestry in Preindustrial Japan* (Berkeley: University of California Press, 1989). For a history of Japan's colonial forestry in general, see Hagino Toshio, *Chōsen, Manshū, Taiwan ringyō hattatsu shiron* (Zaidanhōjin Rin'ya Kōsaikai, 1965). For Japan's colonial forest policy in Taiwan in particular, see Li Wen-liang, "Rizhi shiqi Taiwan zongtufu de linye zhipei yu suoyouquan: Yi yuangu guanxi wei zhongxin," *Taiwanshi yanjiu* 5, no. 2 (2000): 35–54; Hung Kuang-chi, "Linxue, ziben zhuyi yu bianqu tongzhi: Rizhi shiqi linye diaocha yu zhengli shiye di zaisikao," *Taiwanshi yanjiu* 11, no. 2 (2005): 77–144; Antonio Tavares, "The

Japanese Colonial State and the Dissolution of the Late Imperial Frontier Economy in Taiwan, 1886–1909," *Journal of Asian Studies* 64, no. 2 (2005): 361–85.

2 Recent studies in environmental history have urged historians not to consider the colonizing class as a monolithic bloc with coherent interests and views about nature. Regarding the intricate relationship between modern or scientific forestry and colonialism, see, for example, Ramachandra Guha, *The Unquiet Woods: Ecological Change and Peasant Resistance in the Himalaya* (Berkeley: University of California Press, 1990); Richard H. Grove, *Green Imperialism: Colonial Expansion, Tropical Island Edens, and the Origins of Environmentalism, 1600–1860*, Studies in Environment and History (Cambridge: Cambridge University Press, 1995); K. Sivaramakrishnan, *Modern Forests: Statemaking and Environmental Change in Colonial Eastern India* (Stanford, Calif.: Stanford University Press, 1999); Arun Agrawal, *Environmentality: Technologies of Government and the Making of Subjects* (Durham, N.C.: Duke University Press, 2005); Ravi Rajan, *Modernizing Nature: Forestry and Imperial Eco-development 1800–1950* (Oxford: Oxford University Press, 2006); Peter Vandergeest and Nancy Lee Peluso, "Empires of Forestry: Professional Forestry and State Power in Southeast Asia, Part 1," *Environment and History* 12 (2006): 31–64; Gregory A. Barton, "Empire Forestry and American Environmentalism," *Environment and History* 6, no. 2 (2000): 187–203; Caroline Ford, "Reforestation, Landscape Conservation, and the Anxieties of Empire in French Colonial Algeria," *American Historical Review* 113, no. 2 (2008): 341–62.

3 For a detailed analysis of the contents, contexts, and influences of this particular forest law, see Li, "Rizhi shiqi," 38–40, and Hung, "Linxue," 83–92.

4 This particular view of Taiwan's forests and forestry can be found in voluminous publications issued by the *Shokusankyoku*. See, for example, Taiwan Sōtokufu Shokusankyoku, *Taiwan no rin'ya* (Taipei: Shokusankyoku, 1911).

5 Regarding Taiwanese peoples' resistance against Japan's colonial rule vis-à-vis Japan's land and forest policies, see Paul Katz's remarkable studies on the Ta-pa-ni Incident: Paul Katz, "Governmentality and Its Consequences in Colonial Taiwan: A Case Study of the Ta-pa-ni Incident of 1915," *Journal of Asian Studies* 64, no. 2 (2005): 387–424. Regarding the tension between Taiwan's indigenous peoples and the colonial government over forest uses, a succinct account can be found in Yamaji Katsuhiko, *Taiwan no shokuminchi tōchi: "Mushu no yabanjin" to iu gensetsu no tenkai* (Nihon Tosho Sentā, 2004).

6 One forester's recollections about his colleagues' motivation to come to Taiwan can be found in Dainippon Sanrinkai, *Meiji ringyō isshi zokuhen* (Dainippon Sanrinkai, 1931), 291–93.

7 The results of this particular forest survey were published in Taiwan Sōtokufu Minseibu Shokusanka, *Taiwan Sōtokufu minseikyoku shokusanbu hōbun* (Shokusanka, 1898–99). For a detailed account of how colonial foresters deployed the normal forest model to analyze Taiwan's forest condition, see Hung, "Linxue," 88–90.

8 The text of this particular regulation can be found in Taiwan Sōtokufu Eirinkyoku, *Taiwan rin'ya hōki* (Taipei: Eirinkyoku, 1919), 70. Regarding the colonial government's effort to establish a class of private forest managers, see Hung, "Linxue," 102–11.

9 Hung, "Linxue," 112–18.

10 Relevant information about the Five-Year Project for Controlling the Barbarians can be found in the reports issued by the Keimukyoku and Banmu Honsho. See, for example, Taiwan Sōtokufu Ribanka, *Riban gaiyō* (Taipei: Banmu Honsho, 1913); Taiwan Sōtokufu Keimukyoku, *Ribanshi kō* (Taipei: Keimukyoku, 1918–1938). For relevant research, see Fujii Shizue, *Lifan: Riben zhili Taiwan de jice* (Taipei: Wenyingtang, 1997), and Yamaji, *Taiwan no shokuminchi tōchi.*

11   Taiwan Sōtokufu Naimukyoku, *Taiwan kanyū rin'ya seiri jigyō hōkokusho* (Taipei: Naimukyoku, 1926), 277–78, 280.

12   Takekoshi Yosaburō, *Taiwan tōchishi* (Hakubunkan, 1905).

13   Dainippon Sanrinkai, *Meiji ringyō isshi* (Tokyo: Dainippon Sanrinkai, 1931), 458–77.

14   *Meiji ringyō isshi*, 438–40; Magata Sei, "Sara ni Arizan mondai o kōkyū seyo," *Dainippon sanrin kaihō* (hereafter cited as *DSK*) 324 (1909): 26; Taiwan Sōtokufu Eirinkyoku, *Ringyō ippan* (Taipei: Eirinkyoku, 1919), 84–91.

15   Kawai Shitarō, "Taiwan ringyō ni tsuite," *DSK* 392 (1915): 68–83.

16   "Beizai to honpōsan mokuzai no hikaku kenkyū," *DSK* 484 (1923): 69–71.

17   Michael Williams, *Americans and Their Forests: A Historical Geography* (Cambridge: Cambridge University Press, 1989), 301–302.

18   Okamoto Sei, "Tokushu rinboku hinoki, benihi no jukyū oyobi torihiki ni tsuite: Hontō ni okeru mokuzai no jukyū to hinoki, benihi," *Taiwan no sanrin* (hereafter cited as *TNS*) 85 (1933): 128; Miyase Hiroshi, "Eirinjozai to sono shijō," *TNS* 85 (1933): 132–33.

19   Taiwan Sanrinkai, *Taiwan no ringyō* (Taipei: Taiwan Sanrinkai, 1933), 130–31. A German forester traveled to Taiwan at that time and was much impressed by Japan's operations on Ari Mountain: Amerigo Hofmann, *Aus den Waldungen des Fernen Ostens; Forstliche Reisen und Studien in Japan, Formosa, Korea und den angrenzenden gebieten Ostasiens* (Wien and Leipzig: W. Frick, 1913), 66–74.

20   "Arizan sagyō shinkeikaku," *Taiwan nichinichi shinpō* (hereafter cited as *TWNS*), September 5, 1914; "Hachisenzan batsuboku no keikaku," *TWNS*, September 24, 1914; "Arizan kinkyō," *TWNS*, October 13, 1914; "Hinokizai tōitsu keikaku," *TWNS*, November 26, 1914.

21   Masuzawa Shinji, "Taiwansan katsuyōjyu no shiyō kachi kōjōsaku ni tsuite," *Taiwan sanrin kaihō* (hereafter cited as *TSK*) 35 (1929): 82; Nagayama Kikuo, "Taiwansan katsuyōjyu no riyō kachi zōshin ni tsuite," *TSK* 62 (1931): 38, 43.

22   Fukaya Ryūzō, "Taiwan ni okeru shiyūringyō no shinkō to shinrin kumiai," *TNS* 34 (1928): 2–11; Hiraga Shōji, "Rinsanbutsu haraisageru hōhō no kaizen to mokuzai no kinyūka," *TSK* 35 (1929): 51; Lai Yunxiang, "Taiwan ni okeru minrin keiei no taiken," *TNS* 79 (1932): 12–27.

23   Taiwan Sanrinkai, *Taiwan no ringyō*, 85–86.

24   "Taiwan zōrin jigyō no enkaku oyobi genjō I," *Taiwan nōjihō* (hereafter cited as *TWN*) 123 (1917): 17–25; Azebu Sadakuma, "Shinbunshi o tsūji kantaru bokoku rinsei no shinkeikō," *TSK* 8 (1924): 10–11; Ara Usaburō, "Kusunoki zōrin ni kansuru shokan," *TNS* 108 (1935): 4–5.

25   "Beimatsu yunyū gekizō," *DSK* 446 (1920): 37; Watanabe Isao, "Gaikokuzai no yunyū," *DSK* 477 (1922): 33; Azebu Yoshihiro, "Mokuzai kanzei mondai no keii," *TSK* 37 (1929): 13–50.

26   Yamazaki Yoshio, "Taiwan ni okeru mokuzai jukyū ni tsuite," *TSK* 3 (1923): 6.

27   "Arizan keiei shippai," *DSK* 387 (1915): 52–53; Saji Takanori, "Taiwan no eirin jigyō ni tsuite," *TSK* 79 (1932): 55–56.

28   A detailed analysis of this subject can be found in Watanabe Isao, *Gaizai yunyū no kyokusei to sono taisaku* (Teikoku Sanrinkai, 1925).

29   Yatagai Masayoshi, "Taiwan no shinrin to sono kaihatsu," *TNS* 79 (1932): 28–33.

30   "Rinmu kaigi ni okeru shokusankyokuchō kunji," *TSK* 7 (1924): 2.

31   Asano Yasukichi, "Taiwan ringyō no kaiko," *TNS* 78 (1932): 15–18.

32   Kamiyama's own account of his forestry experience can be found in *Meiji ringyō isshi zokuhen*, 347–61, under the title "Sanrinkyoku jidai no omoide."

33   Azebu Sadakuma, "Shinrin keikaku jigyō no chakusō subekaraku endainare," *TSK* 33 (1928): 3–7.

34   Taiwan Sōtokufu Shokusankyoku, *Shinrin keikaku jigyō hōkokusho II* (Taipei: Shokusankyoku, 1937), 66–68, 572–73, 577.

35   Taiwan Sōtokufu Shokusankyoku, ed., *Taiwan shuyō rinboku seichōryō chōsasho* (Taipei: Shokusankyoku, 1922); Taiwan Sōtokufu Shokusankyoku, *Shinrin keikaku*, 72–73.

36   Taiwan Sōtokufu Shokusankyoku, *Taiwan ringyō no kihon chōsasho* (Taipei: Shokusankyoku, 1931), 172–73; Okada Shin, "Taiwan ni okeru mokuzai jukyū zōrin," *TSK* 72 (1932): 9–17; "Honpō rinsanbutsu no juyō nami kyōkyūryō," *DSK* 484 (1923): 63–69.

37   Taiwan Sōtokufu Shokusankyoku, *Shinrin keikaku*, 593, 597, 607; Aoki Shigeru, *Taiwan ringyō-jyō no kiso chishiki* (Taipei: Shinkōdō, 1925), 553–54.

38   By far the most comprehensive account of Japan's colonial forestry during the 1930s and 1940s is Hagino, *Chōsen, Manshū, Taiwan*.

39   Daniel R. Headrick, *The Tools of Empire: Technology and European Imperialism in the Nineteenth Century* (New York: Oxford University Press, 1981); James Scott, *Seeing Like a State: How Certain Schemes to Improve the Human Condition Have Failed* (New Haven, Conn.: Yale University Press, 1998).

40   I would like to thank Professor Philip Brown for reminding me of this point. Indeed, environmental historians have recently pointed out the importance of maintaining a balance between a locally grounded approach and a more globally scaled analysis. See, for example, Kenneth Pomeranz, "Introduction," in *The Environment and World History*, ed. Edmund Burke III and Kenneth Pomeranz (Berkeley: University of California Press, 2009), 3–32. Regarding Japan's environmental history, exemplar case studies in this regard are Brett L. Walker, "Meiji Modernization, Scientific Agriculture, and the Destruction of Japan's Hokkaido Wolf," *Environmental History* 9, no. 2 (2004): 248–74; and Brett L. Walker, *The Lost Wolves of Japan*, Weyerhaeuser Environmental Books (Seattle: University of Washington Press, 2005).

41   Discussions about how Japan's traditional forestry may be relevant to contemporary environmental issues can be found in Margaret A. McKean, Clark Gibson, and Elinor Ostrom, eds., *People and Forests: Communities, Institutions, and Governance* (Cambridge, Mass.: MIT Press, 2000); and Jared M. Diamond, *Collapse: How Societies Choose to Fail or Succeed* (New York: Viking, 2005).

PART IV

Climate

10

A History of Climatic Change in Japan

A Reconstruction of Meteorological Trends from Documentary Evidence

TAKEHIKO MIKAMI, MASUMI ZAIKI, AND JUNPEI HIRANO

Among the global climatic variations known from historic times are the Medieval Warm Period (ninth–fourteenth centuries) and the Little Ice Age (sixteenth–nineteenth centuries). Although there were important regional differences in climatic variation during both eras, global mean temperatures in the Medieval Warm Period were perhaps 1 to 2 degrees Celsius warmer than during the Little Ice Age. However, as is discussed in the Fourth Assessment Report of the Intergovernmental Panel on Climate Change, to define medieval warmth in a way that has more relevance for exploring the magnitude and causes of recent global warming, widespread and continuous paleoclimatic evidence must be reconciled and scaled against recent measured temperatures to allow a meaningful quantitative comparison against twentieth-century warmth.[1]

In Japan, the earliest official meteorological data come from the Hakodate observatory in Hokkaido in 1872. However, we have several kinds of documentary sources for reconstructing climatic variations in earlier centuries, including records of lake-freezing dates going back to the fifteenth century and weather diary records. Also, several kinds of meteorological observational data from the early nineteenth century are available from Nagasaki, Tokyo, and other cities (Tokyo was known as Edo prior to 1868, but we use the modern name throughout this essay for purposes of clarity and consistency). In this paper, we discuss the characteristics of these valuable proxies and instrumental data and describe the climatic changes in Japan that they indicate.

Reconstruction of Winter Temperatures Based on Lake-Freezing Records

Continuous records of lake-freezing dates since the fifteenth century come from central Japan, where a small lake is known for a mysterious winter phenomenon. When Lake Suwa freezes, shrinkage and expansion of the ice due to diurnal temperature variations cause an unusual type of ice cracking known as Omiwatari, said to resemble "a bridge crossing the lake" (figure 10.1). The ancient village people apparently believed it to be the track of a god visiting a goddess on the opposite shore (and indeed, Omiwatari literally translates as "divine crossing").

Since the fifteenth century, the formation of Omiwatari has been celebrated in a ceremony two or three days after its occurrence. The dates of complete freezing of the lake and Omiwatari have been recorded at the Suwa Shrine since the fifteenth century (figure 10.2), and also independently by the Suwa Meteorological Observatory since 1951. The motivation for recording lake-freezing dates was probably their usefulness for predicting summer climatic conditions in connection with rice harvests. During a cold winter, Omiwatari would occur before mid-December, whereas in a warm winter, it would be delayed until the end of February, or no Omiwatari would occur at all.

We sought to reconstruct winter temperatures based on the Lake Suwa freezing records. A linear regression of the freezing dates with the instrumental series of the Suwa Meteorological Observatory revealed that December–January mean temperatures are highly correlated ($r = 0.80$, significant at the 1 percent level) with the Lake Suwa series over the calibration period 1945–1990. Based on the regression equation, winter temperatures were estimated for central Japan from 1444 to 2010 (plate 8). Although the lake-freezing records are not continuous from the late seventeenth century to the nineteenth century, a clear warming trend stands out during the final stage of the Little Ice Age, from the 1750s to the 1850s. On the other hand, the coldest period since the fifteenth century was the early 1600s, when reconstructed mean winter temperatures were about 1 to 1.5 degrees Celsius lower than at present (1981–2010). To verify the reliability of this estimated temperature series, we made comparisons with other climate reconstructions on the basis of different proxy records, such as tree rings (see figure 10.9).[2] The results show a relatively good agreement with our winter temperature reconstruction.

As for the long-term freezing records of Lake Suwa, the reliability of yearly records is still not assured for some periods. More effort should be made in the verification and calibration of these valuable documentary records. Comparison with instrumentally observed data during the overlapping period is a prerequisite to producing a robust reconstruction.[3]

FIGURE 10.1. Ice-cracking phenomenon known as Omiwatari at Lake Suwa.
Photograph by Takehiko Mikami, January 31, 1998.

XII: December; I: January; II: February; III: March; — no description;
* approximate date, not later than; no: did not freeze or did not crack;
did: freezed or cracked, missing dates; open: "Akiumi" (open sea), except
land-fast ices which form along the shore; freezing date: full freez day.

Winter	Freezing dates	"Omiwatari"	Re-marks	Winter	Freezing dates	"Omiwatari"	Re-marks
1397–98	18 XII 1397	21 XII 1397		1480–81	3 I 1481	6 I 1481	
				81–82	15 I 1482	18 I 1482	
				82–83	24 I 1483	26 I 1483	
1443–44	7 I 1444	9 I 1444		83–84	31 XII 1483	1 I 1484	
44–45	23 XII 1444	26 XII 1444		84–85	1 I 1485	3 I 1485	
45–46	31 XII 1445	3 I 1446		85–86	10 I 1486	12 I 1486	
46–47	1 I 1447	4 I 1447		86–87	11 I 1487	14 I 1487	
47–48	30 XII 1447	2 I 1448		87–88	5 II 1488	7 II 1488	
48–49	7 I 1449	10 I 1449		88–89	20 I 1489	23 I 1489	
49–50	12 I 1450	15 I 1450		89–90	24 XII 1489	25 XII 1489	
1450–51	7 I 1451	11 I 1451		1490–91	9 II 1491	12 II 1491	
51–52	22 I 1452	25 I 1452		91–92	5 I 1492	7 I 1492	
52–53	28 XII 1452	31 XII 1452		92–93	1 I 1493	3 I 1493	
53–54	2 I 1454	4 I 1454		93–94	5 I 1494	7 I 1494	
54–55	4 I 1455	7 I 1455		94–95	8 I 1495	10 I 1495	
55–56	31 XII 1455	3 I 1456		95–96	3 I 1496	6 I 1496	
56–57	4 I 1457	7 I 1457		96–97	22 I 1497	25 I 1497	
57–58	no ?	no ?	open ?	97–98	11 XII 1497	13 XII 1497	
58–59	5 I 1459	8 I 1459		98–99	13 I 1499	14 I 1499	
59–60	20 XII 1459	23 XII 1459		99–1500	18 XII 1499	20 XII 1499	

FIGURE 10.2. Example of Lake Suwa freezing date records

SOURCE: Hidetoshi Arakawa, "Fujiwhara on Five Centuries of Freezing Dates of Lake Suwa in the
Central Japan," Archiv für Meteorologie, Geophysik und Bioklimatologie, Series B 6 (1954): 153,
table 2.

Reconstruction of Summer Temperatures Based on Weather Diaries

Many weather diaries from most parts of Japan are preserved in local libraries and museums. Figure 10.3 shows an example of a weather diary from Edo (modern Tokyo) with very detailed weather descriptions for November 1, 1834. The boxed area reads, "1st day [of the tenth month], fine weather during the day, cloudy in the evening and began to rain at night." Some official diaries have been kept continuously since the seventeenth century. Over the past few decades, most of the known daily weather records have been digitized and added to the Historical Weather Database of Japan.[4] Descriptions of daily weather differ from diary to diary. In some diaries, simple expressions such as "fine," "cloudy," or "rain" were used; others feature relatively detailed information about the duration of particular weather phenomena or the wind direction and intensity. Descriptions of precipitation are the most important in quantitative climate reconstruction. For instance, descriptions such as "rainy all day," "occasionally rainy," "heavy rainfall," "showery," or "light rain in the morning" were frequently used in old diaries. Based on these daily weather records, it should be possible to estimate the amount of precipitation and the seasonal mean temperature and even to reconstruct typical pressure patterns that are relevant to regional precipitation distributions.[5]

Here we show an example of summer temperature reconstruction based on daily weather records since the early eighteenth century from an old diary. To discuss the decadal to century time-scale variations in the summer temperatures of Tokyo, meteorological data for the past 136 years would be insufficient. Thus we made an attempt to reconstruct summer temperatures in Tokyo based on the weather records of the Ishikawa family, who lived in the western suburbs of Tokyo and kept diaries from generation to generation, beginning in 1721.

In general, Japan experiences hot summers under the influence of strong subtropical highs, which bring dry and sunny weather conditions, and cool summers under the influence of stagnant polar fronts and passing extra-tropical cyclones, which bring cloudy and rainy weather situations. This suggests that there may be a high correlation between the number of rainy days and the mean temperature in summer months. Fortunately, historical weather records in the Ishikawa diaries and observed meteorological data of the Japan Meteorological Agency (JMA) from Tokyo overlap for sixty-five years, from 1876 to 1940, allowing us to cross-check the two sources of information.

We first compared the number of rainy days in the Ishikawa diaries and the JMA data. For the JMA data set, a rainy day was defined as a day when rainfall exceeded 1 millimeter. During the sixty-five-year period, the average number of rainy days in the Ishikawa diaries for June, July, and August were 9.2, 8.4, and 7.7,

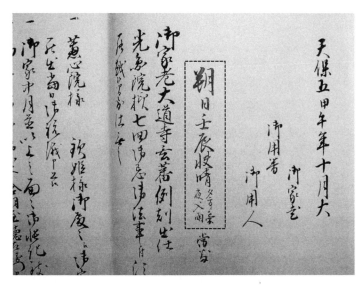

FIGURE 10.3. Example of daily weather records (November 1, 1834) in the official diary of the Hirosaki Domain, modern Aomori Prefecture

SOURCE: Hirosaki Municipal Library at Hirosaki, Japan. Author's photocopy.

respectively, while those in the JMA data for the same months were 10.8, 8.7, and 7.1, respectively. Although June rainy days were underestimated in the Ishikawa diaries, the July and August rainy days demonstrated good agreement between the diaries and the official observations.

To confirm the temperature and rainy day relationships, we calculated correlation coefficients, using the JMA data, between the number of rainy days and the mean temperatures in summer months for the sixty-five-year period. Correlation coefficients for June, July, and August were −0.41, −0.70, and −0.45, respectively. Because July mean temperatures are most highly correlated with the number of rainy days in a month, it should be possible to reconstruct July temperatures in Tokyo for 1721–1940 based on the weather records in the Ishikawa diaries.

We applied a simple least-square regression method to the July mean temperatures and the number of rainy days in Tokyo (JMA data) (figure 10.4). We obtained the linear regression equation $Y = 26.46 - 0.235X$, where X (the predictor) is the number of rainy days and Y (the predictant) is the monthly mean temperature in July. The coefficient of determination R^2 equals 0.49, and the standard error of Y is 0.98. Based on this regression equation, we estimated July temperatures in Tokyo for each year from 1721 to 1940, a period during which both proxy (diary) and observed (meteorological) data were available from 1876 to 1940.

Combining the reconstructed temperature time series for 1721–1940 with

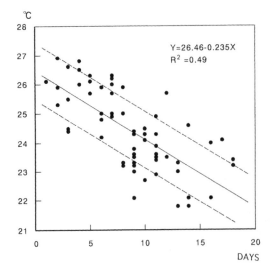

FIGURE 10.4. Relationship between the number of rainy days and the monthly mean temperature of Tokyo in July, 1876–1940

SOURCE: Takehiko Mikami, "Long Term Variations of Summer Temperatures in Tokyo since 1721," *Geographical Reports of Tokyo Metropolitan University* 31 (1996): 161, fig. 2.

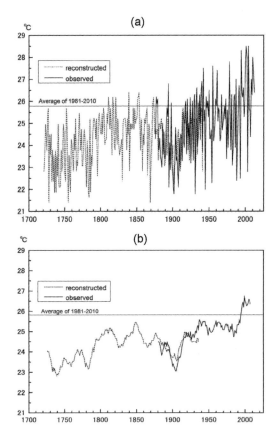

FIGURE 10.5. Combined time series of reconstructed (broken line) and observed (solid line) July temperatures in Tokyo, 1721–2012. (a) Year-to-year variations; (b) Smoothed curve (eleven-year running means).

SOURCE: Based on Takehiko Mikami, "Long Term Variations of Summer Temperatures in Tokyo since 1721," *Geographical Reports of Tokyo Metropolitan University* 31 (1996): 162, fig. 3, with additional data.

the observed temperatures from JMA data for 1876–2012 reveals the long-term variations in July temperatures for 1721–2012. Figure 10.5(a) shows year-to-year variations, with both reconstructed (dashed line) and observed (solid line) temperatures. Figure 10.5(b) displays smoothed temperature curves with eleven-year running means; the horizontal line indicates the average temperature for 1981–2010.

Estimated temperature series show several cooler and warmer periods. From 1721 to 1790, temperatures are estimated to have been around 1.5 to 2 degrees Celsius lower than at present. During this period, July temperatures show large year-to-year variability, with the lower values below 22 degrees in 1728, 1736, 1738, 1755, 1758, 1783, 1784, and 1786. It should be noted that the temperatures in the 1780s were often extremely low and displayed large interannual variations. The summer of 1783, for example, had an extremely poor rice harvest because of exceedingly cool and wet conditions, and this unusual weather caused a historic severe famine in Japan.[6]

On the other hand, it was rather warm in the nineteenth century, especially in the 1810s and the early 1850s, with the higher values above 26 degrees Celsius in 1811, 1817, 1821, 1851, 1852, and 1853. During this generally warm century, the 1830s, late 1860s, and late 1890s were relatively cool decades, and great famines recurred in the 1830s, just as in the 1780s. July temperatures reached their lowest level around 1900, when eleven-year mean temperatures were the same as those around 1740. It is interesting to note that there may be 160-year periodicity in summer temperature variations, which should be examined in more detail. Of course, Tokyo has one of the strongest urban heat islands in the world, and this is likely to have influenced temperature trends, at least in the modern period.

The observed temperatures allow us to verify the reliability of our reconstructed temperature time series based on the diaries' weather records. As indicated in figure 10.5, both time series are well correlated during the overlapping period. Observed temperatures show the lowest level around 1900; this minimum is also clearly evident in the reconstructed temperature series.

Winter Climate Variations since 1700 Based on Official Records

We estimated winter temperatures based on daily weather records in the official diary of the Isahaya fief (1700–1872) at Nagasaki. Although Nagasaki is located in southwestern Japan, snow sometimes falls in colder winters. Therefore winter temperatures could be estimated from the frequency of snow, indicating colder days, versus the frequency of rain, indicating warmer days. Because the Isahaya diaries had recorded the number of days with snow and rain for nearly

200 years, beginning in 1700, we applied a simple linear regression analysis to estimate winter temperatures at Nagasaki, using meteorological data from the Nagasaki JMA Observatory.

The snowfall ratio is highly correlated with January mean temperatures at Nagasaki. Thus we applied the equation of linear regression ($Y = 7.895 - 3.365X$) to estimate January mean temperatures at Nagasaki for 1700–1872. We combined the estimated temperature series with the meteorological data since 1879 (plate 9). Observed instrumental temperature data are also available for the early nineteenth century (figure 10.14).

During the past three hundred years, variations in January mean temperatures at Nagasaki are characterized by decadal oscillations and large year-to-year variability (plate 9). Estimated temperatures were relatively high in the early eighteenth century, and then showed a gradual decreasing trend toward the early nineteenth century, which was the coldest period since 1700. However, January temperatures again show an upward trend from the 1840s to the 1860s, which was the final stage of the Little Ice Age.

Observed temperature series also indicate decadal and year-to-year variations with warmer periods (1901–1920, 1950s, and 1990s) and colder periods (1871–1890 and 1940s). Although reconstructed temperatures (red line) seems to agree with the observed temperatures (green line) during the overlapping period in the early to middle nineteenth century, statistical standard error (about 1 degree Celsius) and the upper-lower limit values peculiar to a linear regression analysis should be considered.

◼ Reconstruction of Winter Climate Variations in the Nineteenth Century

We sought to reconstruct winter climate conditions in Japan for 1810–11 to 1858–59 from daily weather records documented in old diaries. Daily weather maps for each winter were drawn using nineteenth-century weather records collected by our research group. Maps were divided into five types by classifying daily snowfall and rainfall distributions (two major types are shown in plate 10), and the occurrence frequencies of each weather pattern for the period were analyzed (figure 10.6). The frequencies of typical winter monsoon weather patterns were high from the late 1820s to the early 1840s. This period almost coincided with a summer cold period in the nineteenth century. The result implies that strengthening of a cold air mass around Japan occurred in the late 1820s, not only in summer but also in winter.

The frequencies of the typical winter monsoon patterns correspond with the freezing dates of Lake Suwa, which have been used as an indicator of winter

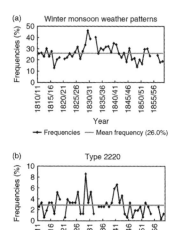

(a) Winter monsoon weather patterns

Year
Frequencies — Mean frequency (26.0%)

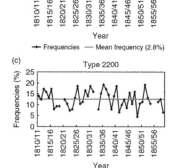

(b) Type 2220

Year
Frequencies — Mean frequency (2.8%)

(c) Type 2200

Year
Frequencies — Mean frequency (12.5%)

FIGURE 10.6. Occurrence frequencies for each weather pattern, 1810–11 to 1858–59 (a) Winter monsoon weather patterns; (b) Type 2220; (c) Type 2200. Gray line indicates mean frequency.

SOURCE: Junpei Hirano and Takehiko Mikami, "Reconstruction of Winter Climate Variations during the 19th Century in Japan," *International Journal of Climatology* 28 (2008): 1431, Fig. 7.

coldness in previous studies (figure 10.7). On the basis of the frequencies of the winter monsoon weather patterns, we estimated mean January temperatures for western Japan (figure 10.8). The time series of estimated temperatures reveal a cooling period from the late 1820s to the early 1830s. By contrast, a warming trend can be detected during 1840s and 1850s, which corresponds to the final stage of the Little Ice Age.

We have collected a tremendous number of daily weather records in old diaries from all over Japan, most of which have been digitized and are available as computer-readable files. As for the long-term freezing records of Lake Suwa, there remains some problem with the reliability of yearly records during particular periods. More effort should be made to verify and calibrate these valuable documentary records in Japan during historical periods. Comparison with other independent proxies, such as the tree-ring records described by Sweda and Takeda (figure 10.9) and the cherry flowering date records described by Aono and Kazui (figures 10.10 and 10.11), may also be important for the accuracy of climate reconstructions.[7]

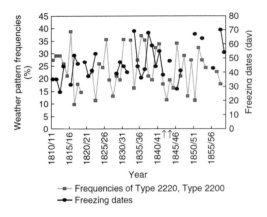

FIGURE 10.7. Time series for the frequencies of Type 2220 and Type 2200 weather patterns in December–January and the freezing dates of Lake Suwa, 1810–11 to 1858–59 Squares indicate frequencies of weather patterns, circles indicate freezing dates, and arrows indicate years without freezing (open lake).

SOURCE: Junpei Hirano and Takehiko Mikami, "Reconstruction of Winter Climate Variations during the 19th Century in Japan," *International Journal of Climatology* 28 (2008): 1432, fig. 10.

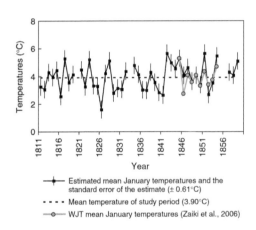

FIGURE 10.8. Time series of mean January temperatures for western Japan, 1811–1859 Squares indicate estimated temperatures, and circles show the West Japan temperature series. Error bar for estimated temperatures indicates the standard error (±0.61 degrees C).

SOURCE: Junpei Hirano and Takehiko Mikami, "Reconstruction of Winter Climate Variations during the 19th Century in Japan," *International Journal of Climatology* 28 (2008): 1433, Fig. 11, p. 1433 (WJT data are from M. Zaiki, G.P. Können, T. Tsukahara, P.D. Jones, T. Mikami, and K. Matsumoto, "Recovery of Nineteenth-Century Tokyo/Osaka Meteorological Data in Japan," *International Journal of Climatology* 26 (2006): 418–21, Appendix A1).

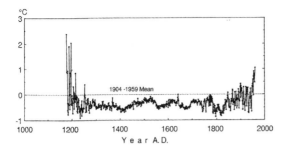

FIGURE 10.9. Reconstruction of winter temperatures in central Japan based on tree rings

SOURCE: Tatsuo Sweda and Shinichi Takeda, "Construction of an 800-Year-Long *Chamaecyparis* Dendrochronology for Central Japan," *Dendrochronologia* 11 (1994): 83, fig. 2.

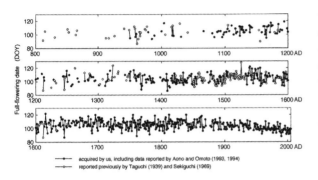

acquired by us, including data reported by Aono and Omoto (1993, 1994)
reported previously by Taguchi (1939) and Sekiguchi (1969)

FIGURE 10.10. Interannual variation of the full-flowering dates of the cherry tree, *P. jamasakura*, at Kyoto, acquired from old diaries and chronicles

SOURCE: Y. Aono and K. Kazui, "Phenological Data Series of Cherry Tree Flowering in Kyoto, Japan, and Its Application to Reconstruction of Springtime Temperatures since the 9th Century," *International Journal of Climatology* 28, no. 7 (2008): 908, fig. 2.

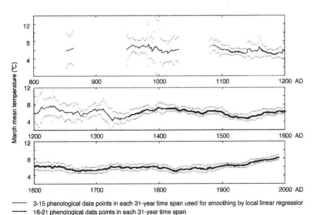

3-15 phenological data points in each 31-year time span used for smoothing by local linear regression
16-21 phenological data points in each 31-year time span
22-31 phenological data points in each 31-year time span
95% confidence intervals in smoothing procedure

FIGURE 10.11. Reconstruction of March mean temperatures at Kyoto in the historical period, since the ninth century.

The thicknesses of the curves indicate differences in the number of phenological data points in each thirty-one-year span used by the local linear regression procedure. Confidence intervals (95 percent) on the smoothed value are shown in dotted lines.

SOURCE: Y. Aono and K. Kazui, "Phenological Data Series of Cherry Tree Flowering," 911, fig. 6.

Recovery of Nineteenth-Century Meteorological Data

Japan's climate is regarded as one of the many blank spots in the pre-1900 world because the country's official meteorological network started only in the 1870s. Prior to 1872, no Japanese meteorological records were thought to exist apart from the notations in the diaries of administrators at many places in Japan.[8] It was believed that the only pre-1872 instrumental data on Japanese climate were taken by the Dutch in the settlement of Dejima in Nagasaki (figure 10.12).

Recently, however, subdaily weather records taken routinely by Japanese scientists for a few places in Japan, most notably Edo (Tokyo) and Osaka, have been discovered (figure 10.13).[9] These Japanese meteorological observations were associated with the development of astronomical research by the so-called Dutch Studies scholars, as a result of Japan's modernization and the introduction

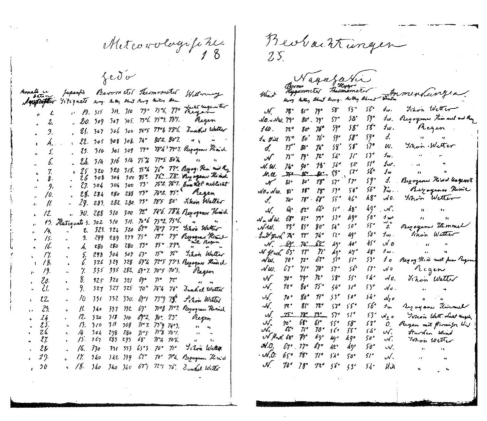

FIGURE 10.12. Example of meteorological observation data by von Siebold
SOURCE: Library of Ruhr University, Bochum, Germany. Author's photocopy.

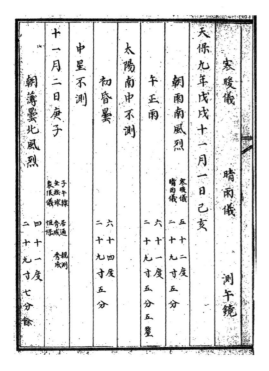

FIGURE 10.13. Example of old Japanese instrumental data from the compilation *Reiken Kōbo*.
Detailed descriptions of daily weather, temperature, and air-pressure observations are shown for December 17 and 18, 1838.

SOURCE: National Archives of Japan. Author's photocopy.

of modern, Western instruments. Because these data partly overlap with the Dejima data and also fill some of the gaps, we began a search to recover them from Japanese archives and elsewhere.

As a follow-up to the work of evaluating and reconciling early-nineteenth-century Japanese observations taken at Dejima, we present here an evaluation of the recovered nineteenth-century instrumental data from Tokyo, Yokohama, Osaka, and Kobe (figure 10.14).[10] The availability of the 1819–1878 temperature and pressure observations taken at all five locations holds the potential for constructing a nineteenth-century instrumental temperature and pressure series for western Japan.

The rediscovery of these weather data have greater significance than just extending the Japanese instrumental record back in time. First, the recovered series happens to be in a region of the world that is poorly covered by instrumental data; second, it overlaps with the long daily series of weather reports from 1700 to 1868 documented in administrative diaries from many locations in Japan.[11]

We have recovered instrumental temperature and pressure observations from Tokyo for 1825–1828, 1839–1855, and 1872–1875; from Yokohama for 1860–1871

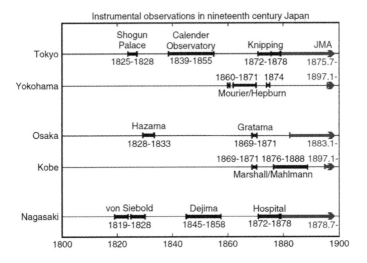

FIGURE 10.14. Availability of meteorological data in Japan before 1900
Gray represents data from official stations of the Japan Meteorological Agency; black indicates earlier data. The subseries are named by their observer or location. The Nagasaki data are discussed in G. P. Können, M. Zaiki, A. P. M. Baede, T. Mikami, P. D. Jones, and T. Tsukahara, "Pre-1872 Extension of the Japanese Instrumental Meteorological Observations Series Back to 1819," *Journal of Climate* 16 (2003): 118–31; and Zaiki et al., "Recovery of Nineteenth-Century Tokyo/Osaka Meteorological Data in Japan," *International Journal of Climatology* 26 (2006): 399–423.
SOURCE: Zaiki et al., "Recovery," 401, fig. 2.

and 1874; from Osaka for 1828–1833 and 1869–1871; and from Kobe for 1869–1871 and 1875–1888. The newly recovered records contain data before the 1870s—decades in the nineteenth century for which no instrumental data were believed to exist. This extension of Japanese records, along with the recently recovered intermittent Dejima-Nagasaki series for 1819–1878, implies that for the nineteenth century, the Japanese instrumental record no longer contains major temporal gaps.

The recovered data were used for a preliminary calculation of the West Japan temperature series, which is a representative temperature series for the area (figure 10.14). The existence of a warm epoch in the 1850s in western Japan and a downward temperature trend until the early twentieth century, as previously inferred from documentary records, is confirmed by the data. It should be noted that the Central England temperature series also shows similar climatic variations since the early nineteenth century (figure 10.15).

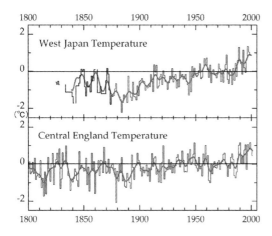

FIGURE 10.15. The West Japan temperature series and the Central England temperature series.

SOURCES: Data for Japan from Zaiki et al., "Recovery"; data for England from D.E. Parker, T.P. Legg, and C.K. Folland, "A New Daily Central England Temperature Series, 1772-1991," *International Journal of Climatology* 12 (1992): 317-42.

Conclusions

Climatic changes in Japan since the eighteenth century were analyzed using various kinds of reconstructed data and meteorological data. The results obtained are summarized as follows:

- Winter temperature was relatively higher during the early eighteenth century, becoming lower from the late eighteenth century to the early nineteenth century. The 1810s and 1820s were particularly cold. Rapid warming commenced in the 1830s, probably in conjunction with the termination of the Little Ice Age. The 1850s and 1860s were particularly warm.
- Summer temperature shows somewhat different changes from winter temperature variations, especially during the eighteenth century. Estimated summer temperatures were rather lower in the mid-eighteenth century, which was followed by a warming trend until the mid-nineteenth century. Summer temperatures reached their peak around 1850 and thereafter gradually fell to a low around 1900.
- Comparison of annual mean temperatures in West Japan temperature series and Central England temperature series shows similar trends, in accordance with the changes in global circulation patterns.

NOTES

1   Susan Solomon, Dahe Qin, Martin Manning, Melinda Marquis, Kristen Averyt, Melinda M. B. Tignor, Henry LeRoy Miller, Jr., and Zhenlin Chen, eds., *Climate Change 2007: The*

Physical Science Basis (Contribution of Working Group I to the Fourth Assessment Report of the Intergovernmental Panel on Climate Change) (Cambridge: Cambridge University Press, 2007).

2   Tatsuo Sweda and Shinichi Takeda, "Construction of an 800-Year-Long *Chamaecyparis* Dendrochronology for Central Japan," *Dendrochronologia* 11 (1994): 79–86.

3   G. P. Können, M. Zaiki, A. P. M. Baede, T. Mikami, P. D. Jones, and T. Tsukahara, "Pre-1872 Extension of the Japanese Instrumental Meteorological Observations Series Back to 1819," *Journal of Climate* 16 (2003): 118–31; M. Zaiki, G. P. Können, T. Tsukahara, P. D. Jones, T. Mikami, and K. Matsumoto, "Recovery of Nineteenth-Century Tokyo/Osaka Meteorological Data in Japan," *International Journal of Climatology* 26 (2006): 399–423.

4   Takehiko Mikami, "Climatic Reconstruction in Historical Times Based on Weather Records," *Geographical Review of Japan (Series B)* 61, no. 1 (1988): 14–22; Takehiko Mikami, "Quantitative Climatic Reconstruction in Japan Based on Historical Documents," *Bulletin of the National Museum of Japanese History* 81 (1999): 41–50.

5   E.g., Takehiko Mikami, "Climatic Variations in Japan Reconstructed from Historical Documents," *Weather* 63, no. 7 (2008): 190–93; Junpei Hirano and Takehiko Mikami, "Reconstruction of Winter Climate Variations during the 19th Century in Japan," *International Journal of Climatology* 28 (2008): 1423–34.

6   Takehiko Mikami, "Climate of Japan during 1781–90 in Comparison with That of China," in *The Climate of China and Global Climate: Proceedings of the Beijing International Symposium on Climate,* Beijing, October 30–November 3, 1984, ed. Duzheng Ye, Congbing Fu, Jiping Chao, and M. Yoshino (Beijing: China Ocean Press, 1987), 63–75; Takehiko Mikami, "Climate Variations in Japan during the Little Ice Age—Summer Temperature Reconstructions since 1771—," in *Proceedings of the International Symposium on the Little Ice Age Climate,* ed. T. Mikami (Tokyo: Tokyo Metropolitan University, 1992), 176–81.

7   Sweda and Takeda, "Construction of an 800-Year-Long *Chamaecyparis* Dendrochronology"; Yasuyuki Aono and Keiko Kazui, "Phenological Data Series of Cherry Tree Flowering in Kyoto, Japan, and Its Application to Reconstruction of Springtime Temperatures since the 9th Century," *International Journal of Climatology* 28, no. 7 (2008): 905–14.

8   Takehiko Mikami, "Climatic Reconstruction in Historical Times"; T. Mikami, M. Zaiki, G. P. Können, and P. D. Jones, "Winter Temperature Reconstruction at Dejima, Nagasaki Based on Historical Meteorological Documents during the Last 300 Years," in *Proceedings of the International Conference on Climate Change and Variability—Past, Present and Future,* ed. Takehiko Mikami (Tokyo: Tokyo Metropolitan University, 2000), 103–106; Mikami, "Climatic Variations in Japan."

9   Amano Ryūji, "1830 nen zengo ni okeru Ōsaka no kishō, *Shōsen Daigaku kenkyū hōkoku* A3 (1952): 137–57; Amano Ryūji, "1838 nen yori 1855 nen ni itaru Tōkyō no kishō," *Shōsen Daigaku kenkyū hōkoku* A4 (1953): 167–94; Tsukahara Tōgo, Zaiki Masumi, Matsumoto Keiko, and Mikami Takehiko, "Nihon no kiki kansoku no hajimari: Dare ga, dono yōna jōkyō de hajimeta no ka," *Gekkan chikyū* 27, no. 9 (2005): 713–20.

10 Können et al., "Pre-1872 Extension."

11 Mikami, "Climatic Reconstruction in Historical Times"; Mikami et al., "Winter Temperature Reconstruction at Dejima, Nagasaki"; Mikami, "Climatic Variations in Japan Reconstructed from Historical Documents."

11

Climate, Famine, and Population in Japanese History

A Long-Term Perspective

OSAMU SAITO

Famine is a phenomenon of mass starvation caused primarily by poor harvests, typically triggered by bad weather, such as drought, excessive rainfall, or cold temperatures in summer. Since the time of T. R. Malthus, who asserted that when all other checks on population growth had failed, famine emerged to bring population levels in line with food supply, scholars have maintained that the frequency of famines was one of the most powerful determinants of mortality in the past and have assumed that there were significant correlations between long-run changes in climate and population.[1] In particular, paleoclimatologists have identified a very long spell of cold years beginning in the thirteenth century, and this period of cooling, known as the Little Ice Age, is thought to explain why there were so many severe famines in various parts of the world from late medieval to early modern times. In Japanese history, the Little Ice Age corresponds to the Kamakura through the Tokugawa eras. The question I would like to address is whether there was any significant change in the relationship between climate and famine during this long period. Some might argue that the adverse consequences of climate continued to be felt even in the late Tokugawa period, citing the various famines that took place at that time. However, consensus is that a major break came with the establishment of the Tokugawa regime. According to this interpretation, population growth took off in the seventeenth century as the country's food supply began to expand.

In this chapter I first set out a revised chronological table of famines in the Japanese past, then examine to what extent global cooling and warming were related to changing frequencies of famine, and finally, identify exactly when the nature of the relationship between climate and famine started to change. The new famine data set is expected to provide historians with clues to how population changed in the transition from medieval to early modern times in Japan. Akira Hayami's thesis that Japan's population was as low as twelve million in 1600 but grew rapidly during the seventeenth century has so far been accepted widely.[2] Although Hayami did not explicitly discuss the implications of this thesis, one possible scenario is Malthusian: with population pressure on land mounting, susceptibility to changing climatic conditions—and hence, the frequency of famine—may have increased over time. However, there are competing interpretations, and William Wayne Farris's recent synthesis, based on those revisionist estimates as well as a survey of other evidence, concludes that population must have been in the range of fifteen million to seventeen million in 1600, much larger than Hayami's estimate.[3] This chapter will lend support to the revisionist claims. It suggests that the real break with the medieval past took place earlier, during a period of civil war in which climatic conditions were consistently unfavorable, and hence that the rate of population increase in the period after 1600 must have been lower than previously thought.

The finding that the number of serious famines decreased despite worsening climatic conditions may be taken to imply that factors other than climate gained weight well before the Tokugawa came to power. This raises an issue concerning how human resilience was achieved in the late medieval setting. Yet we will also see that even in the post-1700 phase of gradual upturn, the country could not escape from disasters, such as the so-called Tenmei famine in the 1780s and the Tenpō famine in the 1830s, both of which stretched over a period of years. How to interpret these contrasting outcomes in a consistent manner is another issue. The Malthusian framework suggests that agrarian progress is crucial to escaping from the mortality trap. On the other hand, politics and institutional structure may matter for the timing of change. It is expected that the general level of resilience will rise as a result of long-run growth in agrarian productivity, but whether political and institutional changes will give rise to positive or negative changes probably hinges more on factors specific to the time period in question.

▓ Famine Records

Historical famines in Japan and their causes can be identified with the help of several data books on historical natural disasters.[4] A recent addition to this genre is a fascinating database compiled by Fujiki Hisashi and covering the tenth to mid-seventeenth centuries, a period for which records have long been scanty.[5] Based on these data sets, I have now identified a total of 281 famine years between 567 and 1869, inclusive. This is 47 more than my previous count of 234 such years.[6]

In identifying a case of famine, it is important to make two distinctions in the historical records. The first distinction is between famine-related deaths and epidemic-related deaths, and the second is between phenomena associated with "hunger," "starvation," and "crisis" and those described merely as "crop failure" or "disastrous harvest," since not all harvest failures resulted in famines.

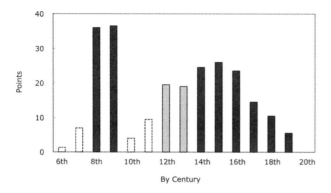

FIGURE 11.1. Japanese famine points, by century

SOURCES: Famine chronology table compiled by the author from Ogashima Minoru, *Nihon saiishi* (Nihon Kōgyōkai, 1894); Nishimura Makoto and Yoshikawa Ichirō, eds., *Nihon kyōkōshi kō* (Maruzen, 1936); and Fujiki Hisashi, comp., *Nihon chūsei kishō saigaishi nenpyō kō* (Kōshi Shoin, 2007).

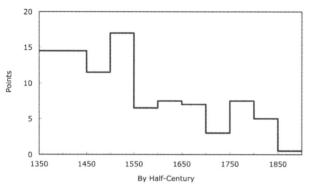

FIGURE 11.2. Japanese famine points, by half-century, 1350–1900

SOURCES: See fig. 11.1.

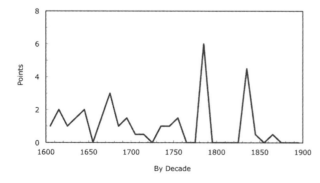

FIGURE 11.3. Japanese famine points, by decade, 1601–1900

SOURCES: See fig. 11.1.

Next we have to consider the intensity, extent, and duration of the "hunger" or "starvation" in question. The best measure of the intensity of a famine is probably the death toll or the rate of excess mortality, but given the nature of the records we have for earlier centuries, there is little hope of recovering such data. It is often possible, however, to determine how geographically widespread the famine was and how long it lasted. Admittedly this is not always an easy task, but here I attempt to identify the severity of famines by assigning 1 "famine point" to cross-regional occurrences (those for which records exist for more than five or six provinces), 0.5 point to famines apparently confined to a single province, and 0 to genuinely local famines.[7] Note that the mere reference to unusual weather conditions, such as "extremely hot" or "unprecedentedly cold," does not necessarily imply that there was a famine unless the reference is accompanied by observations that "everybody starved" or "fields were covered with dead bodies." Even when starvation did occur, it is frequently difficult to determine whether the phenomenon was cross-regional and countrywide or confined to a single region. For example, when a local source from the late medieval period states, "The whole country starved," caution is in order, since there may be no evidence of mass starvation in other regions. In cases like this, I regard the description as an exaggeration and conclude that the famine was only regional.

Famines from the Eighth to the Nineteenth Centuries

Figure 11.1 shows the distribution of all 281 famine years by century. Records for the sixth and seventh centuries are scanty, but those for the eighth and ninth centuries are surprisingly numerous. Then there is a long period of data scarcity between the ninth and the fourteenth centuries. The reason is the disintegration in the beginning of the Heian period of the ancient, Chinese-style *ritsuryō* state system, which created a political vacuum and put an end to effective record-keeping. After the establishment of the Kamakura government in 1192, the number of records, both official and private, began to increase gradually, but it is not until the fourteenth century that the data recover to the point that we can be reasonably confident about the frequency of famine. From 1350 onward, it is possible to show the changing frequency of famine by half-century (figure 11.2), and from 1600 onward, when the number of written records multiplied under Tokugawa rule, by decade (figure 11.3).

It is evident from these graphs that famine was very frequent in ancient times but became less frequent in subsequent periods. In the eighth and ninth centuries, famine occurred on average once every three years. In the medieval period, or more precisely, in the period from ca. 1300 to 1550, the frequency declined a little, to an average of once every four years. Then came an unexpected, substantial fall in the number of famine years between the first and second halves of the sixteenth century, as a result of which the late sixteenth to seventeenth centuries averaged one famine every seven years. From about 1700 onward, the incidence was further reduced: for the eighteenth century we have a total of 10.5 famine points, in the nineteenth century a total of only 5.5, and in the twentieth century a complete absence of famine, whether regional or countrywide.

Famine in the Context of Climatic Change

Historians of global climate have paid special attention to the Little Ice Age. In Europe, for example, a bitterly cold period began with the great rains of 1315, and similar or even harsher climatic conditions came back at the height of the Little Ice Age, in the seventeenth century.[8] For Japan, a variety of data sources have been used to map the long-term chronology of climatic change; these include fossil pollen, tree rings, and other scientific data, as well as literary records.[9] While some studies focus on subperiods such as the Little Ice Age, others cover a much longer span, sometimes going back to ancient times. Particularly useful are temperature reconstructions based on data from (1) pollen collected from Ozegahara in the mountains of north Kanto; (2) tree rings of Japanese cedars (*sugi*) on Yaku Island, off Kyushu; (3) boreholes at a site on the Lake Biwa shore; (4) tree rings of Japanese cypresses (*hinoki*) in southern Shinshū, central Japan; and records of (5) the freezing of Lake Suwa in central Japan and (6) the flowering of cherry trees in Kyoto.[10] There are also studies based on (7) categorical data, such as records of weather hazards. Although it is difficult to translate such records into estimates of actual temperatures, long-term series of weather records are nonetheless valuable, particularly if they provide season-specific information.[11] Data series (1), (2), (3), and (7) are relatively deficient in this sense, but series (4), (5), and (6) generally provide information about temperatures in winter to springtime.

The long-term temperature series derived from the Yaku tree rings and the Biwa boreholes (2 and 3) cover two millennia, and the Ozegahara data series (1) goes back to the fifth millennium BCE. The three series agree that the warmest period was reached in the twelfth or thirteenth century, but there is disagreement with respect to the coldest: the former two series indicate that it comes in the seventeenth century, whereas the Ozegahara data show the lowest temperatures in the fourteenth century. The Yaku average temperature series also shows a trough in the fourteenth century, although not as deep as that in the seventeenth, but in the Biwa series the fourteenth century does not seem to have been particularly cold. There is another disagreement between this long-term Biwa series and the shorter series of tree ring data from Kiso (4): in the latter there was a brief but deep trough at the end of the eighteenth century, but in the former the deepest trough came in the seventeenth century. At this stage I am not in a position to account for these differences among the data series. Considering the global picture of a long period of cooling between the thirteenth-century peak and the seventeenth-century trough,[12] I surmise that in Japan, too, the seventeenth century was probably the coldest overall. With this periodization, what we have found from the famine data can be summarized in table 11.1.

TABLE 11.1. Climate change, drought, cold summer, and the frequency of famine

Period	Climate	Causes of famine: drought vs. cold summer	Famine points per century
7th–12th	Warm period	6 : 4	31*
13th–early 16th	Cooling	7 : 3	22
Late 16th–17th		5 : 5	12
18th–19th	Warming	0 : 10	8

* Average calculated with the seventh, tenth, and eleventh centuries excluded.

SOURCES: See text and figure 11.1.

It is clear that the frequency of famine fell, in the long run, with a 10-point decline in the middle of the cooling period. Also, drought as a cause of famine became less and less important over time. It would appear that these famine-related trends have nothing to do with the changing levels of temperature, but to test this hypothesis, we must take a closer look at temperature fluctuations during the age of cooling. According to the estimates derived from the tree ring and cherry-tree flowering data (series 4 and 6), for example, there were two more turning points in the series of temperatures toward the end of the period.[13] Specifically, there was another trough in the mid-sixteenth century, followed by a short but full cycle of warming and cooling that ended in the 1670s, which suggests that the period around 1600 may have been relatively warm rather than cold. If this were really the case, then one could argue that the 10-point decline in the frequency of famines shown in table 11.1 was triggered by this warming spell in the late sixteenth century. However, the data from cherry blossoming and tree rings reflect mean temperatures in winter and spring and may not be a good guide to those in other seasons. In a rice-growing country like Japan, moreover, a successful harvest depends not only on abundant rainfall in spring and early summer but also on sufficiently high temperatures in midsummer.

It is interesting in this respect to look at a diagram compiled by two Japanese paleoclimatologists from qualitative data to show both summer and winter distributions of climatic hazards over a fourteen-hundred-year period (data series 7).[14] According to the diagram, from the mid-sixteenth century, winters became warmer but summers became cooler, which suggests that the century after about 1550 may have been one of the most disruptive for rice cultivators in the Little Ice Age. Moreover, the same diagram shows that even after the thirteenth century, summer temperatures remained generally high until the late sixteenth century, whereas cool summers were more frequent in the period after 1600. It is not unlikely that the diagram overstates the degree of warmth prior to 1550 because the authors base their judgment on the number of droughts rather than on summer temperatures per se, but it is probably safe to say that as far as the summer season is concerned, cooling over the period 1200–1550 was a slower process than previously imagined. According to a more detailed account of summer temperatures in the eighteenth and nineteenth centuries, it was only after the early eighteenth century that average summer temperatures began to rise gradually.[15] This suggests that not just the decades around 1600 but as much as a century and a half beginning in the mid-1500s were climatically unfavorable for the rice economy.

TABLE 11.2. The phase of climate, drought, cold summer, and the frequency of famine by half-century

Period	Trend in summer temperature	Drought vs. cold summer	Famine points
Late 14th		7 : 3	14.5
Early 15th	Stable, or cooling gradually	8 : 2	14.5
Late 15th		8 : 2	11.5
Early 16th		7 : 3	17
Late 16th		6 : 4	7
Early 17th	Cooling	4 : 6	7.5
Late 17th		3 : 7	7
Early 18th		0 : 10	3
Late 18th	Warming with cool summers	0 : 10	7.5
Early 19th	in 1750s, 1780s, and 1830s	0 : 10	5
Late 19th		0 : 10	0.5

sources: Figure 11.1 and, for summer temperature, Maejima and Tagami, "Climatic Change" and Mikami, "Long Term Variations."

Given the above chronology of climatic change in the summer season, table 11.2 sets out the same information as in table 11.1, this time by half-century. Although short-term fluctuations in both climatic and famine series can sometimes be extremely wide, the observations based on the half-century periods confirm that famine became less frequent and drought less problematic over the long run. However, more significant is the finding that the break with the past in the famine frequency series came earlier than that in the causes-of-famine column. The major decrease in frequency occurred between the first and the second halves of the sixteenth century, with a 10-point magnitude of decline. One may argue that the early-sixteenth-century figure of 17 points is overstated and that an average figure of 14 points should be taken to represent the predecline level; on the other hand, given the nature of pre-Tokugawa records, it is equally likely that the earlier the period in question, the more incomplete the data. Moreover, close scrutiny of the chronological data reveals that for the timing of decline, this is a fairly robust observation. Famine suddenly became less frequent in the mid-sixteenth century: the magnitude of that decline was in the range of 7–10 points.

The finding implies, first, that the mid-sixteenth-century decline in famine frequency took place despite global cooling, and second, that this major decline was not necessarily brought about by a spurt in agricultural progress since, as indicated above, the period was climatically unfavorable for rice culture. Not surprisingly, the causes-of-famine column in table 11.2 indicates that earlier famines had more to do with drought that affected the early growth of the rice plant, whereas those after 1700 were triggered by cold summer temperatures that affected its final growth. There is evidence that the introduction of Champa rice, an *indica* variety resistant to both flood and drought, and the implementation of better irrigation facilities had begun by the so-called Muromachi Optimum of 1370–1450, and also that by the 1540s and 1550s, hunger had become more or less localized as the age-old "spring hungers" pattern of death seasonality began to disappear.[16] However, it took another 200 years for Japanese society to become completely safe from famines caused by drought, as table 11.2 suggests. Agricultural progress was a long, gradual process, but the medieval decline in famine frequency was more articulated. Thus, we have to turn to other resilience factors to account for the momentous decline, an issue discussed in the next section.

▓ Discussion

The evidence we now have in the form of famine chronology suggests that there were two broad tendencies. The first was a long-term decline in the frequency of famines. This was a very slow process that began in ancient times and continued even after the beginning of the Little Ice Age. Then came an unexpectedly substantial decline in the sixteenth century—unexpected because the decline took place in a period of global cooling, and also because it has long been assumed that, as in the Book of Jeremiah, the sword and famine went hand in hand.[17] In Japan, however, it seems that the specter of famine diminished during a period of civil war in the sixteenth century, not in the postunification period of peace.

This finding, therefore, should be interpreted as indicating growing human resilience since, although crop failures may not have decreased so much as one would expect from the famine table, the probability that any poor harvest would cause widespread starvation must have diminished significantly. The reduction in this probability was associated with another tendency—that is, for drought to have become less problematic as a cause of famine. In earlier centuries a majority of famines were triggered by drought, but from the early sixteenth century on, the number of famines caused by either extreme warmness or short rainfall began to decline. Crucial for this shift, as noted earlier, were agrarian changes that took place in rice farming—reclamation of fertile lowland deltas, varietal selection, and investment in irrigation and other forms of infrastructure. All these increased the carrying capacity of the land; the transition had begun in late medieval times but became more apparent in the seventeenth century and after. The link between climate and famine must have become looser all the way through the later Little Ice Age.

Whether this is what actually happened from ca. 1500 to 1900 is open to question. Before addressing this question, however, I would like to explore the demographic implications of the late medieval decline in famine frequency. All other things being equal, such a decline should result in population growth; the question is how much.[18] Elsewhere I have attempted to measure the demographic effects of a decline in famine by regressing population totals on famine points and epidemic frequency for 1721 to 1846, controlling for the time trend.[19] This analysis showed, first, that one occurrence of famine meant a reduction in population by 325,400; second, that the effects of epidemics were smaller than those of famine; and third, that the growth potential of the late Tokugawa population was low but positive and its level increased gradually over time. Since the context is different, the entire regression model cannot be used to simulate late medieval population change; instead, I would like to confine the discussion to the effects of famine alone. I apply the coefficient of 325,400 per famine point to an 8.5-point decline in famines (the average of 7 and 10 points) over the fifty-year period from the first to the second half of the sixteenth century (i.e., from 1525 to 1575). The coefficient is an estimate derived from the early modern—not medieval—situation. Of course, the population at risk from famine in the late medieval period was half the size of that in the late Tokugawa period, which implies that a coefficient of 325,400 might be a gross exaggeration. On the other hand, literary accounts of medieval famines, especially those that stretched over two or more years, suggest that the level of excess mortality must have been far higher than that in the early modern period. According to my own calculations, the early modern famines on average reduced the country's population by 1.4 percent; demographers believe that excess mortality of a two-year famine in the past could have been more serious, resulting in a 9 percent decline in population.[20] On balance, therefore, the 325,400 figure is probably not too far off the mark.

The aim of the exercise is to calculate the implied rate of increase in population over 1525 to 1600. For this we have to know the initial size of population in 1525. Accepting William Wayne Farris's view that population in 1450 was 9.6 million to 10.5 million, I assume that the population in 1525 was 10.5 million or 13 million.[21] Also assumed is that there was no change in the frequency of epidemic outbreaks from 1525 to 1600, and that the underlying rate of population growth remained zero until 1575 but increased by 0.1 point from 1575 to 1600 as unification was under way (given intensified warfare and the relatively high levels of epidemic disease after the Muromachi Optimum, this supposition is presumably not unrealistic). All this enables us to estimate Japan's population in 1600. The results of the exercise are set out in table 11.3.

TABLE 11.3. Demographic implications of a decline of famine frequency in the range of 7-10 points

Assumed population in 1525	Estimated population in 1600	Implied rate of increase (% p.a.)		
		1525–75	1575–1600	1525–1600
Case (1): 10.5 m	15.4 m	0.49	0.59	0.53
Case (2): 13.0 m	17.8 m	0.39	0.49	0.42

Method and assumptions: See text.

The demographic consequence of the great decline in the frequency of famine during the sixteenth century was phenomenal. Whether we accept the lower or upper figure for population in 1525, we see a substantial increase in population during the period, such that Japan's population in 1600 must have been substantially larger than Hayami's estimate of twelve million. This in turn suggests that the seventeenth-century population growth was not as strong as Hayami's estimate implies. Probably there was no acceleration in the tempo of population increase after 1600, in which case per capita output levels may well have remained more or less stable during the century, leaving no room for a strictly Malthusian interpretation.

Turning to the determinants of the late medieval decline in famines, I have argued in this chapter that agrarian innovations cannot explain its onset. Although it is undoubtedly true that human resilience rose with agricultural growth, the whole process was a slow and prolonged one, and much of the agricultural progress was made in the seventeenth rather than the sixteenth century. Also, it is difficult to link the decline in famines to the return of peace, since the decline had already begun in the middle of the Warring States period. In this respect, it is interesting to note that the mid-sixteenth century witnessed the localization of famine, on the one hand, and the consolidation of warring daimyo domains, on the other.[22] This suggests that reforms introduced by daimyo may have been crucially important in preventing a poor harvest from developing into a regional or cross-regional famine. No doubt, most of the innovations made by daimyo administrations were motivated by military and political imperatives, but some of those reforms may well have led to further changes in systems of governance and spatial organization. The warlord paved the way to the consolidation of power over his territory by paying greater attention to trade, as well as to land clearance, than his predecessors, which may have meant a higher level of public action—and also a greater reliance on market forces—in response to climatic hazards and disasters.[23] In other words, it is not unlikely that the sudden decline in the frequency of famines was an unintended consequence of state formation efforts by these warring overlords.

With the establishment of Tokugawa rule, Confucian influence grew, and the notion that the ruling class ought to exercise "benevolence" *(jinsei)* by giving relief *(o-sukui)* to the populace was explicitly accepted by both the central government (the bakufu) and the regional daimyo. The Confucian notion often led the samurai administrations to establish state-run granary reserves. In 1654 Aizu Domain introduced a granary system. The initial stocking was made by the domainal government, but storage was managed by local magistrates, with occasional soft loans to needy peasants. In 1671 a similar scheme was adopted by the Okayama Domain.[24] The underlying *jinsei* concept was also shared by village officials, who during times of difficulty often requested governmental relief in the form of provisions and tax reductions so that they could continue to make a living "as peasants" (hyakushō naritachi).[25] However, political and institutional reforms carried out under the Tokugawa regime did not always have positive effects: in some cases political or organizational reforms actually increased the likelihood or the severity of famines.

As an example of the unintended negative consequences of government policy, let us turn to another case in Japan's history to ask why the country was struck abruptly by two prolonged famines despite the general warming trend late in the Tokugawa period. Cold summers have been blamed for the Tenmei and Tenpō disasters. However, whereas cooling in the 1780s was global, the abnormally cold weather of the 1830s seems to have been a phenomenon limited to northern Japan. Indeed, a comparative analysis of weather diary data for two places, Ikeda in the Osaka area and Hirosaki in the Tohoku, for 1714 to 1864 reveals that cool summers did not invariably lead to mass hunger and starvation. In both places, only one in four cool summers (defined as summers in which the share of fair days in July and August was 25 percent or lower) resulted in a regional or countrywide famine, confirming the positive influence of agricultural trends since the late sixteenth century.[26] According to readily available macroeconomic estimates, farm output grew a little faster than arable land from 1650 to 1850: in other words, land productivity kept growing throughout the two-hundred-year period.[27]

On the other hand, it has been suggested that a subtle but important shift took place between the ruling and the ruled. The shift concerns the issue of who was responsible for systems of disaster relief and prevention, especially grain procurement and storage. The eighteenth and early nineteenth centuries saw the bakufu and, to a lesser extent, daimyo governments becoming less interested in maintaining granaries and rice storage. Daimyo-sponsored granaries became difficult to maintain properly as government finances grew increasingly chaotic. There is evidence that the granary reserves were often cashed in when the daimyo government faced financial trouble.[28] In the Tenmei and Tenpō famine periods in particular, the bakufu's focus is said to have been riveted on Edo, the capital city, in contrast with the earlier response to the Kyōhō famine of 1732.[29] It is well known that in the wake of Tenmei disasters, Matsudaira Sadanobu, the bakufu's chief counselor of the late 1780s and early 1790s, issued edicts urging village and township offices in bakufu lands and also daimyo administrations to store grain as a precaution against famine. However, unlike their seventeenth-century predecessors, most of the granaries set up in this period were not fully government run. Local people were asked to contribute, and requests for contributions were often regarded as a temporary surtax on cultivated land. In the Akita Domain of northern Dewa, for example, the daimyo government had to withdraw a 1784 proposal to create domain-wide grain reserves, and in 1834 peasants in a district where some form of grain reserves had been introduced staged a protest and demanded, among other things, the abolition of the granary system.[30] In 1841, moreover, when a new granary scheme was planned for another daimyo territory in southern Dewa, the local magistrate had to include in the document a clause reading, "even a grain of barley or millet *[zakkoku hitotsubu mo]* should not be used for the sake of government affairs."[31] All this shows growing distrust among the peasantry in the late eighteenth and early nineteenth centuries. On the other hand, many daimyo governments allowed village and regional authorities to have a bigger say in disaster prevention and relief in exchange for assuming greater financial burdens. The change, however gradual and subtle, must have affected the Confucian notion of benevolence and the long-held relationship between the ruling and the ruled based on that notion.

According to that interpretation, the Tenmei and Tenpō cases were famine disasters aggravated by a coordination failure between the central and regional authorities that resulted from a multilevel change in the structure of governance. This account of famines in late Tokugawa Japan suggests that not all political and institutional changes contributed to socioenvironmental resilience.

▨ Conclusion

This chapter has explored the relationship between estimated famine frequencies and climatic cycles of cooling and warming over the very long run. Cold conditions in historic periods are often supposed to have been associated with high frequencies of famine and also with years of disturbance of conflict, but the major conclusion of this analysis is that such a supposition does not hold. A case in point is the late medieval decline in famine frequencies: this unexpected decline took place in a period of global cooling, which in the context of Japanese history coincided with the age of Warring States. Moreover, two of the three most serious famines in the Tokugawa era occurred in the late eighteenth and early nineteenth centuries. Both were triggered by cold summers. Climatically, however, this century-long period saw a gradual upturn in global temperature, but in the Japanese context, the two famine periods corresponded to deviations from the generally decreasing trend in the frequency of famines.

The finding has a direct bearing on the issue of the balance between nature and human resilience. In the Japanese islands it was in the mid-sixteenth century, not in the seventeenth, that the societal ability to withstand a crop failure caused by natural hazards started to increase (and hence, population took off into sustained growth). From this, however, one should not conclude hastily that the major driving force behind the scene was agricultural progress. It has been suggested that some progress was made even in medieval circumstances: particularly important is the introduction of a non-*japonica* rice variety, which played a part in lowland, river delta areas. But the climatic condition of the sixteenth and early seventeenth centuries was so unfavorable that much of the fruit of progress must have been harvested after that cooling phase. In fact, as we have seen above, farm output per unit of land kept growing throughout the period after 1650. On the other hand, the post-1700 phase of gradual upturn in global temperatures saw Japanese population stagnate, with the Tenmei and Tenpō famines taking heavy death tolls in the 1780s and the 1830s. All this cannot be accounted for by any Malthusian interpretation of the history of nature-population interactions.

What I have proposed instead is to turn to the political side of the story. In the case of the mid-sixteenth-century decline in famines, it is important to ask why some climatically induced poor harvests did not develop into famines under regional daimyo rule. Across the warring states of this period, daimyo's energy was devoted to military campaigns. But militarily motivated reform efforts led in most cases to territorial consolidation and administrative centralization, which may well have made it easier for daimyo to take action within their territories in response to climatic hazards. The sudden decline in the number of famines was thus an unintended consequence of state formation efforts made by those warring daimyo. As for the Tenmei and Tenpō famines in the latter half of the Tokugawa era, similarly, we should ask why just two of the cool summers in years that happened to be deviations from the general trend of warming resulted in such devastating famines. Here, too, my argument involves famine prevention and relief. Under the pax Tokugawa, the ruling samurai's attention was turned to the Confucian concerns for people's welfare. One of their policy-making efforts was directed to famine prevention and relief, centered on the system of granary reserve management. As time went on, however, the management of state-run granaries led to official abuses and the government placed the onus on the people themselves in the form of voluntary contributions and village-level management. The process of this multilevel change in the structure of governance was prolonged and not always articulated; as a result, any coordination failure between the central and regional authorities may well have had disastrous consequences in times of poor harvest. The lesson to be learned, therefore, is that politics and institutions mattered even in the history of climate, famine, and population.

*   This chapter is a substantially revised version of Saito Osamu, "Climate and Famine in Historic Japan: A Very Long-Term Perspective," in *Demographic Responses to Economic and Environmental Crises*, ed. Satomi Kurosu, Tommy Bengtsson, and Cameron Campbell (Reitaku University, 2010), 272–81, http://www.archive-iussp.org/members/restricted /publications/Kashiwa09/Chapter15.pdf. In revising the paper, I benefited greatly from comments, suggestions, and bibliographical help offered by Bruce Batten, Wayne Farris, and Kitō Hiroshi, to whom I am most grateful.

1   T. R. Malthus, *An Essay on the Principle of Population* (London: Murray, 1798), chap. 7. For a demonstration of the postulated relationship, see Patrick R. Galloway, "Long-Term Fluctuations in Climate and Population in the Preindustrial Era," *Population and Development Review* 12, no. 1 (1986): 1–24.

2   Akira Hayami, "The Population at the Beginning of the Tokugawa Period," *Keio Economic Studies* 4 (1967): 1–28. See also Hayami's *The Historical Demography of Pre-Modern Japan* (Tokyo: University of Tokyo Press, 2001), 43–46; and Matao Miyamoto, "Quantitative Aspects of Tokugawa Economy," in *Emergence of Economic Society in Japan, 1600–1859*, ed. Akira Hayami, Osamu Saito, and Ronald P. Toby (Oxford: Oxford University Press, 2004), 36–41.

3   William Wayne Farris, *Japan's Medieval Population: Famine, Fertility, and Warfare in a Transformative Age* (Honolulu: University of Hawai'i Press, 2006). The revisionist estimates include my own unpublished one, of 17 million for 1600.

4   Ogashima Minoru, *Nihon saii-shi* (Nihon Kōgyōkai, 1894); Nishimura Makoto and Yoshikawa Ichirō, eds., *Nihon kyōkōshi kō* (Maruzen, 1936).

5   Fujiki Hisashi, comp., *Nihon chūsei kishō saigaishi nenpyō kō* (Kōshi Shoin, 2007).

6   Osamu Saito, "The Frequency of Famines as Demographic Correctives in the Japanese Past," in *Famine Demography: Perspectives from the Past and Present*, ed. T. Dyson and C. Ó Gráda (Oxford: Oxford University Press, 2002), 223.

7   The risk of overstating the frequency of famines would increase by counting such a case, since records of local famines prior to the seventeenth century are sporadic and survive purely by chance.

8   See, for example, Emmanuel Le Roy Ladurie, *Times of Feast, Times of Famine: A History of Climate since the Year 1000* (New York: Doubleday, 1971); and Andrew B. Appleby, "Epidemics and Famine in the Little Ice Age," *Journal of Interdisciplinary History* 10, no. 4 (1980): 643–63. For the situation in East Asia, see William S. Atwell, "Volcanism and Short-Term Climatic Change in East Asian and World History, ca. 1200–1699," *Journal of World History* 12, no. 1 (2001): 29–98.

9   For an excellent survey of literature in this field, see Bruce L. Batten, "Climate Change in Japanese History and Prehistory: A Comparative Overview," Occasional Papers in Japanese Studies Number 2009-01 (Cambridge, Mass.: Edwin O. Reischauer Institute of Japanese Studies, Harvard University, 2009), http://rijs.fas.harvard.edu/pdfs/batten.pdf. See also Takehiko Mikami, "Climatic Variations in Japan Reconstructed from Historical Documents," *Weather* 63, no. 7 (2008): 190–93, and chapter 10 of this volume.

10 On pollen, cedar tree rings, and boreholes, see Yutaka Sakaguchi, "Warm and Cold Stages in the Past 7600 Years in Japan and Their Global Correlation: Especially on Climatic Impacts to the Global Sea Level Changes and the Ancient Japanese History," *Bulletin of the Department of Geography, University of Tokyo* 15 (1983): 1–31; Hiroyuki Kitagawa and Eiji Matsumoto, "Climatic Implications of $\delta^{13}C$ Variations in a Japanese Cedar *(Cryptomeria japonica)* during the Last Two Millennia," *Geophysical Research Letters* 22, no. 6 (1995): 2155–2158; and

Shunsaku Goto, Hideki Hamamoto, and Makoto Yamano, "Climatic and Environmental Changes at Southeastern Coast of Lake Biwa over Past 3000 Years, Inferred from Borehole Temperature Data," *Physics of the Earth and Planetary Interiors* 152, no. 4 (2005): 314–25. On cypress tree rings, see Tatsuo Sweda and Shinichi Takeda, "Construction of an 800-Year-Long *Chamaecyparis* Dendrochronology for Central Japan," *Dendrochronologia* 11 (1994): 79–86. On freezing of Lake Suwa, see Takehiko Mikami, "Quantitative Climatic Reconstruction in Japan Based on Historical Documents," *Bulletin of the National Museum of Japanese History* 81 (1999): 41–50. On the flowering of cherry trees, see Yasuyuki Aono and Keiko Kazui, "Phenological Data Series of Cherry Tree Flowering in Kyoto, Japan, and Its Application to Reconstruction of Springtime Temperatures since the 9th Century," *International Journal of Climatology* 28, no. 7 (2008): 905–14; and Yasuyuki Aono and Shizuka Saito, "Clarifying Springtime Temperature Reconstructions of the Medieval Period by Gap-Filling the Cherry Blossom Phenological Data Series at Kyoto, Japan," *International Journal of Biometeorology* 54, no. 2 (2010): 211–19.

11 Ikuo Maejima and Yoshio Tagami, "Climatic Change during Historical Times in Japan: Reconstruction from Climatic Hazard Records," *Geographical Reports of Tokyo Metropolitan University* 21 (1986): 157–71.

12 See Anders Moberg, Dmitry M. Sonechkin, Karin Holmgren, Nina M. Datsenko, and Wibjörn Karlén, "Highly Variable Northern Hemisphere Temperatures Reconstructed from Low- and High-Resolution Proxy Data," *Nature* 433 (2005): 613–17.

13 Sweda and Takeda, "Construction of an 800-Year-Long *Chamaecyparis* Dendrochronology"; Aono and Kazui, "Phenological Data Series"; and Aono and Saito, "Clarifying Springtime Temperature Reconstructions" (Sweda and Takeda's diagram is reproduced as figure 10.9 and Aono and Kazui's as figure 10.10 in chapter 10 of this volume). In the winter temperature series reconstructed from freezing dates of Lake Suwa, however, it is difficult to discern cycles because the graph shows incessant fluctuations between ca. 1450 and 1690 (Mikami, "Quantitative Climatic Reconstruction" and figure 10.5 in chapter 10 of this volume).

14 Maejima and Tagami, "Climatic Change," 162–63.

15 Takehiko Mikami, "Long Term Variations of Summer Temperatures in Tokyo since 1721," *Geographical Reports of Tokyo Metropolitan University* 31 (1996): 157–65; and chapter 10 of this volume.

16 Saito, "The Frequency of Famines," 225–26; and Farris, *Japan's Medieval Population*, 132–36, 185–90.

17 Cormac Ó Gráda, *Famine: A Short History* (Princeton, N.J.: Princeton University Press, 2009), 12.

18 This is because the growth in population, if strong enough, will exert a negative feedback within the Malthusian system of climatic and societal interactions.

19 The following exercise is based on equation (2) of table 11.5 in Saito, "The Frequency of Famines," 234. The estimated "famine effect" includes not only the death toll but the decline in fertility as well.

20 Susan Cotts Watkins and Jane Menken, "Famines in Historical Perspective," *Population and Development Review* 11, no. 4 (1985), 661.

21 Farris, *Japan's Medieval Population*, 95–100.

22 Farris, *Japan's Medieval Population*, 186.

23 Farris, *Japan's Medieval Population*, 222–47.

24 Kikuchi Isao, *Kinsei no kikin* (Yoshikawa Kōbunkan, 1997), 181.

25 Irwin Scheiner, "Benevolent Lords and Honorable Peasants: Rebellion and Peasant

Consciousness in Tokugawa Japan," in *Japanese Thought in the Tokugawa Period, 1600–1868*, ed. Tetsuo Najita and Irwin Scheiner (Chicago: University of Chicago Press, 1978), 39–62; and Fukaya Katsumi, *Hyakushō naritachi* (Hanawa Shobō, 1993), 15–66.

26 Saito, "The Frequency of Famines," 228–30.

27 Miyamoto, "Quantitative Aspects of Tokugawa Economy," 38. The annual average growth rate in farm output per unit of land was 0.06 percent for the initial period of 1600–1650 versus 0.14 percent for the 1650–1850 period. The percentage increased to 0.49 in 1872, but much of this increase is probably accounted for by technical problems associated with the linking of Tokugawa to Meiji data series.

28 Kikuchi, *Kinsei no kikin*, 181–82; and Kikuchi Isao, "Tokugawa Nihon no kikin taisaku," in *Shakai keizai shigaku no kadai to tenbō*, ed. Shakai Keizaishi Gakkai (Yūhikaku, 2002), 375–85.

29 See Conrad Totman, *Early Modern Japan* (Berkeley: University of California Press, 1993), 308, 243, 512–14.

30 Kikuchi, *Kinsei no kikin*, 187–88.

31 Kikuchi, *Kinsei no kikin*, 191–92.

12

The Climatic Dilemmas of Built Environments

Tokyo, Heat Islands, and Urban Adaptation

SCOTT O'BRYAN

Cities tend to concentrate environmental burdens spatially.[1] From their struggles with the local problems associated with periods of high-speed urbanization (water access, sanitation, sewerage, adequate housing) to the forms of air, water, and ground pollution that especially tend to emerge as modern modes of material accumulation ("economic growth") expand, modern cities have been environmental hotspots. It was staggering levels of city-scale pollution that eventually fueled the mid-twentieth-century rise of the ecology movement in the United States, Europe, and Japan and the classic regulatory responses of the 1960s and 1970s by the states in those regions that went some distance toward ameliorating the worst degradations.

At just the same time, however, environmental scientists were discovering that what had been seen as the "externalities" of industrial production and mass consumption had unimagined free-ranging, nonlocal consequences as well. Recognition of the perils of global "population bombs," resource exhaustion, ozone depletion, and finally climate change itself signaled the rise of a new sort of systemic environmentalism, one that began warning that the scale of problems threatened the very viability of modern human society. The "green" challenges that systemic environmentalism identified, moreover, were often disproportionately associated with cities the richer these became. By the end of the twentieth century, an emerging field of urban climate studies was even revealing that, beyond the frightening global climate effects of our reliance on fossil fuels, such very large cities as Tokyo can themselves quite directly and dramatically reshape the local climatic environments in which they stand.

By examining the contemporary history of Tokyo as one of the world's premier examples of the urban heat island (UHI) effect—and of scientific and official attempts to understand this climatic phenomenon and devise adaptive responses—this paper seeks to suggest the ways in which our built environments reveal the very "built" nature of the climate itself. We are by now used to thinking of the malign climatic effects of high-growth societies as global in reach. Yet urban climate researchers are now focused on understanding the ways that large cities can produce sufficient additional heat to create sustained temperature "islands" many degrees warmer than their surrounding rural terrains, a thermal phenomenon capable of altering weather on specifically local yet nevertheless significant scales.

The physical landscapes in which cities are situated—their geographical relationships to mountains, valleys, or seas and the hydrological cycles these terrains produce—can influence the heat island effect. But design matters, too. Climate experts, engineers, architects, and officials alike have been scrambling over the past two decades to understand the ways in which urban planning approaches, economic development, changing lifestyles, and architecture have conspired in Tokyo to affect the most apparently natural elements—the temperature, winds, and rain—of the city's environment. Heat generated by the engines and air-conditioners of the city, miles of concrete largely unbroken by green spaces, and high-rise buildings that block cooling flows of ocean air have all helped create a climatic island in which sustained rises in city temperature have gone far beyond those attributable to global climate change alone.

Heat Island Tokyo helps underscore the tangled feedback relations between the local (urban and periurban) and the systemic (global). As with the problem of air pollution, we are once again reminded that climatic dilemmas exist at overlapping geographic scales. Overheated Tokyo, moreover, shows that large, rich cities that seem to have grappled with many older pollution challenges and even begun to institute green environmental responses (solid waste recycling, carbon dioxide cap-and-trade programs) continue to produce unintended consequences that, like those of any dynamic system, seem always to lie just outside our ability to predict. Moreover, they produce such unforeseen consequences, in the phrase of Fisher, Hill, and Feinman, "at multiple timescales." The temporal lags separating problem, response, and consequence lie at the heart of the thermal challenge in Tokyo, a city that in becoming large, rich, and structurally safe, unintentionally contributed to its own long-term climate change. As part of the larger struggle to create socionatural environments of "resilience," able "to absorb" and adapt "to unforeseen and unintended consequences in rapid fashion," the history of urban heat in Tokyo forces its millions of residents to confront the problem of exactly what they and their predecessors have built.[2]

The "Aberrant" Climate of Tokyo

The city of Tokyo has indeed been heating up. The average temperature of the city has increased by 3 degrees Celsius over the course of the twentieth century, a rate that was five times larger than the estimated 0.6-degree rate of increase in general global warming over the same period.[3] Tokyo is not alone in its heat problem. Many other large cities in Japan—Nagoya, Kyoto, Fukuoka, Sapporo, and Sendai, among others—are being studied in regard to the phenomenon. Nor is this a problem in Japan alone. Such large cities as Atlanta, Phoenix, Los Angeles, and Mexico City in the Americas also are the focus of heat island concern, and NASA tracks their temperature changes using satellite imaging.

The scale of temperature increases in Tokyo, however, has led to the unfortunate recognition of the Japanese capital city as exhibiting "some of the most dramatic consequences" of UHI anywhere in the world.[4] The average temperature in such smaller Japanese cities as Mito increased by just 1 degree Celsius during roughly the same hundred-year period.[5] New York's significant 1.5-degree increase during the past century is still only half that of Tokyo. The European Union has also sounded the alarm over urban heating in its major cities. The challenge posed by the average 1-degree rises there are surely significant, but again pale next to the warming problem of Tokyo.[6]

The rise in the average temperatures of Tokyo and other large and medium cities in Japan is in fact considered so much a climatic phenomenon in its own right, standing outside the "normal" thermal spatial topography of the country as a whole, that when meteorologists calculate the mean seasonal temperature of the nation (as, for example, when they declared summer 2010 the hottest on record in Japan going back to 1898), they routinely exclude temperature data from the many weather stations believed to be the most affected by city heat islands. They instead use readings taken from seventeen stations around the archipelago in largely nonurban areas.[7] The intent is to attain a sense, independent of local warming effects, of the overall shifts of the larger weather universe in which Japan sits, yet one cannot help but note that one perverse result of setting aside data that would "skew" the overall climate shift record is that researchers thereby exclude climate trends in the urban regions where most Japanese actually live.

Beyond the simple calculation of increasing average highs in the city, many other measures contribute to the statistical picture of thermal change cited by weather scientists to further illuminate the many layers of the atmospheric challenge facing Tokyo. Between 1931 and 1980, there was only one year with more than thirty so-called tropical nights, when temperatures never fall below 25 degrees Celsius, an important indicator of overall warming and the additional heat burden it places on city inhabitants. Between 1981 and the end of the century, there were suddenly seven such years.[8] When data from the first years of the new century are included, the number of tropical nights has doubled in forty years.[9] Not only has the average city temperature increased, so too has the length of time that high temperatures must be endured: the total average annual summer hours at 30 degrees or over experienced by the city between 1980 and 2000 more than doubled, from 168 to 357.[10]

The heat litany continues. More dramatic even than the increase in the average temperature has been the rise of 3.8 degrees Celsius in the average daily low temperature over the century ending in 2000.[11] Temperatures are higher not only in summer, but most dramatically of all, during winter, too, rising a full 5 degrees on average since the beginning of the twentieth century. There are consequently fewer freezing days and snowfalls are now quite rare.[12]

The total surface area of the city and surrounding region now experiencing high temperatures has also grown far larger in recent years. At the beginning of the 1980s, only Nerima ward in Tokyo and parts of Saitama and Gunma prefectures logged 190 or more annual hours at temperatures over 30 degrees Celsius. By the end of the century, however, most of Tokyo and the contiguous areas subsuming much of Saitama, Gunma, Kanagawa, and Chiba prefectures experienced such temperatures, and they did so for twice as much time or more per annum.[13]

Rainmaker: The Weather Effects of Heating and the Built Environment

The extreme warming trend has engendered wide public discourse over the past decade on what is increasingly conceived as Tokyo's aberrant climate.[14] Indeed, such popular references to climatic change seem highly appropriate. The regional urban warming of Tokyo affects not just background temperature but by extension the full range of regional atmospheric phenomena: wind speed and daily vector changes, cloud formation, and precipitation levels, intensity, and timing. Even the patterns of *yūdachi*, cooling rain showers so closely associated with summertime late afternoons and early evenings in Tokyo, now seem to have been disrupted, one link in a series of feedback loops that exacerbate the problem. Such showers form less frequently along the same pattern as before, and when they do, they come generally much later at night.[15]

At the same time, new types of torrential rain squalls, often focused intensely in small areas at any one time, are occurring in new patterns in densely populated, inner-city precincts of Tokyo, increasing the overall rainfall levels of the city as a whole and leading to more frequent flooding (in subways, most dramatically). In the past twenty years, rainfall amounts greater than 50 millimeters in an hour, something that "had only been seen rarely before," have become more frequent.[16]

The overheated western downtown district of Suginami-ku is one of the severest hit by these "urban-pattern localized downpours" (a meteorological term rendered in Japanese as *toshi-gata shūchū gōu*).[17] A particularly torrential urban-pattern downpour on the night of September 4, 2005, infamously produced more total rainfall (258 millimeters) in the district than seven of the eight typhoons that had caused flooding during the prior twenty years. Moreover, the September 2005 urban rainstorm released far more rain per hour (112 millimeters, 20 more per hour than the next most drenching storm) than any weather event, whether urban downpour or typhoon, during that same period. Unprecedented levels of flooding resulted, especially in the dense neighborhoods along nearly the entire stretch of the Zenpukuji River in the ward.[18]

Data from the Japanese Aerospace Exploration Agency and NASA satellites have climate scientists convinced that the very presence of cities and the heating of the local atmosphere for which they are responsible influence cloud formation and rainfall.[19] Researchers believe that the particular signature of Tokyo as a built landscape does not by itself drive intense localized rainstorms. Yet they increasingly find evidence that effects of the built environment do play a significant role in exacerbating the squall-like nature of the storms, in rearranging their timing, and in making them more frequent.

As the city temperature rises during the fair, hot days of summer, classic onshore sea breeze systems form, in which the air over the land rises as it becomes heated, and higher-pressure, cool air from the sea flows in to fill the low-pressure areas created on shore by the rising warm air mass.[20] The geography of the Kanto land mass on which the metropolis of Tokyo stands affects the dynamics of storm systems that develop under these basic conditions. The megalopolis is surrounded on three sides by water—Sagami Bay to the southwest, Tokyo Bay, and the Kashima Sea to the northeast—and such onshore breeze fronts from all three directions can often converge over west-central Tokyo, centered roughly on the Nerima, Suginami, and Nakano areas of the city. This complex air convergence causes more lift, pulling in from the ocean more moist air that becomes more unstable as it clashes with cooler air in the upper atmosphere. Rainstorms result.[21]

The hotter Tokyo becomes, the more these processes are strengthened. The heat island effect energizes the hot air in and over Tokyo, which rises to break through the superheated boundary layer over the city into cooler air above, becoming steadily more unstable as the warm and cool air masses interact, generating cumulonimbus storm systems and increasing the possibility of lightning.[22] Mikami Takehiko and other investigators, such as Fumiaki Kobayashi and his colleagues, all argue that the UHI by itself may not be sufficient to generate Tokyo's new patterns of urban squall-like downpours, but that higher levels of overheating function as a "trigger," in Mikami's word, setting off powerful storm dynamics.[23]

The built environmental characteristics of Tokyo have a significant effect on the heat trigger. The increase in the built-up, heat-absorbing surface area since the late nineteenth century has been large, as the city stretched westward and spawned exurb metropolitan areas in all surrounding prefectures. Between 1875 and 1935 the population of the formerly designated Tokyo Prefecture (Tōkyō-fu) rose from 1.5 million to 6.4 million, a 329 percent increase.[24] By the 1960s, with a population then exceeding 10 million, Tokyo was identified as one of a class of "world cities," rapidly urbanizing capital centers of politics, economics, and culture.[25] In 2010, Tokyo had a population of just under 13 million. If one includes the metropolitan area 50 kilometers in circumference from city hall, the population soars to 32 million, exactly 25 percent of the total 128 million population of Japan. Tokyo prefecture is seventeen times more densely populated than the national average.[26]

With more people have come more intensive and extensive building of human-made heat-absorbing surfaces and more sources of thermal waste exhaust. Successive building and infrastructure booms from the late nineteenth century kept pace with the influx of people and natural population increase. During the feverish expansions of the postwar period, highways covered cooling surface waterways, and drainage systems redirected them fully underground; green space shrank dramatically. One of the most significant changes shaping the heat island signature of Tokyo was the new dominance of concrete in the post–World War II reconstruction and growth of the city. The heating effects of the postwar shift away from thermally cooler wood-based construction seem borne out in the longitudinal temperature data, which reveal rates of temperature increase in Tokyo during the first half of the century similar to those in New York and Paris, but greater acceleration in the Japanese city beginning after World War II. Such accelerating thermal effects of building material change have been mirrored in other rapidly growing East Asian cities as well in recent decades.[27]

Even in centuries past, when cities were smaller than now, some temperature increases occurred.[28] However, modern urban forms have resulted in built environments so massive in density, heat generation, surface area and, often, upward thrust that they now exert significant thermal and atmospheric effects on the ocean of air above them. The city, we are discovering, acts as its own form of landscape, one of specific shape, height, geographic area, and surface characteristics, influencing weather in its own ways, much like any natural set of land or water forms, but one erected exceedingly quickly in geological time on top of existing geographic forms of much longer standing.

Requisite to Adaptation: Knowledge Formations and Urban Climatology

Requisite to adapting to change in socionatural environments is the need for institutions of knowledge production that identify and, simply but powerfully, name the challenges in the first place. The now startling evidence for the rapidity of urban thermal change has become known through a field of urban climatology that grew up in Japan and around the world over the course of the twentieth century. The increasing relevance of the UHI problem has fueled contemporary expansion of urban climate studies, which until recent decades was seen at best as a niche research area, one with little prominence in the larger discourses of either environmentalism or city planning.

Japanese researchers during the second half of the twentieth century have arguably been, along with their German counterparts, among the most active in the development of the modern and contemporary forms of urban climatology.[29] A small cohort of Japanese researchers applied themselves to urban thermal research beginning in the 1920s up through the wartime 1940s. Most prominent among them was Fukui Eiichirō, who used an automobile-based observation system to produce the first graphical display of the heat island effect in downtown Tokyo, publishing a simple isothermal map in 1941.[30] In the 1950s, government research funding for urban weather studies rose, and prewar researchers like Fukui and others led the way, investigating the effects of new building materials in Japan's rebuilt cities and rapidly rising postwar urban populations. In the 1960s the still relatively small field continued to examine what the explosive expansion of the city and the sprawling bedroom suburbs would mean for the atmospheric environment.[31] By the beginning of the 1970s, Japanese researchers, along with their colleagues in industrial nations elsewhere, were employing the term heat island effect *(hīto airando genshō)* to denote the thermal change they observed.

Dramatic improvement in computing power and innovations in modeling technologies revolutionized the fields of industrial system controls, stock forecasting, demographics, and general weather prediction during the 1960s and 1970s, accounting for the "circular, interlocking" relations within dynamic systems.[32] These new modeling powers, fueled by the data generated within what Paul N. Edwards has described as a rising "informational globalism" of the weather, played a fundamental role over the past thirty years in identifying the unfolding challenge of planetary climate change.[33] Climate science at both global and local levels recast heat as no longer a neutral variable in weather prediction but now the critical element in a very serious systemic problem. Those studying the urban climate realized that what they were dealing with was itself an incurred environmental "cost" like any other form of waste.

One of the standard and purposefully more assertive terms that thus emerged for the overwarming of cities among urban weather researchers beginning in the 1990s was "thermal pollution" *(netsu osen)*, signifying a product of the human environment that degraded other social and living systems. In 2001, the Japanese Ministry of Environment followed the lead of climatologists and officially defined UHI as "air pollution," inaugurating a new age of political attention and fueling the eventual development over the subsequent decade of official adaptive responses in the form of mitigation guidelines and regulations.[34]

Rearticulated now in the older language of pollution, this new rendering of urban temperature phenomena recalls the discourses of earlier postwar decades during which Tokyo had represented a different sort of climatic dystopia. Particulate and gaseous air pollutants from coal-burning and oil-burning industries were a primary environmental fixation of the high-growth 1950s and 1960s. Rising car usage only worsened atmospheric conditions. This was the age when pollution of the air was so dire that major street intersections featured electric signs registering real-time street levels of sulfur dioxide and carbon monoxide.[35] The official institutions created to inform the adaptive approaches of public policy on such "traditional" air pollution problems, those like the Tokyo Metropolitan Research Institute for Environmental Protection (Tōkyō-to Kankyō Kagaku Kenkyūjo), which tellingly began its operations in 1968 as the Tokyo Metropolitan Pollution Research Institute (Tōkyō-to Kōgai Kenkyūjo), are now at the forefront of investigating UHI in the twenty-first century.

The term thermal pollution, when used to denote urban heating, is itself a direct appropriation from an older usage. It originally referred to the discharge into rivers and lakes of warmed water after it had been used to cool power plants or other industrial processes—a problem, along with chemical waste discharge into water, contemporaneous with air quality problems of the early postwar decades and at least partly addressed by water regulations of the 1970s. The rubric of thermal pollution is thus connected discursively to an older era of high-industrial pollution dilemmas of air and water, only now repurposed for an age of climate change and a perhaps increasingly postindustrial Tokyo. Neither air pollution, whether primary (particulates, sulfur dioxide) or photochemical (ozone), nor water pollution has ever, of course, gone away entirely. The point here rather is the recognition of an entirely new category of environmental dilemma, not disconnected from the older category (airborne effluents can influence relative amounts of atmospheric heating, for example) but now pointing outward as well, linking urban localities to the ominous climatic changes unfolding on a planetary scale. Here the problem has been finally recognized as one with real consequences, a lamentably "radiant" effect of modern urbanity now given a name with discursive power, ideally, to shape adaptive responses in the forms of public policy and design practices.

Urban Design, Wind, and the Time-Scale Challenges of Adaptation

The prior section of this essay was predicated on the simple suggestion that adaptation requires knowledge. Yet despite the international professionalization of urban meteorological studies and its increasing relevance to cities like Tokyo, international assessment and adaptation planning for global climate change have yet to account well for heat islands.[36] With the exception of recent pushes for energy efficiencies and greener building material, urban planning around the world has also continued in almost complete disconnect from the knowledge generated by urban climate studies.

In Tokyo, this gap between thermal conditions, assessment of the causes, and the nature of actual building has now begun to change—but only, it must be emphasized, in the first decade of the new century. The cases where change is beginning to take place are fascinating, but many urban design and construction project decisions of longer standing (and even some from surprisingly recently) reveal the historical and time-scale challenges of adaptation.

On the heels of the declaration in 2001 that excess heat is a pollution issue, national and city officials have scrambled to erect a mitigation regime. UHI is also now a focus of the environmental bureau of Tokyo's metropolitan government.[37] Some efforts, such as the government promotion of "Cool Biz" professional dress styles, with an emphasis on short sleeves and the forgoing of ties, seem pathetically inadequate to the problems at hand. Yet others outlined in the metropolitan government "Guidelines for Heat Island Control Measures," promulgated in final form in 2005, seem to express a range of adaption and mitigation strategies that can have good effect and are easy to implement within the near-to-medium term. Those aspects of the city's new anti-UHI efforts that seem to hold good promise for early and fruitful implementation are those that relate especially to the surface properties of the built landscape, in terms of both building materials and new sorts of "heat shielding" approaches. Under the new UHI guidelines, campaigns to increase green space in the city have become more visible, and these were further enunciated in Tokyo's bid for the 2016 Olympics.[38]

A familiar aspect of hopes for urban reform for the better part of a century, perennial calls for the "greening" (ryokka) of Tokyo have now been imbued with new meaning. Greenery is no longer necessary on purely aesthetic or lifestyle grounds, but now as an urgent means to cool the city. The much-touted move toward green building roofs indeed goes back half a decade prior to the UHI guidelines. More recently, the metropolitan government, in coordination with the Ministry of Education, has initiated a program to sod all school playing fields, replacing the sand and clay surfaces so common in Japanese public spaces.[39] Though taking some time to roll out over the massive landscape, other effective city programs to cool the surfaces of the built environment—changing the albedo (reflectivity versus heat absorption qualities) of streets and instituting the new CASBEE-HI architectural guidelines to reduce the heat footprint of buildings—present few rigid structural dilemmas: they do not require fundamental changes in city layout or large infrastructure.[40]

FIGURE 12.1. Suginami Ward, Tokyo, heat mitigation efforts.
In an attempt to create a model practice for mitigating rising temperatures, Suginami Ward has planted a "green curtain" of trailing vines, mostly *hechima* gourd and cucumber plants, in pots along one of the walls of the local government office to shield it from the summer sun, thus lowering the amount of heat absorbed by the building. The vegetables harvested are given to local schools.

Photograph by author, October 2010.

Yet even when mitigation can occur on relatively short time scales, the challenges to efficacious change are immense. Although new park space has increased in some wards, such as Suginami-ku, the total area of park development "simply cannot keep up" with total green space depredations.[41] There was an estimated overall decline of 1 percent in green space and water surface area in the ward portion of the city between 1998 and 2003, though perhaps the renewed greening guidelines can reverse this long-standing trend.[42] It must be said, however, that plantings around new large construction projects built under the new UHI guidelines—as can be seen, for example, around Ōsaki or in the multitower Shiosaito redevelopment project in Shiodome—seem dwarfed by the magnitude and heat-generating capacity of the structures.

FIGURE 12.2. Small trees along Meguro River, Tokyo. Trees planted alongside new tower projects along the Meguro River in the Ōsaki area of Tokyo certainly do some good and mark a significant departure from past approaches. They do still seem, however, a diminutive response to the massive nature of the adjacent buildings.

Photograph by author, October 2010.

Other categories of urban adaptive strategy also seem to require much longer time scales as they confront the historical legacies of earlier city-building decisions. For many, the key to cooling the city lies in restoring older wind patterns, in part by reconnecting the downtown core of the city with its waterfront. Much as in urban thought elsewhere in the world, concern with airflows through Tokyo has played no sustained role in the development of the city. Yet notionally, at any rate, the wind simulations of urban climate experts are slowly beginning to inform visions of a transformed Tokyo of the future. Such prominent architects as Andō Tadao and government officials alike have articulated a desire to use what researchers are learning about the most salutary building shapes, pathways for winds, and street layouts. Andō's conception, adopted by the Tokyo metropolitan government in its bid for the 2016 Olympics, is to create a "Forest of the Sea" *(Umi no mori)*, whereby new forests planted on landfill islands in the harbor would help cool breezes flowing into the city from the water.[43]

It is relatively more difficult, however, to bend the wind to one's city visions than to change something like thermal reflectivity. This is not simply because the wind is mercurial. Enticing the wind over and through the city to perform the desired cooling function might often require literally reshaping the built forms of the city, its buildings and infrastructures and their physical arrangement.[44]

The thirteen-tower urban redevelopment megaproject, Shiosaito, in Shiodome on the site of the old JR freight train yards, is a notorious case in point. In planning since the late 1980s and built between the end of the 1990s and 2006, the massive business tower and condo project was touted as the jewel of a revamped Tokyo for the twenty-first century. The entire Shiodome project covers more than 30 hectares, and each of its buildings rises taller than 100 meters, several reaching 190 meters or more (plates 11 and 12).

Almost immediately upon completion, however, the project's height and its shoulder-to-shoulder layout of office towers became the object of urban climatological suspicion. Heat researchers have dubbed the Shiodome development the "Tokyo Wall," cutting off as it does a central portion of the downtown area from its waterfront. They have shown that the project impedes the flow of cooling air that used to travel from the harbor waters, over Hamarikyū Park, and into the downtown core.[45] Wind studies were done in the environmental impact statements required by the city for the project, but only on "building wind," the localized acceleration effects of wind along sidewalks abutting tall structures.[46] Consideration of heat island air-flow issues was not required in the environmental assessments. When the project was planned, urban climatological knowledge had not penetrated the regulatory system for city design and construction. Indeed, such UHI wind issues yet remain absent from impact statement requirements.

Calls for new heat-informed approaches to thinking about wind flow have become especially salient in the context of ongoing urban redevelopment in Japan during the past two decades. Despite the steady dirge sounding the supposed end of the Japanese economy, building construction has, after a pause in the late 1990s, continued in Tokyo, and often, as at Shiodome, on extremely large scales. Officials and developers tout big projects as the key to urban revitalization in the new century and frequently target waterfront areas of the city so potentially critical to any future adaptive efforts toward heat island cooling.

▦ Conclusion

The initial adaptive response to the knowledge presented by urban climatologists mirrored the regulatory mitigation approaches of earlier postwar moments of pollution challenges in the city. Such regulation was certainly to be welcomed. The predicament, however, for the city now armed with knowledge of this new kind of pollution was the inherent complexity of its origins: it was not so easily identifiable with its specific points of origin. Responses could no longer be pinpointed at particular smokestacks, as in earlier eras. The downstream effects of thermal pollution were in some ways homologous with those of more familiar forms of air pollution, and yet the newly central discipline of urban climatology was revealing in ever more detail its dynamic, overlapping connections at multiple scales with systemic weather change.

One-off fixes seem increasingly anachronistic. Moreover, although some of the many required adaptations to environmental challenge can be embarked upon in relatively quick order, others, when they encounter structural rigidities or must negotiate the results of design and siting decisions taken years in the past, require very long urban historical time scales. How does one remediate brand-new, multitower megaprojects like Shiosaito, the very mass, height, and density of which are primary drivers of the thermal challenge they pose? On what time line would the dismantling and rearranging of old street patterns set in place in decades and even centuries past occur so as to better enable the flow of cooling air? What is the calculation by which one determines the cost-effectiveness of extensive built environmental restructuring versus continued suffering under the ill effects of significant climate change?

The Shiodome tower project reveals the disassociations between science and city building policies, between short-term economic imperatives and the challenge of environmental adaptation. And it underscores the long temporal lags in urban design. Building projects are expensive, and because their development pathways are often set early on in one context, it is difficult later to make the significant changes to built environments that a new order of green environmental challenge may be demanding in another. Urban structures erected in one time can come to stand later as confoundingly "permanent" rigidities within changed socionatural contexts. Such dilemmas suggest an entirely different take on what Fisher, Hill, and Feinman call the "resilience of built capital."[47] Rather than considering "resilience" a positive attribute of anthropogenic landscapes facing change or catastrophe, we might use the phrase in the context of city-level climate change to refer to the "blockages" that can be caused by the immobile, built environmental forms of the overheated urban landscape.

NOTES

1   Paraphrase of Peter J. Marcotullio and Gordon McGranahan, "Scaling the Urban Environmental Challenge," in *Scaling Urban Environmental Challenges: From Local to Global and Back*, ed. Peter J. Marcotullio and Gordon McGranahan (London: Earthscan, 2007), 2.

2   Christopher T. Fisher, J. Brett Hill, and Gary M. Feinman, eds., *The Archaeology of Environmental Change: Socionatural Legacies of Degradation and Resilience* (Tucson: University of Arizona Press, 2009), 3, 11; Roderick McIntosh, Joseph A. Tainter, and Susan Keech McIntosh, "Climate, History, and Human Action," in *The Way the Wind Blows: Climate, History, and Human Action*, ed. Roderick McIntosh, Joseph A. Tainter, and Susan Keech, McIntosh Historical Ecology Series (New York: Columbia University Press, 2000), 13.

3   Tōkyō-to Kankyōkyoku, "Hīto airando genshō o saguru: Atsukunaru Tōkyō," 1, http://www2.kankyo.metro.tokyo.jp/heat2/index.htm (accessed September 21, 2010); Intergovernmental Panel on Climate Change, *Third Assessment Report, Climate Change 2001: Working Group 1, The Scientific Basis, Summary for Policymakers*, online edition (GRID-Arendal, 2003), http://www.grida.no/publications/other/ipcc_tar/?src=/climate/ipcc_tar/.

4   Bjorn Lomborg, *Cool It: The Skeptical Environmentalist's Guide to Global Warming* (New York: Alfred A. Knopf, 2008), 19.

5   Tōkyō-to, "Hīto airando genshō," 1; Eriko Arita, "Tokyo's Heat Island Keeps Breezes at Bay," *The Japan Times Online*, August 18, 2004, http://search.japantimes.co.jp/cgi-bin/nn20040818f1.html (accessed March 1, 2011).

6   Tokyo Metropolitan Government, *Environmental Whitepaper 2006*, "Feature Story: Towards Sustainable Cities," chap. 1, 3, http://www2.kankyo.metro.tokyo.jp/kouhou/env/eng_2006/index.html (accessed February 11, 2011).

7   Japan Meteorological Agency, "Japan Hit by Hottest Summer in More Than 100 Years," news release, September 10, 2010, and "Primary Factors of Extremely Hot Summer 2010 in Japan," news release, September 16, 2010, http://www.jma.go.jp/jma/en/News/indexe_news.html (accessed April 17, 2012).

8   Kankyō Jōhō Kagaku Sentā, *Hīto airando genshō no jittai kaiseki to taisaku no arikata ni tsuite: Heisei 12 nendo hōkokusho Kankyoshō ukeoi gyōmu hōkokusho* (Kankyō Jōhō Kagaku Sentā, 2001), 12.

9   Tokyo Metropolitan Government, *Environmental Whitepaper 2006*, "Feature Story: Towards Sustainable Cities," chaps. 1, 3.

10 Kankyō Jōhō Kagaku Sentā, "Hīto airando genshō," 8.

11 Mikami Takehiko, "Fueru Tōkyō no manatsubi to nettaiya," in *Tōkyō ijō kishō*, ed. Mikami Takehiko (Yōsensha, 2006), 9–10.

12 Mikami, "Fueru Tōkyō no manatsubi," 11; Eric Talmadge, "Overpopulated Tokyo Becoming a Heat Island," *Seattle Times*, August 26, 2002, http://community.seattletimes.nwsource.com /archive/?date=20020826&slug=heatisland26 (accessed September 22, 2010).

13 Kankyō Jōhō Kagaku Sentā, "Hīto airando genshō," 8, 10.

14 The term often used is *ijō kishō*, "anomalous" or "aberrant" weather.

15 Mikami Takehiko, "Pinpointo de Tōkyō o osou toshi-gata shūchū gōu," in *Tōkyō ijō kishō*, ed. Mikami Takehiko (Yōsensha, 2006), 15.

16 Suginami-ku Toshi-gata Suigai Taisaku Kentō Senmonka Iinkai, *Arata na toshi-gata suigai no gensai ni idomu* (Suginami-ku Seisaku Keieibu Kiki Kanrishitsu, 2006), 11.

17 Yoshino Minoru, Suginami Kuyakusho, Midori Kōen Kachō, interview with author, October 8, 2010.

18 Suginami-ku Toshi-gata Suigai Taisaku Kentō Senmonka Iinkai, *Arata na toshi-gata suigai*, 2–4, 13–15, fig. 2-12, fig. 3-11, fig. 3-12.

19 Holli Riebeek, NASA Earth Observatory, "Urban Rain," December 8, 2006, http://earthobservatory.nasa.gov/Features/UrbanRain/ (accessed February 17, 2011); Japan Aerospace Exploration Agency, "Missions," http://www.jaxa.jp/projects/util/eos/index _e.html (accessed February 17, 2011).

20 Fumiaki Kobayashi, Maki Imai, Hirofumi Sugawara, Manabu Kanda, and Hitoshi Yokoyama, "Generation of Cumulonimbus First Echoes in the Tokyo Metropolitan Region on Mid-Summer Days" (paper presented at the Seventh International Conference on Urban Climate, Yokohama, Japan, June 29–July 3, 2009), http://www.ide.titech.ac.jp/~icuc7/extended _abstracts/pdf/375801-1-090512171643-003.pdf (accessed September 2010).

21 Suginami-ku Toshi-gata Suigai Taisaku Kentō Senmonka Iinkai, *Arata na toshi-gata suigai*, 12; Mikami, "Pinpointo," 14–15.

22 Kobayashi et al., "Generation of Cumulonimbus," section 4.

23 Suginami-ku Toshi-gata Suigai Taisaku Kentō Senmonka Iinkai, *Arata na toshi-gata suigai*, 12; Kobayashi et al., "Generation of Cumulonimbus," section 4.

24 Hachiro Nakamura, "Urban Growth in Prewar Japan," in *Japanese Cities in the World Economy*, ed. Kuniko Fujita and Richard Child Hill (Philadelphia: Temple University Press, 1993), 45, table 2.4.

25 Peter Hall, *The World Cities* (London: Weidenfeld and Nicolson, 1966).

26 Sōmushō Tōkeikyoku, *Statistical Handbook of Japan 2010*, http://www.stat.go.jp/data /handbook/index.htm (accessed February 20, 2010); Tokyo Metropolitan Government, "City Profile," http://www.metro.tokyo.jp/ENGLISH/PROFILE/overview03.htm (accessed February 20, 2010).

27 I thank Mikami Takehiko for this insight on accelerating rates of heating.

28 Nihon Kenchiku Gakkai, *Hīto airando to kenchiku, toshi: Taisaku no bijon to kadai* (Nihon Kenchiku Gakkai, 2007), 35–38.

29 On history of the field in Europe, Vladimir Janković and Michael Hebbert, "Hidden Climate Change: Urban Meteorology and the Scales of Real Weather," *Climate Change* 113 (2012): 23–33.

30 Nihon kenchiku gakkai, *Hīto airando*, 38–41; Fukui Eiichirō and Wada Sadao, "Honpō no dai-toshi ni okeru kion bunpu," *Chirigaku hyōron* 17, no. 5 (1941): 354–72.

31 Nihon kenchiku gakkai, *Hīto airando*, 44.

32 Donella H. Meadows, Dennis L. Meadows, Jørgen Randers, and William W. Behrens III, *The Limits to Growth: A Report for the Club of Rome's Project on the Predicament of Mankind* (Washington: Potomac Associates, 1972), 31.

33   Paul N. Edwards, *A Vast Machine: Computer Models, Climate Data, and the Politics of Global Warming* (Cambridge, Mass.: MIT Press, 2010), 23–25.

34   Yoshika Yamamoto, "Measures to Mitigate Urban Heat Islands," *Science and Technology Trends Quarterly Review* (NISTEP) 18 (2006): 65, www.nistep.go.jp/achiev/ftx/eng/stfc/stt018e/.../ STTqr1806.pdf (accessed September 10, 2010).

35   Tokyo Metropolitan Government, *Tokyo Fights Pollution: An Urgent Appeal for Reform*, Municipal Library, No. 4 (1971), 32 (photo).

36   Mark McCarthy, "Including Cities in Climate Models," in *City Weathers: Meteorology and Urban Design, 1950–2010*, ed. Michael Hebbert, Vladimir Janković, and Brian Webb (Manchester: University of Manchester, Manchester Architecture Research Centre, 2011), http://www.academia.edu/3434577/City_Weathers_Meteorology_and_Urban_ Design_1950-2010 (accessed October 28, 2013).

37   Tokyo Metropolitan Government Bureau of the Environment, "Guidelines for Heat Island Control Measures [Summary Edition]" (Tokyo Metropolitan Government, 2005), 2013, http:// www.kankyo.metro.tokyo.jp/en/attachement/heat_island.pdf (accessed October 28, 2013); Tokyo Metropolitan Government Bureau of the Environment, "Thermal Environment Map of Tokyo" (Tokyo Metropolitan Government, 2005), http://www.metro.tokyo.jp/ENGLISH/ TOPICS/2005/ftf56100.htm (accessed October 28, 2013).

38   Tokyo Metropolitan Government Bureau of the Environment, *Basic Policies for the Ten-Year Project to Green Tokyo* (Tokyo Metropolitan Government, 2006); Tōkyō-to, *2016 nen Tōkyō Orinpikku Pararinpikku kankyō gaidorainu*, pamphlet (Tōkyō-to, 2006); Tokyo Metropolitan Government Building, lobby display, October 2010.

39   Kine Ruriko, Suginami Kuyakusho Kankyōbu Shisuishin kachō, interview with author, October 6, 2010; Suginami-ku Toshi Seibibu Midori Kōenka, ed., *Suginami-ku midori no kihon keikaku* (Suginami-ku, 2010), 33; Suginami-ku Kyōiku Iinkai, untitled Japanese-language environmental model school pamphlet for Ogikubo Elementary School (Suginami-ku Kyōiku Iinkai, 2009).

40   Tōkyō-to Kankyōkyoku, *Tōkyō-to kenchikubutsu shōenerugī seinō hyōkasho no aramashi* (2010).

41   Yoshino, interview with author, October 8, 2010.

42   Tokyo Metropolitan Government, *Environmental Whitepaper 2010*, http://www.kankyo.metro. tokyo.jp/en/documents/white_paper_2010.html (accessed July 15, 2012).

43   "Andō Tadao ga kataru Tōkyō kaizō ron: Tōkyōwan ni midori no shima o, soshite natsu ha suzukaze ga fukinukeru machi ni," *Zaikai* No. 2, special ed. (Summer, 2008): 32–37; Tōkyō-to Midori no Tōkyō Bokin Jikkō Iinkai, *Midori no Tōkyō bokin*, pamphlet (Tōkyō-to, 2006).

44   On "built capital," see Fisher et al., eds., *Archaeology*, 11–12.

45   Nihon kenchiku gakkai, *Hīto airando*, 122–26.

46   See Tōkyō-to Kankyōkyoku, *Shiodome kankyō eikyō hyōkasho: Shiryō hen* (Tōkyō-to, 1999).

47   Fisher et al., eds., *Archaeology*, 11–12.

Concluding Thoughts

In the Shadow of 3.11

BRUCE L. BATTEN AND PHILIP C. BROWN

Although certainly not by design, "Environment and Society in the Japanese Islands" turned out to be one of the timeliest conferences ever held in the field of Japanese history. As organizers, we had decided nearly a year in advance to hold our meetings on March 28–29, 2011, in Honolulu. The timing and location were chosen to take advantage of the fact that many of our participants would be going to Hawaii to attend the annual meeting of the Association for Asian Studies, scheduled to begin as our program ended. What none of us knew, of course, was that just a few weeks before our conference, Japan would be struck by the most powerful earthquake in that nation's records of seismic magnitude—the event now referred to as the Great East Japan Earthquake or, more simply, 3.11.

One of us, Batten, lives in Tokyo, and remembers the events of that day with great clarity. The violent shaking went on for what seemed like minutes, but that was only the beginning. There followed power outages, shortages of essentials such as food, bottled water, and gasoline, and aftershocks that went on, day after day, month after month. Then there was the ongoing fear of invisible radiation. It was clear from the daily newscasts that a disaster of immense magnitude had occurred, whether measured in loss of human life, physical and psychological suffering, environmental impact, or economic cost. Even now, several years later, it seems likely that the affected region of northern Japan will not fully recover from the events of 3.11 for decades, if not longer.

The earthquake and its aftermath notwithstanding, our Hawaii conference ran as scheduled; somewhat miraculously, all the Japanese participants who signed on to attend were able to do so. That said, it was a subdued group of academics who met on the nineteenth floor of Tokai University Pacific Center on the morning of March 28; all of us were still in shock at the magnitude of the calamity, particularly the large number of casualties reported in the media. (At the time of this writing, January 2014, the numbers of deaths and disappearances directly attributable to the disaster are officially listed as 15,884 and 2,640, respectively.[1])

Yet it was also clear even then that our research was directly relevant to 3.11, and vice versa. Although only two of our presenters—Barnes and Smits—had prepared papers specifically related to earthquakes, none of the participants could avoid being struck by how elements and themes of the studies each of us had undertaken were playing out in real time. The disaster provided striking illustrations of the interconnectedness of human society and the natural environment, and of how historical processes are influenced by complex feedback cycles operating over diverse geographic and chronological scales. Rather than summarize the work presented in this volume, it thus seems appropriate to reflect on 3.11 in relation to the broader issues raised by environmental history and the essays we have presented.[2]

The Great East Japan Earthquake was, at its core, a natural event. The magnitude (M) 9.0 temblor was a natural result of geologic forces and would have occurred whether or not anyone was living in the region. The Japanese islands' location at the conjunction of multiple plate boundaries makes such events inevitable. The devastating tsunami that followed the earthquake were also purely natural phenomena, their configuration and extent determined by the characteristics of the earthquake, including its location, magnitude, and direction of movement, by the topography (both undersea and coastal) of the affected region, and to a lesser extent, by the tides.

To make an obvious point, it was the presence of people that turned the earthquake and tsunami into what is commonly referred to as a natural disaster. Although that term leaves the human dimension unspoken, what it presumes is that the natural occurrence wreaks havoc on humans' lives and activities. Had an M9.0 earthquake occurred in the same location a million years ago, it would not have constituted a "disaster" for the simple reason that no *Homo sapiens* lived in Japan—or for that matter, anywhere else that might have been affected by a Pacific tsunami.[3]

In general, it seems fair to say that all other factors being equal, the greater the population of a given area, the greater the potential human suffering from any given geologic event. Consider the closest historical equivalent to the 2011 temblor, the Jōgan Earthquake of 869, also centered off the Miyagi coast. With an estimated magnitude of 8.6, the Jōgan Earthquake precipitated a deadly tsunami that inundated the Sendai Plain, reportedly drowning one thousand people and causing extensive damage to property and crops.[4] This was a terrible disaster by the standards of the time, but it pales in comparison with the events of 2011—or those of 1896, when a tsunami resulting from an M8.2–8.5 earthquake killed some twenty thousand in Miyagi and Iwate prefectures.[5] One reason for the difference is simply that Tohoku is much more densely populated in modern times than it was in the ninth century.

The effects of a natural disaster are influenced not just by size and density of population, of course, but also by its distribution. In the case of the Great East Japan Earthquake and other historical earthquakes in the region, population was concentrated in coastal areas and low-lying alluvial plains most vulnerable to tsunami. In a pattern evocative of the dilemma of settlement on easily inundated floodplains, such regions were preferentially settled because of their flat topography, easy transportation by road, river, or sea, and availability of food, whether from fishing, farming, or economical transportation networks.

Other elements of human agency, in addition to choice of residence, also play a significant role in shaping the consequences of natural forces. One example is the nature of the local infrastructure. From a historical perspective, infrastructure tends to become more complex over time, but by the same token it becomes potentially more susceptible to disruption. In the worst case, modern technologies can become actual agents of destruction. Contemporary Japanese construction standards are widely and justifiably praised and undoubtedly served to limit certain kinds of damage from the 2011 earthquake. The devastation wrought by the tsunami, however, was magnified by several specific elements of the local infrastructure. Reinforcement of riverbanks with concrete dikes channeled water moving inland, increased its speed and force, and projected damage inland. Embankments constructed to support high-speed rail and road transport functioned to retain water that had overtopped them and prolonged the drainage process, turning many areas into cesspools of chemical and human waste and providing a breeding ground for bacteria and viruses. Last but not least was the presence of four nuclear reactors at the Fukushima Daiichi power plant. The "unforeseen" size of the earthquake and tsunami exceeded planning specifications for the plant; electric power and cooling capability were lost completely, and three of the four reactors in operation suffered catastrophic meltdown. The most obvious damage resulting from this event was the release of large amounts of radiation into the atmosphere and ocean. But energy shortages also resulted, partly because of the loss of the reactors themselves but more importantly because of the public backlash that virtually shut down the entire Japanese nuclear power industry.

Another question, also one with direct policy implications, regards people's individual and collective expectations about the likelihood of disaster and their consequent willingness to undertake preparations. Ultimately, these factors find expression not just in the physical infrastructure described above, but also in social institutions of all kinds. (Although not directly related to earthquakes, examples of infrastructure and institutions created to meet social needs can be seen in the urban construction described by Kawasumi and O'Bryan, the military activities noted by Tyner, and the efforts to manage nature catalogued by Brown, Dinmore, and Sano.) Earthquakes are a fact of life in Japan, one that its residents have struggled to comprehend in many ways since premodern times, as documented by Smits. Over decades and centuries, such understandings have had tangible institutional, social, and technical consequences, with the result that the Japanese are on the whole better prepared for earthquakes and tsunami than are citizens of most other countries: building codes are strict; disaster plans are in place; individuals and families are generally aware of the steps they can take to plan for and mitigate risk.

That said, people may become complacent, whether because of familiarity ("we survived the last earthquake, so we'll manage to get through the next one, too"), forgetfulness, or simple misplaced optimism. Individuals and institutions frequently discount future risks when the alternative is to spend time and money now. Why stock up on water and disaster rations that most likely will never be needed? Why build a seawall to withstand 15- or 20-meter waves when the cost is prohibitive (at least by some assessments) and the likelihood of such a tsunami in our lifetimes remote? Prior to 3.11, Tohoku had not experienced a major tsunami since 1933. On that occasion, there were "only" about fifteen hundred deaths. The relatively small number of fatalities is generally ascribed to lessons taken to heart from the calamitous 1896 tsunami, which was recent enough to remain in people's minds.[6] By contrast, the much longer interval between 1933 and 2011, coupled with the relatively small number of fatalities in 1933, may have resulted in a certain complacency and unwillingness to spend money that might have saved lives and prevented the nuclear meltdowns following the Great East Japan Earthquake.

Related to the issue of planning and preparedness, people's individual and collective reactions to disaster caused by natural phenomena also make a meaningful difference. As was noted by many foreign observers, the affected population was by global standards remarkably stoic and well behaved following 3.11.[7] On the whole, survivors politely moved into temporary shelters, and there were very few reports of looting. Farther away, in urban Tokyo, there was initially some panic buying, contributing to shortages of gasoline, bottled water, and batteries, but people generally went about their business as usual and complied with the government's requests to cut back on energy use, even during the unusually hot summer. With regard to Tohoku, at least, the sober response may suggest that long-standing rural networks, embedded in customary practices such as those discussed by Sano, provided stability and self-help mechanisms.

On the other hand, whereas the response of affected Japanese was generally exemplary, that of the Japanese government was critically slow and inadequate, in part because politicians in an already struggling administration appeared to spend as much time bickering and backbiting as in actually dealing with the problems at hand. A parallel fumbling had characterized the response to the Great Hanshin Earthquake of 1995, but after that disaster, policy changes were made to facilitate the restoration of "lifelines," such as water, electricity, and gas, to affected populations after a major earthquake. The result appeared to be a better response to subsequent temblors, such as those that struck Niigata in 2004 and 2007, and led to a renewed sense of confidence in government preparedness. The catastrophic events of 2011 shattered any such self-assurance by completely overwhelming existing systems. The Japanese government's handling of the disaster raised questions about the degree to which bureaucrats were even capable of understanding the potential harms, human and environmental, of nuclear energy and natural forces. (As the essays by Dinmore and O'Bryan suggest, such doubts can be extended to the bureaucratic handling of other modern environmental issues as well.)

Although perhaps obvious, we also note that these various actions and inactions had consequences not just for human society but also fed back into the "natural" environment. The most obvious example is radiation damage. Radiation from Fukushima Daiichi contaminated agricultural produce as far away as Shizuoka, to the west of Tokyo. Also affected, though to what extent is not clear even today, was marine life off the Tohoku coast and indeed in other parts of the Pacific Ocean.[8] Another example was the vast amount of waste swept into the Pacific and now, as we write, littering the shores of Hawaii and the west coast of the U.S. mainland. These and other environmental effects of the disaster will have social repercussions that will again have environmental consequences, in a spreading and complex web of causation. (Similar if more geographically circumscribed interrelationships between human activities and environmental regimes feature in the case studies in this volume by Hung, Sano, and Tyner.) The most profound effects will be felt in Japan, but residents of other countries will hardly remain untouched. The Fukushima accident fueled antinuclear sentiment throughout the world. A turn away from nuclear energy would certainly have implications for the environment, not only in terms of reducing risks of nuclear contamination but also involving the prospects for alternative, more environmentally friendly energy sources. Human populations today are more attuned and responsive to these kinds of risk than the more subtle ones touched on in several of our essays.

Although the Great East Japan Earthquake and associated tsunami were "natural" and the study of history can do nothing to prevent similar occurrences, human beings are not just passive observers—or victims—of such events. They can act to mitigate or alleviate the effects on human society. As we have argued above, to a large extent, the damage or significance of an earthquake—or any environmental circumstance, for that matter—is determined by our own actions, or lack thereof.

That being the case, an understanding of history is important, even crucial, in seeking to evaluate and minimize environmental risks. Calculation of risk based on modern instrument data or experience is simply too limited to serve as a guide to potential magnitude of the worst flood or tsunami on a centennial or millennial scale. Inexact as the results might be, the kind of historical reconstructions represented in this collection—for example, by Barnes and Smits for geology; by Mikami, Zaiki, and Hirano for climate; and by Brown, Kawasumi, and Sano for environmental risks related to water—must be factored into our calculations.

Current events and recent history lead many to presume that the issues we face today—pollution, resource depletion, global warming, and more—are unprecedented. Indeed, some of our essays do paint a rather depressing picture of the current situation. For example, based on O'Bryan's chapter, it is difficult to imagine that Tokyo will get cooler anytime soon. Yet other studies presented in this volume suggest that humans have not only faced severe problems in the past (including some of their own creation), they have also generally adapted to such dangers. Kawasumi's essay touches on urban pollution in a distinctly premodern context, and Sano's documents the human capacity to adapt natural cycles to our own benefit, at least within certain productive constraints. The essays by Saito, Higuchi, and Hung all provide examples of how technological innovation and imports of resources have helped ameliorate natural conditions (e.g., climatic cooling, access to natural biological fertilizers and timber) and promote "resilience" in some sense. Historical case studies, like that by Tyner, will further reveal ways in which we reshape that which we call nature, something particularly important as Japan (and the world) considers ways to move from growthism to sustainability, a trend mentioned in the conclusion to Dinmore's chapter on Sakuma Dam.

All of these points are relevant to the disaster in northeastern Japan and to other environmental issues we face today. We cannot assume that history will always provide clear answers to present problems, which are often quantitatively or qualitatively different from those humans faced in the past. Nonetheless, to meet the challenges that face us on our shrinking and increasingly interconnected planet, we must take into account what the historical disciplines have to say about past relationships between human society and the natural environment. Such study suggests the range of options available, issues and approaches deserving of attention, and the creativity human society has exercised when confronted with previous challenges. It is our hope that the chapters in the present book will not only deepen readers' understanding of environmental history in Japan but will also stimulate further work in this important field and thus contribute, albeit indirectly, to greater resilience of the global socioenvironmental system that we share.

1   National Police Agency of Japan, "Damage Situation and Police Countermeasures Associated with 2011 Tohoku District—Off the Pacific Ocean Earthquake," http://www.npa.go.jp/archive /keibi/biki/higaijokyo_e.pdf (accessed January 14, 2014).

2   What follows makes no pretense of being a systematic examination of 3.11. Among many other books, see Lucy Birmingham and David McNeill, *Strong in the Rain: Surviving Japan's Earthquake, Tsunami, and Fukushima Nuclear Disaster* (New York: Palgrave Macmillan, 2012); Jeff Kingston, ed., *Natural Disaster and Nuclear Crisis in Japan: Response and Recovery after Japan's 3/11*, The Nissan Institute/Routledge Japanese Studies Series (London: Routledge, 2012); Tom Gill, Brigitte Steger, and David Slater, eds., *Japan Copes with Calamity: Ethnographies of the Earthquake, Tsunami and Nuclear Disasters of March 2011* (New York: Peter Lang, 2013); and Richard Samuels, *3.11: Disaster and Change in Japan* (Ithaca, N.Y.: Cornell University Press, 2013). Also see the "Digital Archive of Japan's 2011 Disasters," created by the Edwin O. Reischauer Institute of Japanese Studies, Harvard University, and partners, http://jdarchive .org/en/home.

3   See Gina Barnes's discussion in chapter 1 of the present volume of proximity as a factor in risk: no proximity, no risk.

4   Kuroita Katsumi and Kokushi Taikei Henshūkai, eds., *Nihon sandai jitsuroku*, vol. 4, Shintei zōho kokushi taikei (Yoshikawa Kōbunkan, 1934), entry for fifth month, twenty-sixth day of Jōgan 11 (869).

5   Shutō Nobuo, "Meiji Sanriku jishin tsunami," in *Nihon rekishi kasai jiten*, ed. Kitahara Itoko, Matsuura Ritsuko, and Kimura Reō (Yoshikawa Kōbunkan, 2012), 379.

6   Shutō Nobuo, "Shōwa Sanriku jishin tsunami," in *Nihon rekishi kasai jiten*, ed. Kitahara Itoko, Matsuura Ritsuko, and Kimura Reō (Yoshikawa Kōbunkan, 2012), 459–67.

7   Individual fortitude and good behavior (the subject of this paragraph) and government inept- itude (that of the next) became well-worn tropes in American news coverage of 3.11, as noted by Theodore C. Bestor, "Disasters, Natural and Unnatural: Reflection on March 11, 2011, and Its Aftermath," *Journal of Asian Studies* 72, no. 4 (2013): 763–82. Bestor takes a jab at the media by noting how it "painstakingly discovered that the Japanese elite and the Japanese everyperson were acting precisely as established stereotypes would have it!" (p. 765). We agree that some of the coverage made heavy use of stereotypes but nonetheless feel that these generalizations have an important core of truth. Bestor would seem to agree (ibid.).

8   As a case in point, elevated levels of radioactive iodine have been detected in kelp off the California coast: Steven L. Manley and Christopher G. Lowe, "Canopy-Forming Kelps as California's Coastal Dosimeter: [131]I from Damaged Japanese Reactor Measured in *Macrocystis pyrifera*," *Environmental Science and Technology* 46 (2012): 3731–36.

Bibliography

Place of publication for all Japanese-language works is Tokyo unless otherwise noted.

Abe Yasunari. "Jishin to hitobito no sōzōryoku." In *1855 Ansei Edo jishin hōkokusho*, edited by Chūō Bōsai Kaigi, 129–91. Fuji Sōgō Kenkyūjo, 2004.

Ackerman, Edward A. *Japan's Natural Resources and Their Relation to Japan's Economic Future*. Chicago: University of Chicago Press, 1953.

Adachi Keiji. "Daizu kasu ryūtsū to Shin dai no shōgyō teki nōgyō." *Tōyōshi kenkyū* 37, no. 3 (1978): 35–63.

Agrawal, Arun. *Environmentality: Technologies of Government and the Making of Subjects*. Durham, N.C.: Duke University Press, 2005.

Aizawa, Koji. "Present Situation of Monitoring, Prediction and Information for Volcanic Disaster Mitigation in Japan." Paper presented at the Sixth Joint Meeting of the UJNR Panel on Earthquake Research, Tokushima, Japan, November 8–11, 2006. http://cais .gsi.go.jp/UJNR/6th/orally/Oo8_Present.pdf (accessed June 24, 2011).

Akahane Sadayuki and Kitahara Itoko, eds. *Zenkōji jishin ni manabu*. Nagano-shi: Shinano Mainichi Shinbunsha, 2003.

Amano Ryūji. "1830 nen zengo ni okeru Ōsaka no kishō." *Shōsen Daigaku kenkyū hōkoku* A3 (1952): 137–57.

Amano Ryūji. "1838 nen yori 1855 nen ni itaru Tōkyō no kishō." *Shōsen Daigaku kenkyū hōkoku* A4 (1953): 167–94.

Amino Yoshihiko. *Nihon chūsei toshi no sekai*. Chikuma Shobō, 1996.

Anderson, Jennifer L. "Nature's Currency: The Atlantic Mahogany Trade and the Commodification of Nature in the Eighteenth Century." *Early American Studies* (2004): 47–80.

"Andō Tadao ga kataru Tōkyō kaizō ron: Tōkyōwan ni midori no shima o, soshite natsu ha suzukaze ga fukinukeru machi ni." *Zaikai* No. 2, special ed. (Summer 2008): 32–37.

Anonymous. *Edo Ōjishin matsudai hanashi no tane*. 1855.

Anyoji, Nobuo. "Technical Efforts to Prepare Volcanic Hazard Maps." *Technical Note of the National Research Institute for Earth Science and Disaster Prevention* 380 (July 2013): 125–28.

Aoki Shigekazu. *Nara-ken kishō saigaishi*. Yōtokusha, 1956.

Aoki Shigeru. *Taiwan ringyōjyō no kiso chishiki*. Taipei: Shinkōdō, 1925.

Aono, Yasuyuki, and Keiko Kazui. "Phenological Data Series of Cherry Tree Flowering in Kyoto, Japan, and Its Application to Reconstruction of Springtime Temperatures since the 9th Century." *International Journal of Climatology* 28, no. 7 (2008): 905–14.

Aono, Yasuyuki, and Shizuka Saito. "Clarifying Springtime Temperature Reconstructions of the Medieval Period by Gap-Filling the Cherry Blossom Phenological Data Series at Kyoto, Japan." *International Journal of Biometeorology* 54, no. 2 (2010): 211–19.

Appleby, Andrew B. "Epidemics and Famine in the Little Ice Age." *Journal of Interdisciplinary History* 10, no. 4 (1980): 643–63.

Ara Usaburō. "Kusunoki zōrin ni kansuru shokan." *Taiwan no sanrin* 108 (1935): 1–11.

Arakawa Hidetoshi. "Fujiwhara on Five Centuries of Freezing Dates of Lake Suwa in the Central Japan." *Archiv für Meteorologie, Geophysik und Bioklimatologie*, Series B 6 (1954): 152-66.

Arakawa Hidetoshi, ed. *Jitsuroku, Ō-Edo kaimetsu no hi: Ansei kenmonroku, Ansei kenmonshi, Ansei fūbunshū.* Kyōikusha, 1982.

Aramaki Shigeo. "Asama Tenmei no funka no suii to mondaiten." In *Kazanbai kōkogaku*, edited by Arai Fusao, 83–110. Kokon Shoin, 1993.

Arima, Midori. "An Ethnographic and Historical Study of Ogasawara/the Bonin Islands, Japan." Ph.D. diss., Stanford University, 1990.

Arita, Eriko. "Tokyo's Heat Island Keeps Breezes at Bay." *Japan Times Online*, August 18, 2004. http://search.japantimes.co.jp/cgi-bin/nn20040818f1.html (accessed March 1, 2011).

"Arizan keiei shippai." *Dainippon sanrin kaihō* 387 (1915): 52–53.

"Arizan kinkyō." *Taiwan nichinichi shinpō*, October 13, 1914.

"Arizan sagyō shinkeikaku." *Taiwan nichinichi shinpō*, September 5, 1914.

Asai Ryōi. *Kaname'ishi* (1662). In *Kanazōshishū*, edited and translated by Taniwaki Masachika, Oka Masahiko, and Inoue Kazuhito, 13–83. Shōgakukan, 1999.

Asano Yasukichi. "Taiwan ringyō no kaiko." *Taiwan no sanrin* 78 (1932), 15–18.

Atwater, Brian F., Musumi-Rokkaku Satoko, Satake Kenji, Tsuji Yoshinobu, Ueda Kazue, and David K. Yamaguchi. *The Orphan Tsunami of 1700: Japanese Clues to a Parent Earthquake in North America.* U.S. Geological Survey Professsional Paper 1707. Reston, Va., and Seattle: United States Geological Survey in association with University of Washington Press, 2005. http://pubs.usgs.gov/pp/pp1707/.

Atwell, William S. "Volcanism and Short-Term Climatic Change in East Asian and World History, ca. 1200–1699." *Journal of World History* 12, no. 1 (2001): 29–98.

Azebu Sadakuma. "Shinbunshi o tsūji kantaru bokoku rinsei no shinkeikō." *Taiwan sanrin kaihō* 8 (1924): 7–13.

Azebu Sadakuma. "Shinrin keikaku jigyō no chakusō subekaraku endainare." *Taiwan sanrin kaihō* 33 (1928), 3–7.

Azebu Yoshihiro. "Mokuzai kanzei mondai no keii," *Taiwan sanrin kaihō* 37 (1929): 13–50.

Ball, Philip. *Critical Mass: How One Thing Leads to Another.* New York: Farrar, Straus and Giroux, 2004.

Barnes, Gina L. "Earthquake Archaeology in Japan: An Introduction." In *Ancient Earthquakes*, edited by Manuel Sintubin, Iain Stewart, Tina Niemi, and Erhan Altunel, 81–96. Boulder, Colo.: Geological Society of America, 2010.

Barton, Gregory A. "Empire Forestry and American Environmentalism." *Environment and History* 6, no. 2 (2000): 187–203.

Batten, Bruce L. "Climate Change in Japanese History and Prehistory: A Comparative Overview." Edwin O. Reischauer Institute of Japanese Studies Occasional Paper 2009–01. Cambridge, Mass.: Harvard University, 2009. http://rijs.fas.harvard.edu /pdfs/batten.pdf.

Beechey, Frederick William. *Narrative of a Voyage to the Pacific and Bering's Strait: To Co-operate with the Polar Expeditions: Performed in His Majesty's Ship* Blossom, *under the Command of Captain F. W. Beechey, R.N. In the Years 1825, 26, 27, 28.* Philadelphia: Carey & Lea, 1832.

Beechey, Frederick William, John Richardson, Nicholas Aylward Vigors, George Tradescant Lay, Edward Turner Bennett, Richard Owen, John Edward Gray, George Brettingham Sowerby, William Buckland, Edward Belcher, and Alexander Collie. *The Zoology of Captain Beechey's Voyage.* London: H. G. Bohn, 1839.

"Beimatsu yunyū gekizō." *Dainippon sanrin kaihō* 446 (1920): 37.

"Beizai to honpōsan mokuzai no hikaku kenkyū." *Dainippon sanrin kaihō* 484 (1923): 69–71.

Bestor, Theodore C. "Disasters, Natural and Unnatural: Reflections on March 11, 2011, and Its Aftermath." *Journal of Asian Studies* 72, no. 4 (2013): 763–82.

Biggs, David. *Quagmire: Nation-Building and Nature in the Mekong Delta.* Seattle: University of Washington Press, 2011.

Bird, Isabella L. *Unbeaten Tracks in Japan.* Virago/Beacon Travelers. Boston: Beacon Press, 1987.

Birmingham, Lucy, and David McNeill. *Strong in the Rain: Surviving Japan's Earthquake, Tsunami, and Fukushima Nuclear Disaster.* New York: Palgrave Macmillan, 2012.

Biwako Hakubutsukan. *Nihon chūsei gyokairui shōhi no kenkyū.* Kusatsu: Biwako Hakubutsukan, 2010.

Bray, Francesca. *The Rice Economies: Technology and Development in Asian Societies.* Oxford: Basil Blackwell, 1986.

Brown, Cynthia Stokes. *Big History: From the Big Bang to the Present.* New York: The New Press, 2007.

Brown, Philip C. "A Case of Failed Technology Transfer—Land Survey Technology in Early Modern Japan." *Senri Ethnological Studies* 46 (1998): 83–97.

Brown, Philip C. *Central Authority and Local Autonomy in the Formation of Early Modern Japan: The Case of Kaga Domain.* Stanford, Calif.: Stanford University Press, 1993.

Brown, Philip C. "Constructing Nature." In *Japan at Nature's Edge: The Environmental Context of a Global Power,* edited by Ian Jared Miller, Julia Adeney Thomas, and Brett L. Walker, 90–114. Honolulu: University of Hawai'i Press, 2013.

Brown, Philip C. *Cultivating Commons: Joint Ownership of Arable Land in Early Modern Japan.* Honolulu: University of Hawai'i Press, 2011.

Brown, Philip C. "'Feudal Remnants' and Tenant Power: The Case of Niigata, Japan, in the Nineteenth and Early Twentieth Centuries." *Peasant Studies* 15, no. 1 (1987): 5–26.

Brown, Philip C. "Moving Rivers: Lowland Water Management in Nineteenth Century

Japan." Paper presented at the annual meeting of the Society for the History of Technology, Tacoma, Wash., September 30–October 3, 2010.

Brown, Philip C. "Never the Twain Shall Meet: European Land Survey Techniques in Tokugawa Japan." *Chinese Science* 9 (1989): 53–79.

Brown, Philip C. "Practical Constraints on Early Tokugawa Land Taxation: Annual Versus Fixed Assessments in Kaga Domain." *Journal of Japanese Studies* (1988): 369–401.

Brunton, R. Henry. *Building Japan 1868–1876*. Sandgate, Folkestone, Kent: Japan Library, 1991.

Bullard, Fred M. "Volcanoes and Their Activity." In *Volcanic Activity and Human Ecology*, edited by Payson D. Sheets and Donald K. Grayson, 9–48. New York: Academic Press, 1979.

California Institute of Technology Tectonics Observatory. "What Happened during the 2004 Sumatra Earthquake?" http://www.tectonics.caltech.edu/outreach/highlights /sumatra/what.html (accessed June 25, 2011).

Callon, Michel. "Some Elements of a Sociology of Translation: Domestication of the Scallops and the Fishermen of St. Brieue Bay." In *Power, Action, and Belief: A New Sociology of Knowledge*, edited by John Law, 196–233. London: Routledge & Kegan Paul, 1986.

Cholmondeley, Lionel Berners. *The History of the Bonin Islands from the Year 1827 to the Year 1876, and of Nathaniel Savory, One of the Original Settlers; To Which Is Added a Short Supplement Dealing with the Islands after Their Occupation by the Japanese*. London: Constable, 1915. http://mysite.du.edu/~ttyler/ploughboy/boninchol.htm (accessed November 1, 2013).

Christian, David. *Maps of Time: An Introduction to Big History*. The California World History Library 2. Berkeley: University of California Press, 2004.

Chūō Bōsai Kaigi. *1847 Zenkōji jishin hōkokusho*. Nihon Shisutemu Kaihatsu Kenkyūjo, 2007.

Chūzu-chō Kyōiku Iinkai, ed. Ōmi no kuni Yasu gun Awaji Kuyū monjo mokuroku. Chūzu (Shiga Prefecture): Chūzu-chō Kyōiku Iinkai, 1995.

Clancey, Gregory. *Earthquake Nation: The Cultural Politics of Japanese Seismicity, 1868–1930*. Berkeley: University of California Press, 2006.

Clark, Brett, and John Bellamy Foster. "Ecological Imperialism and the Global Metabolic Rift: Unequal Exchange and the Guano/Nitrates Trade." *International Journal of Comparative Sociology* 50, no. 3–4 (2009): 311–34.

Clark, Nigel. "The Demon-Seed: Bioinvasion as the Unsettling of Environmental Cosmopolitanism." *Theory Culture Society* 19, no. 1–2 (2002): 101–25.

Clark, Nigel. "Wild Life: Ferality and the Frontier with Chaos." In *Quicksands: Foundational Histories in Australia and Aotearoa New Zealand*, edited by Klaus Neumann, Nicholas Thomas, and Hilary Ericksen, 133–52. Sydney, New South Wales: NSW Press, 1999.

Cronon, William. *Changes in the Land: Indians, Colonists, and the Ecology of New England*. New York: Hill & Wang, 1983.

Crosby, Alfred W. *The Columbian Exchange: Biological and Cultural Consequences of 1492*. Westport, Conn.: Greenwood Press, 1972.

Crosby, Alfred W. *Ecological Imperialism: The Biological Expansion of Europe, 900–1900*. Cambridge: Cambridge University Press, 1986.

Culshaw, Martin. "Geological Hazards: How Safe Is Britain?" Shell Lecture at Geological Society of London, December 8, 2010.

Curtin, Philip D. *The Rise and Fall of the Plantation Complex: Essays in Atlantic History*. 2nd ed. Cambridge: Cambridge University Press, 1998.

Cushman, Gregory T. *Guano and the Opening of the Pacific World: A Global Ecological History*. Cambridge: Cambridge University Press, 2013.

Dainippon Sanrinkai. *Meiji ringyō isshi*. Dainippon Sanrinkai, 1931.

Dainippon Sanrinkai. *Meiji ringyō isshi zokuhen*. Dainippon Sanrinkai, 1931.

Dangi Kazuyuki. "Ogasawara shotō iminshi." In *Umi to rettō bunka: Kuroshio no michi*, edited by Amino Yoshihiko, Ōbayashi Taryō, Hasegawa Kenichi, Miyata Noboru, and Mori Kazuo, 238–80. Shōgakukan, 1991.

Demarest, Arthur A., Prudence M. Rice, and Don S. Rice, eds. *The Terminal Classic in the Maya Lowlands: Collapse, Transition, and Transformation*. Boulder: University Press of Colorado, 2004.

Dengen Kaihatsu Kabushiki Kaisha. *Denpatsu 30-nenshi*. Dengen Kaihatsu, 1984.

Diamond, Jared. *Collapse: How Societies Choose to Fail or Succeed*. New York: Viking, 2005.

Diamond, Jared. *Guns, Germs, and Steel: The Fates of Human Societies*. New York and London: W. W. Norton & Company, 1997.

"Digital Archive of Japan's 2011 Disasters." http://jdarchive.org/en/home.

Dinmore, Eric G. "Concrete Results? The TVA and the Appeal of Large Dams in Occupation-Era Japan." *Journal of Japanese Studies* 39, no. 1 (Winter 2013): 1–38.

Dinmore, Eric G. "A Small Island Nation Poor in Resources: Natural and Human Resource Anxieties in Trans-World War II Japan." Ph.D. diss., Princeton University, 2006.

Doboku Gakkai. *Meiji izen Nihon doboku shi*. Doboku Gakkai, 1936.

Edo Iseki Kenkyūkai. *Saigai to Edo jidai*. Yoshikawa Kōbunkan, 2009.

Edwards, Paul N. *A Vast Machine: Computer Models, Climate Data, and the Politics of Global Warming*. Cambridge, Mass.: MIT Press, 2010.

Elvin, Mark. *The Retreat of the Elephants: An Environmental History of China*. New Haven, Conn.: Yale University Press, 2004.

Elvin, Mark, and Ts'ui-jung Liu, eds. *Sediments of Time: Environment and Society in Chinese History*. Studies in Environment and History. New York: Cambridge University Press, 1998.

Fagan, Brian. *The Attacking Ocean: The Past, Present, and Future of Rising Sea Levels*. New York: Bloomsbury Press, 2013.

Fagan, Brian. *Floods, Famines, and Emperors: El Niño and the Fate of Civilizations*. New York: Basic Books, 1999.

Fagan, Brian. *The Great Warming: Climate Change and the Rise and Fall of Civilizations*. New York: Bloomsbury Press, 2008.

Fagan, Brian. *The Little Ice Age: How Climate Made History 1300–1850*. New York: Basic Books, 2000.

Fagan, Brian. *The Long Summer: How Climate Changed Civilization*. New York: Basic Books, 2004.

Fairhead, James, and Melissa Leach. *Misreading the African Landscape*. Cambridge: Cambridge University Press, 1996.

Farris, William Wayne. *Japan's Medieval Population: Famine, Fertility, and Warfare in a Transformative Age*. Honolulu: University of Hawai'i Press, 2006.

Fiege, Mark. *Irrigated Eden: The Making of an Agricultural Landscape in the American West*. Weyerhaeuser Environmental Books. Seattle: University of Washington Press, 1999.

Fisher, Christopher T., J. Brett Hill, and Gary M. Feinman, eds. *The Archaeology of Enviromental Change: Socionatural Legacies of Degradation and Resilience*. Tucson: University of Arizona Press, 2009.

Ford, Caroline. "Reforestation, Landscape Conservation, and the Anxieties of Empire in French Colonial Algeria." *American Historical Review* 113, no. 2 (2008): 341–62.

Fraser, Christian. "Vesuvius Escape Plan 'Insufficient.'" BBC News, January 10, 2007. http://news.bbc.co.uk/1/hi/world/europe/6247573.stm (accessed October 15, 2013).

Friday, Karl F., ed. *Japan Emerging: Premodern History to 1850*. Boulder, Colo.: Westview Press, 2012.

Fujii Shizue. *Lifan: Riben zhili Taiwan de jice*. Taipei: Wenyingtang, 1997.

Fujiki Hisashi, comp. *Nihon chūsei kishō saigaishi nenpyō kō*. Kōshi Shoin, 2007.

Fujino Naoki and Kobayashi Tetsuo. "Eruptive History of Kaimondake Volcano, Southern Kyushu, Japan." *Kazan* 42, no. 3 (1997): 195–211. In Japanese with English title and abstract.

Fukao Yōko. "Baikofu ni sasagu." In *"Manshu" no seiritsu: Shinrin no shōjin to kindai kūkan no keisei*, edited by Yasutomi Ayumu and Fukao Yōko, 1–15. Nagoya: Nagoya Daigaku Shuppankai, 2009.

Fukaya Katsumi. *Hyakushō naritachi*. Hanawa Shobō, 1993.

Fukaya Ryūzō. "Taiwan ni okeru shiyūringyō no shinkō to shinrin kumiai." *Taiwan no sanrin* 34 (1928): 2–11.

Fukui Eiichirō and Wada Sadao. "Honpō no daitoshi ni okeru kion bunpu." *Chirigaku hyōron* 17, no. 5 (1941): 354–72.

Fukushima, Daisuke, and Kazuhiro Ishihara. "Volcanic Disaster Prevention and Community Development: How to Convert the Volcano into a Museum." *Annals of Disaster Prevention Research Institute, Kyoto University* 48C (2005). Unpaginated. http://www.dpri.kyoto-u.ac.jp/nenpo/nenpo_e.html (accessed December 16, 2014).

Funakoshi, Masaki. "Ogasawara ni okeru ginnemu hayashi no seiritsu—In'yū to bunpu no kakudai o meguru oboegaki—Sono 2." *Ogasawara kenkyū nenpō* 11 (1987): 39–55.

Furihata Hiroki. "Zenkōji jishin to saigai jōhō." In *Zenkōji jishin ni manabu*, edited by Akahane Sadayuki and Kitahara Itoko, 139–65. Nagano-shi: Shinano Mainichi Shinbunsha, 2003.

Furushima Toshio. *Toshio Furushima chosakushū*. Vol. 6, *Nihon nōgyō gijutsu shi*. Tōkyō Daigaku Shuppankai, 1975.

Furuta Etsuzō. *Kinsei gyohi ryūtsū no chiiki teki tenkai*. Kokon Shoin, 1996.

Galloway, Patrick R. "Long-Term Fluctuations in Climate and Population in the Preindustrial Era." *Population and Development Review* 12, no. 1 (1986): 1–24.

Gill, Richardson B. *The Great Maya Droughts: Water, Life, and Death*. Albuquerque: University of New Mexico Press, 2000.

Gill, Tom, Brigitte Steger, and David Slater, eds. *Japan Copes with Calamity: Ethnographies of the Earthquake, Tsunami and Nuclear Disasters of March 2011*. New York: Peter Lang, 2013.

Gillis, John R. "Islands in the Making of an Atlantic Oceania, 1400–1800." Paper presented at "Seascapes, Littoral Cultures, and Trans-Oceanic Exchanges," Library of Congress, Washington, D.C., February 12–15, 2003. http://www.historycooperative. org/proceedings/seascapes/gillis.html (accessed September 9, 2013).

Gillis, John R. *Islands of the Mind: How the Human Imagination Created the Atlantic World*. New York: Palgrave Macmillan, 2004.

Gisiger, L. "Die titrimetrisehe Bestimmung der Phosphorsäure auf der Grundlage der Methode von N. v. Lorenz unter Anwendung der Tauchfiltration." *Zeitschrift für analytische Chemie* 115 (1938): 15–29.

Gladwell, Malcolm. *The Tipping Point: How Little Things Can Make a Big Difference*. Boston: Little, Brown, 2000.

Gordon, Andrew. *A Modern History of Japan: From Tokugawa Times to the Present*. New York: Oxford University Press, 2003.

Gordon, H. Scott. "The Economic Theory of a Common Property Resource: The Fishery." *Journal of Political Economy* 62 (1954): 124–42.

Gotō Kunio. "The National Land Comprehensive Development Act." In *High Economic Growth Period 1960–1969*, edited by Nakayama Shigeru, Gotō Kunio, and Yoshioka Hitoshi, 333–46. Vol. 3, *A Social History of Science and Technology in Contemporary Japan*. Melbourne: Trans Pacific Press, 2006.

Goto, Shunsaku, Hideki Hamamoto, and Makoto Yamano. "Climatic and Environmental Changes at Southeastern Coast of Lake Biwa over Past 3000 Years, Inferred from Borehole Temperature Data." *Physics of the Earth and Planetary Interiors* 152, no. 4 (2005): 314–25.

Graham, Wade. "Traffick According to Their Own Caprice: Trade and Biological Exchange in the Making of the Pacific World, 1766–1825." Paper presented at "Seascapes, Littoral Cultures, and Trans-Oceanic Exchanges," Library of Congress, Washington, D.C., February 12–15, 2003. http://www.historycooperative.org /proceedings/seascapes/graham.html (accessed September 9, 2013).

Grove, Richard H. *Green Imperialism: Colonial Expansion, Tropical Island Edens, and the Origins of Environmentalism, 1600–1860*. Studies in Environment and History. Cambridge: Cambridge University Press, 1995.

Guha, Ramachandra. *The Unquiet Woods: Ecological Change and Peasant Resistance in the Himalaya*. Berkeley: University of California Press, 1990.

Gunma Prefecture. "Kofun jidai no yoroi chakushō jinkotsu ni tsuite." Gunma Prefecture, 2013. http://www.pref.gunma.jp/03/x4500038.html (accessed October 29, 2013).

Hachiga Susumu. "Kodai tojō no senchi ni tsuite: Sono chikei teki kankyō." *Gakusō* 1 (1979): 33–58.

Hachiga Susumu. "Tojō no zōei gijutsu." *Tojō no seitai*, edited by Kishi Toshio, 171–216. Vol. 9, *Nihon no kodai*. Chūō Kōronsha, 1996.

"Hachisenzan batsuboku no keikaku." *Taiwan nichinichi shinpō*, September 24, 1914.

Hagino Toshio. *Chōsen, Manshū, Taiwan ringyō hattatsu shiron*. Zaidanhōjin Rin'ya Kōsaikai, 1965.

Hagiwara Tatsuo. *Chūsei saishi soshiki no kenkyū.* Yoshikawa Kōbunkan, 1965.

Hall, Peter. *The World Cities.* London: Weidenfeld and Nicolson, 1966.

Harada Kazuhiko. "Matsushiro-han de sakusei sareta jishinzuerui ni tsuite." In *1847 Zenkōji jishin hōkokusho,* edited by Chūō Bōsai Kaigi, 122–23. Nihon Shisutemu Kaihatsu Kenkyūjo, 2007.

Harada Toshimaru. *Kinsei sonraku no keizai to shakai.* Yamakawa Shuppansha, 1983.

Haraway, Donna J. *When Species Meet.* Minneapolis: University of Minnesota Press, 2008.

Harvey, David. *The Limits to Capital.* New ed. London: Verso, 2006.

Hashimoto Jurō. *Jūkōgyōka to dokusen.* Vol. 2, *Senkanki no sangyō hatten to sangyō soshiki.* Tōkyō Daigaku Shuppankai, 2004.

Hashimoto Manpei. *Jishingaku kotohajime: Kaitakusha Sekiya Seikei no shōgai.* Asahi Shinbunsha, 1983.

Hattori Yasunari (text) and Utagawa Yoshiharu (illustrations). *Ansei kenmonroku.* 3 vols. 1856.

Hawks, Francis L. *Narrative of the Expedition of an American Squadron to the China Seas and Japan, Performed in the Years 1852, 1853, and 1854 under the Command of Commodore M. C. Perry, United States Navy.* New York: D. Appleton and Company, 1856.

Hayami Akira. "Kinsei Nihon no keizai hatten to Industrious Revolution." In *Tokugawa shakai kara no tenbō: Hatten, kōzō, kokusai kankei,* edited by Hayami Akira, Saitō Osamu, and Sugiyama Shin'ya, 19–32. Dōbunkan, 1989.

Hayami Akira. *Population and Family in Early-Modern Central Japan.* Nichibunken Monograph Series 11. Kyoto: International Research Center for Japanese Studies, 2010.

Hayami, Akira. *The Historical Demography of Pre-modern Japan.* Tokyo: University of Tokyo Press, 2001.

Hayami, Akira. "The Population at the Beginning of the Tokugawa Period." *Keio Economic Studies* 4 (1967): 1–28.

Hazama Gumi Hyakunenshi Hensan Iinkai. *Hazama Gumi hyakunenshi.* Vol. 2. Hazama Gumi, 1989.

Headrick, Daniel R. *The Tools of Empire: Technology and European Imperialism in the Nineteenth Century.* New York: Oxford University Press, 1981.

Hein, Laura. *Fueling Growth: Energy and Economic Policy in Postwar Japan, 1945–1960.* Cambridge, Mass.: Harvard University Council on East Asian Studies, 1990.

Henke, Christopher R. "Situation Normal? Repairing a Risky Ecology." *Social Studies of Science* 37, no. 1 (2007): 135–42.

Henke, Christopher R. *Cultivating Science, Harvesting Power: Science and Industrial Agriculture in California.* Cambridge, Mass.: MIT Press, 2008.

"Hinokizai tōitsu keikaku." *Taiwan nichinichi shinpō,* November 26, 1914.

Hiraga Shōji. "Rinsanbutsu haraisageru hōhō no kaizen to mokuzai no kinyūka." *Taiwan sanrin kaihō* 35 (1929): 50–52.

Hirakawa Minami, Miyake Kazuo, Ihara Kesao, Mizumoto Kunihiko, and Torigoe Hiroyuki, eds. *Kankyō no Nihonshi.* 5 vols. Yoshikawa Kōbunkan, 2012–13.

Hirano, Junpei, and Takehiko Mikami. "Reconstruction of Winter Climate Variations during the 19th Century in Japan." *International Journal of Climatology* 28 (2008): 1423–34.

Hiratsuka Jun'ichi. "1960 nen izen no nakaumi ni okeru hiryōmo saishūgyō no jittai: Satoko to shite no katako no yakuwari." *EcoSophia (Ekosofia)* 13 (2004): 97–112.

"Hitokage no taeta mura—Dengen Kaihatsu būmu no ato ni." *Asahi jānaru* (March 3, 1972): 31–35.

Hofmann, Amerigo. *Aus den Waldungen des Fernen Ostens; Forstliche Reisen und Studien in Japan, Formosa, Korea und den angrenzenden gebieten Ostasiens.* Wien and Leipzig: W. Frick, 1913.

Hokkaidō Suisanbu. *Hokkaidō gyogyō shi.* Vol. 1. Sapporo: Hokkaidō Suisanbu, 1957.

Hokusui Kyōkai. *Hokusui kyōkai hōkokusho*, no. 32 (March 1888).

Hokusui Kyōkai. *Hokusui kyōkai hyakunenshi.* Sapporo: Hokusui Kyōkai, 1984.

Holling, C. S., Lance H. Gunderson, and Garry D. Peterson. "Sustainability and Panarchies." In *Panarchy: Understanding Transformations in Human and Natural Systems*, edited by Lance H. Gunderson and C. S. Holling, 63–102. Washington, D.C.: Island Press, 2002.

"Honpō rinsanbutsu no juyō nami kyōkyūryō." *Dainippon sanrin kaihō* 484 (1923): 63–69.

Hoshikawa Yutaka, Tajima Ken'ichiro, and Kawai Tadashi. "Nishin sanransho no keisei ni oyobosu shokusei to chikei no eikyō." *Hokkaidō suisan shikenjō hōkoku* 62 (2002): 105–11.

Howell, David L. *Capitalism from Within: Economy, Society, and the State in a Japanese Fishery.* Berkeley: University of California Press, 1995.

Howell, David L. "Fecal Matters: Prolegomenon to a History of Shit in Japan." In *Japan at Nature's Edge: The Environmental Context of a Global Power*, edited by Ian Miller, Julia Adeney Thomas, and Brett L. Walker, 137–51. Honolulu: University of Hawai'i Press, 2013.

Hughes, J. Donald. *An Environmental History of the World: Humankind's Changing Role in the Community of Life.* Routledge Studies in Physical Geography and Environment 2. London and New York: Routledge, 2001.

Hughes, J. Donald. *What Is Environmental History?* Cambridge: Polity Press, 2006.

Hung Kuang-chi. "Linxue, ziben zhuyi yu bianqu tongzhi: Rizhi shiqi linye diaocha yu zhengli shiye di zaisikao." *Taiwanshi yanjiu* 11, no. 2 (2005): 77–144.

Iga Toshirō, ed. *Shiga-ken gyogyōshi.* Vol. 1. Ōtsu: Shiga-ken Gyogyō Kyōdō Kumiai Rengōkai, 1954.

Ikawa Katsuhiko. *Kindai Nihon seishigyō to mayu seisan.* Tōkyō Keizai Jōhō Shuppan, 1998.

Inoue, Masami, and Hajime Sakaguchi. "Estimating the Withers Height of the Ancient Japanese Horse from Hoof Prints." *Anthropozoologica* 25–26 (1996): 119–30.

Inoue Motoyuki. "Promotion of Field-Verified Studies on Sediment Transport Systems Covering Mountains, Rivers, and Coasts." *Science and Technology Trends Quarterly Review* 33 (October 2009).

Intergovernmental Panel on Climate Change. *Third Assessment Report, Climate Change 2001: Working Group 1, The Scientific Basis, Summary for Policymakers.* On-line edition. GRID-Arendal, 2003. http://www.grida.no/publications/other/ipcc_tar/?src= /climate/ipcc_tar/.

International Research Center for Japanese Studies. Kojiruien (Dictionary of Historical Terms) Database. Kyoto: International Research Center for Japanese Studies, 2007. http://www.nichibun.ac.jp/graphicversion/dbase/kojirui_e.html (accessed December 21, 2011).

Irie Hisao. *Manshū kanjin shokumin chiiki.* Dairen (Dalian): Minami Manshū Tetsudō Kabushiki Kaisha, 1937.

Isett, Christopher M. *State, Peasant, and Merchant in Qing Manchuria, 1644–1862.* Stanford, Calif.: Stanford University Press, 2007.

Ishibashi, Katsuhiko. "Status of Historical Seismology in Japan." *Annals of Geophysics* 47, no. 2–3 (2004): 339–68.

Ishihara Shun. *Kindai Nihon to Ogasawara shotō: Idōmin no shimajima to teikoku.* Heibonsha, 2007.

Isomura Teikichi and Tsuda Sen. *Ogasawara-jima yoran.* Ben'ekisha, 1888.

Itō Kazuaki. *Jishin to funka no Nihonshi.* Iwanami Shoten, 2002.

Iwamoto Yoshiteru. *Tōhoku kaihatsu 120-nen.* Tōsui Shobō, 1994.

Janković, Vladimir, and Michael Hebbert. "Hidden Climate Change: Urban Meteorology and the Scales of Real Weather." *Climate Change* 113 (2012): 22–33.

Japan Aerospace Exploration Agency. "Missions." http://www.jaxa.jp/projects/util/eos/ index_e.html (accessed February 17, 2011).

Japan Commission on Large Dams. *Dams in Japan: Past, Present, and Future.* London: CRC Press, 2009.

Japan Meteorological Agency. "Japan Hit by Hottest Summer in More Than 100 Years." News release, September 10, 2010. http://www.jma.go.jp/jma/en/News/indexe _news.html (accessed April 17, 2012).

Japan Meteorological Agency. "Primary Factors of Extremely Hot Summer 2010 in Japan." News release, September 16, 2010. http://ds.data.jma.go.jp/tcc/tcc/news /press_20100916.pdf (accessed April 17, 2012).

Japan Meteorological Agency. "Volcanic Warnings." http://www.jma.go.jp/en/volcano/ (accessed December 16, 2014).

Japan Meteorological Agency. "Weather and Earthquakes: Volcanic Warnings and Volcanic Alert Levels." http://www.seisvol.kishou.go.jp/tokyo/STOCK/kaisetsu/ English/level.html (accessed December 16, 2014).

"Jishin o kizukau mono ari." *Yomiuri shinbun,* special ed. (November 4, 1891): 2.

Johnson, Chalmers. *MITI and the Japanese Miracle.* Stanford, Calif.: Stanford University Press, 1982.

Ka Chih-ming. *Fan toujia: Qingdai Taiwan zuqun zhengzhi yu shufan diquan.* Taipei: Zhongyang Yanjiuyuan Shehuixue Yanjiusuo, 2001.

Kaitakushi. *Kaitakushi jigyō hōkoku.* Vol. 3. Ōkurashō, 1885.

Kamiyama Ikumi. "Sakuma damu kaihatsu to chi'iki fujinkai katsudō—sengo Nihon ni okeru minshuka to josei." In *Kaihatsu no jikan, kaihatsu no kūkan—Sakuma damu to chi'iki shakai no hanseiki*, edited by Machimura Takashi, 133–52. Tōkyō Daigaku Shuppankai, 2006.

Kanehara Masaaki. *Chūsei kōkogaku to shizen kagaku ni okeru gaku yūgō no kanōsei: Chūsei sōgō shiryōgaku no kanōsei*. Shinjinbutsu Ōraisha, 2004.

Kaneko Fumio. *Kindai Nihon ni okeru tai Manshū tōshi no kenkyū*. Kondō Shuppansha, 1991.

"'Kankōchi' Sakuma damu." *Asahi shinbun*, March 3, 1957.

Kankyō Jōhō Kagaku Sentā. *Hīto airando genshō no jittai kaiseki to taisaku no arikata ni tsuite: Heisei 12 nendo hōkokusho Kankyoshō ukeoi gyōmu hōkokusho*. Kankyō Jōhō Kagaku Sentā, 2001.

Kantō Totokufu Minseibu Shomuka. *Manshū daizu ni kansuru chōsa*. Dairen (Dalian): Kantō Totokufu Minseibu Shomuka, 1912.

Katz, Paul. "Governmentality and Its Consequences in Colonial Taiwan: A Case Study of the Ta-pa-ni Incident of 1915." *Journal of Asian Studies* 64, no. 2 (2005): 387–424.

Kawai Shitarō. "Taiwan ringyō ni tsuite." *Dainippon sanrin kaihō* 392 (1915): 68–83.

Kawamura Masami. "Damu kensetsu to iu 'kaihatsu pakkēji.'" In *Kaihatsu no jikan, kaihatsu no kūkan—Sakuma damu to chi'iki shakai no hanseiki*, edited by Machimura Takashi, 73–92. Tōkyō Daigaku Shuppankai, 2006.

Kawanabe, Hiroya. "Biological and Cultural Diversities in Lake Biwa." In *Ancient Lakes: Their Culture and Biological Diversity*, edited by Hiroya Kawanabe, G. W. Coulter, and Anna Curtenius Roosevelt, 17–41. Ghent: Kenobi Productions, 1999.

Kawasumi Tatsunori. "Heian-kyō ni okeru chikei kankyō henka to toshiteki tochi riyō no hensen." *Kōkogaku to shizen kagaku* 42 (2001): 35–54.

Kawasumi Tatsunori. "Rekishi jidai ni okeru Kyōto no kōzui to hanrangen no chikei henka: Iseki ni kirokusareta saigai jōhō o mochiita suigaishi no saikōchiku." *Kyōto rekishi saigai kenkyū* 1 (2004): 13–23.

Kawasumi Tatsunori, Harasawa Ryōta, and Yoshikoshi Akihisa. "Chūsei Kyōto no chikei kankyō henka." In *Chūsei no naka no "Kyōto"* (*Chūsei toshi kenkyū*), edited by Takahashi Yasuo and Chūsei Toshi Kenkyūkai, 151–79. Shinjinbutsu Ōraisha, 2006.

Kazaki Hideo. "Ogasawara kenkyū no igi." *Ogasawara kenkyū nenpō* 1 (1977): 1–4.

Kelly, William W. *Irrigation Management in Japan: A Critical Review of Japanese Social Science Research*. Cornell University East Asia Papers 30. Ithaca, N.Y.: China-Japan Program Rural Development Committee, Cornell University, 1982.

Kelly, William W. *Water Control in Tokugawa Japan: Irrigation Organization in a Japanese River Basin, 1600–1870*. Cornell University East Asia Papers 31. Ithaca, N.Y.: China-Japan Program, Cornell University, 1982.

Kensetsushō Kokudo Chiri'in. *Nihon kokusei zu*. Kensetsushō Kokudo Chiri'in, 1997. CD-ROM.

Kerr, Alex. *Dogs and Demons: Tales from the Dark Side of Japan*. New York: Hill and Wang, 2002.

Kikuchi Isao. *Kinsei no kikin*. Yoshikawa Kōbunkan, 1997.

Kikuchi Isao. "Tokugawa Nihon no kikin taisaku." In *Shakai keizai shigaku no kadai to tenbō*, edited by Shakai Keizaishi Gakkai, 375–85. Yūhikaku, 2002.

Kikuchi Koichirō, Muranaga Mineo, and Itagusu Katsukuni. "Sakuma damu no taisha jōkyō to taisaku." *Denryoku doboku* 291 (2001): 41–45.

Kimura Jun. "Ogasawara no ginnemu hayashi." *Ogasawara kenkyū nenpō* 2 (1978): 19–28.

Kinda Akihiro. *Bichikei to chūsei sonraku.* Yoshikawa Kōbunkan, 1993.

Kingston, Jeff, ed. *Natural Disaster and Nuclear Crisis in Japan: Response and Recovery after Japan's 3/11.* Nissan Institute–Routledge Japanese Studies Series. London: Routledge, 2012.

Kishi Toshio. *Tojō no seitai.* Vol. 9, *Nihon no kodai.* Chūō Kōronsha, 1987.

Kishō-chō [Japan Meteorological Agency]. "Kazan: Genzai no funka keikai." http://www.seisvol.kishou.go.jp/tokyo/keikailevel.html (accessed June 24, 2011).

Kishō-chō [Japan Meteorological Agency]. "Nihon no katsu-kazan bunpu." http://www.jma.go.jp/jma/kishou/intro/gyomu/index95zu.html (accessed June 25, 2011).

Kitagawa, Hiroyuki, and Eiji Matsumoto. "Climatic Implications of δ^{13}C Variations in a Japanese Cedar *(Cryptomeria japonica)* during the Last Two Millennia." *Geophysical Research Letters* 22, no. 6 (1995): 2155–58.

Kitahara Itoko. *Jishin no shakaishi: Ansei daijishin to minshū.* Kōdansha, 2000.

Kitahara Itoko. "Saigai to jōhō." In *Nihon saigaishi,* edited by Kitahara Itoko, 230–60. Yoshikawa Kōbunkan, 2006.

Kitahara Itoko, ed. *Nihon saigaishi.* Yoshikawa Kōbunkan, 2006.

Kitajima Masamoto, ed. *Tochi seidoshi.* Vol. 2. Taikei Nihonshi sōsho 7. Yamakawa Shuppansha, 1975.

Kitamura Toshio. *Ōmi keizaishi ronkō.* Taigadō, 1946.

Kitani Makoto. *Namazue shinkō: Saigai no kosumorojī.* Tsuchiura-shi: Tsukuba shorin, 1984.

Kobayashi, Fumiaki, Maki Imai, Hirofumi Sugawara, Manabu Kanda, and Hitoshi Yokoyama. "Generation of Cumulonimbus First Echoes in the Tokyo Metropolitan Region on Mid-Summer Days." Paper presented at the Seventh International Conference on Urban Climate, June 29–July 3, 2009, Yokohama, Japan. http://www.ide.titech.ac.jp/~icuc7/extended_abstracts/pdf/375801-1-090512171643-003.pdf (accessed September 10, 2010).

Kobayashi Takehiro. *Kindai Nihon to kōshū eisei: Toshi shakaishi no kokoromi.* Yūzankaku Shuppan, 2001.

Koizumi Kinmei. "Chōshin hiroku." In *"Nihon no rekishi jishin shiryō" shūi, Saikō 2-nen yori Shōwa 21-nen ni itaru,* vol. 3, edited by Usami Tatsuo, 212–41. Watanabe Tansa Gijutsu Kenkyūjo, 2005.

Kojima Tōzan and Tōrōan-shujin. *Jishinkō.* Kyoto: Saiseikan, 1830.

"Kojireta Tagokura no hoshō mondai—damu kōji no miokuri mo." *Asahi shinbun,* May 19, 1954, 4.

Kokudo Kōtsūshō Kokudo Chiriin, Kokudo Kōtsūshō Hokuriku Chihō Seibikyoku. *Kochiri ni kansuru chōsa: Kochiri de saguru Echigo no hensen—Arakawa, Aganogawa, Shinanogawa, Himekawa.* Niigata: Kokudo Kōtsūshō Hokuriku Chihō Seibikyoku, 2004.

Komai Tokuzō. *Manshū daizu ron.* Sendai: Tōhoku Teikoku Daigaku Nōka Daigaku Nai Kamera Kai, 1912.

Komine Kazuo. *Manshū: Kigen, shokumin, haken.* Ochanomizu Shobō, 1991.

Können, G. P., M. Zaiki, A. P. M. Baede, T. Mikami, P. D. Jones, and T. Tsukahara. "Pre-1872 Extension of the Japanese Instrumental Meteorological Observations Series Back to 1819." *Journal of Climate* 16 (2003): 118–31.

Kozák, Jan, and Vladimir Cermák. *The Illustrated History of Natural Disasters.* Dordrecht: Springer, 2010.

Kuroita Katsumi and Kokushi Taikei Henshūkai, eds. *Nihon sandai jitsuroku.* Vol. 4, Shintei zōho kokushi taikei. Yoshikawa Kōbunkan, 1934.

Kurokawa Kazue. *Nihon ni okeru Meiji ikō no dojō hiryō kō.* 3 vols. Nihon ni okeru Meiji Ikō no Dojō Hiryō Kō Kankōkai, 1975–1982.

Kurosaki Tadashi. *Suisen toire wa kodai ni mo atta: Toire kōkogaku nyūmon.* Yōshikawa Kōbunkan, 2009.

Kurosaki-chō. *Kurosaki-chō shi.* 8 vols. Niigata: Kurosaki-chō, 1994–2000.

Lai Yunxiang. "Taiwan ni okeru minrin keiei no taiken." *Taiwan no sanrin* 79 (1932): 12–27.

Le Roy Ladurie, Emmanuel. *Times of Feast, Times of Famine: A History of Climate since the Year 1000.* New York: Doubleday, 1971.

Lewis, Michael. *Becoming Apart: National Power and Local Politics in Toyama, 1868–1945.* Cambridge, Mass.: Harvard University Asia Center, 2000.

Li Wen-liang. "Rizhi shiqi Taiwan zongtufu de linye zhipei yu suoyuquan: Yi yuangu guanxi wei zhongxin." *Taiwanshi yanjiu* 5, no. 2 (2000): 35–54.

Lilienthal, David E. *TVA: Minshushugi wa shinten suru.* Translated by Wada Koroku. Iwanami Shoten, 1949.

Lilienthal, David E. *The Venturesome Years, 1950–1955.* Vol. 3, *The Journals of David Lilienthal.* New York: Harper & Row, 1966.

Limerick, Patricia Nelson. *Something in the Soil: Legacies and Reckonings in the New West.* New York: W. W. Norton & Company, 2000.

Lomborg, Bjorn. *Cool It: The Skeptical Environmentalist's Guide to Global Warming.* New York: Alfred A. Knopf, 2008.

Machida Hiroshi and Arai Fusao. *Kazanbai atorasu—Nihon retto to sono shūhen.* Tōkyō Daigaku Shuppankai, 1992.

Machida, Hiroshi and Shinji Sugiyama. "The Impact of the Kikai-Akahoya Explosive Eruptions on Human Societies." In *Natural Disasters and Cultural Change*, edited by Robin Torrence and John Grattan, 313–25. London: Routledge, 2002.

Machida Hiroshi, Sugiyama Shinji, and Moriwaki Hiroshi. *Chisō no chishiki: Daiyonki o saguru.* Tōkyō Bijutsu, 1986.

Machimura Takashi. "Chi'iki shakai ni okeru 'kaihatsu' no juyō—dōin to shutaika no jūsōteki katei." In *Kaihatsu no jikan, kaihatsu no kūkan—Sakuma damu to chi'iki shakai no hanseiki*, edited by Machimura Takashi, 93–132. Tōkyō Daigaku Shuppankai, 2006.

Machimura Takashi. *Kaihatsushugi no kōzō to shinshō—sengo Nihon ga damu de mita yume to genjitsu.* Ochanomizu Shobō, 2011.

Machimura Takashi. "Kioku no naka no Sakuma damu—'bure' to 'nigori' no sōhatsuryoku." In *Kaihatsu no jikan, kaihatsu no kūkan—Sakuma damu to chi'iki shakai no hanseiki*, edited by Machimura Takashi, 233–58. Tōkyō Daigaku Shuppankai, 2006.

Machimura Takashi. "Posuto-damu kaihatsu no hanseiki—chi'iki shakai ni kizamareru

Sakuma damu kensetsu no inpakuto." In *Kaihatsu no jikan, kaihatsu no kūkan—Sakuma damu to chi'iki shakai no hanseiki*, edited by Machimura Takashi, 171–94. Tōkyō Daigaku Shuppankai, 2006.

Machimura Takashi, ed. *Kaihatsu no jikan, kaihatsu no kūkan—Sakuma damu to chi'iki shakai no hanseiki*. Tōkyō Daigaku Shuppankai, 2006.

Maejima, Ikuo, and Yoshio Tagami. "Climatic Change during Historical Times in Japan: Reconstruction from Climatic Hazard Records." *Geographical Reports of Tokyo Metropolitan University* 21 (1986): 157–71.

Magata Sei. "Sara ni Arizan mondai o kōkyū seyo." *Dainippon sanrin kaihō* 324 (1909): 22–28.

Malthus, Thomas Robert. *An Essay on the Principle of Population*. London: Murray, 1798.

Manley, Steven L., and Christopher G. Lowe. "Canopy-Forming Kelps as California's Coastal Dosimeter: [131]I from Damaged Japanese Reactor Measured in *Macrocystis pyrifera*." *Environmental Science and Technology* 46 (2012): 3731–36.

Marcotullio, Peter J., and Gordon McGranahan. "Scaling the Urban Environmental Challenge." In *Scaling Urban Environmental Challenges: From Local to Global and Back*, edited by Peter J. Marcotullio and Gordon McGranahan, 1–17. London: Earthscan, 2007.

Marks, Robert B. *China: Its Environment and History*. Plymouth, U.K.: Rowman & Littlefield, 2012.

Masuzawa Shinji. "Taiwansan katsuyōjyu no shiyō kachi kōjōsaku ni tsuite." *Taiwan sanrin kaihō* 35 (1929): 82.

Matsui Akira. *Kankyō kōkogaku*. Nihon no bijutsu 423. Shibundō, 2001.

Matsukata, Haru. "They're Taming the Heavenly Dragon." *Saturday Evening Post*, November 19, 1955, 34–36.

Mauch, Christof, and Thomas Zeller. *Rivers in History: Perspective on Waterways in Europe and North America*. Pittsburgh: University of Pittsburgh Press, 2008.

McAnany, Patricia A., and Norman Yoffee, eds. *Questioning Collapse: Human Resilience, Ecological Vulnerability, and the Aftermath of Empire*. Cambridge: Cambridge University Press, 2010.

McCarthy, Mark. "Including Cities in Climate Models." In *City Weathers: Meteorology and Urban Design, 1950–2010*, edited by Michael Hebbert, Vladimir Janković, and Brian Webb. Manchester: University of Manchester, Manchester Architecture Research Centre, 2011. http://www.academia.edu/3434577/City_Weathers _Meteorology_and_Urban_Design_1950-2010 (accessed October 28, 2013).

McCormack, Gavan. *The Emptiness of Japanese Affluence*. Armonk, N.Y.: M. E. Sharpe, 2001.

McCormack, Gavan, and Nanyan Guo. "Coming to Terms with Nature: Development Dilemmas on the Ogasawara Islands." *Japan Forum* 13, no. 12 (2001): 177–93.

McDonald, Michael J., and John Muldowny. *TVA and the Dispossessed: The Resettlement of Population in the Norris Dam Area*. Knoxville: University of Tennessee Press, 1982.

McIntosh, Roderick, Joseph A. Tainter, and Susan Keech McIntosh. "Climate, History, and Human Action." In *The Way the Wind Blows: Climate, History, and Human*

Action, edited by Roderick McIntosh, Joseph A. Tainter, and Susan Keech, 1–42. McIntosh Historical Ecology Series. New York: Columbia University Press, 2000.

McKean, Margaret A., Clark Gibson, and Elinor Ostrom, eds. *People and Forests: Communities, Institutions, and Governance.* Cambridge, Mass.: MIT Press, 2000.

McNeill, John Robert. *Mosquito Empires: Ecology and War in the Greater Caribbean.* Cambridge: Cambridge University Press, 2010.

McNeill, John Robert. *Something New Under the Sun: An Environmental History of the Twentieth-Century World.* New York: W. W. Norton & Company, 2000.

Mead, Albert R. *The Giant African Snail: A Problem in Economic Malacology.* Chicago: University of Chicago Press, 1961.

Meadows, H. Donella, Dennis L. Meadows, Jørgen Randers, and William W. Behrens III. *The Limits to Growth: A Report for the Club of Rome's Project on the Predicament of Mankind.* Washington, D.C.: Potomac Associates, 1972.

Melillo, Edward. "The First Green Revolution: Debt Peonage and the Making of the Nitrogen Fertilizer Trade, 1840–1930." *American Historical Review* 117, no. 4 (2012): 1028–60.

Memorandum for Record: Study Abroad of Japanese Technicians Connected with the Resources Committee and Students Interested in Resources Utilization and Planning, October 7, 1948. National Archives and Records Administration, College Park, Md., RG 331 (Supreme Commander for the Allied Powers), Box 9132, Folder #7.

Mikami Takehiko. "Fueru Tōkyō no manatsubi to nettaiya." In *Tōkyō ijō kishō*, edited by Mikami Takehiko, 8–12. Yōsensha, 2006.

Mikami Takehiko. "Pinpointo de Tōkyō o osou toshi-gata shūchū gōu." In *Tōkyō ijō kishō*, edited by Mikami Takehiko, 14–16. Yōsensha, 2006.

Mikami, Takehiko. "Climate of Japan during 1781–90 in Comparison with That of China." In *The Climate of China and Global Climate: Proceedings of the Beijing International Symposium on Climate*, Oct. 30–Nov. 3, 1984, Beijing, China, edited by Duzheng Ye, Congbing Fu, Jiping Chao, and M. Yoshino, 63–75. Beijing: China Ocean Press, 1987.

Mikami, Takehiko. "Climate Variations in Japan during the Little Ice Age—Summer Temperature Reconstructions since 1771," in *Proceedings of the International Symposium on the Little Ice Age Climate*, edited by T. Mikami, 176–81. Tokyo: Tokyo Metropolitan University, 1992).

Mikami, Takehiko. "Climatic Reconstruction in Historical Times Based on Weather Records." *Geographical Review of Japan* (Series B) 61, no. 1 (1988): 14–22.

Mikami, Takehiko. "Climatic Variations in Japan Reconstructed from Historical Documents." *Weather* 63, no. 7 (2008): 190–93.

Mikami, Takehiko. "Long Term Variations of Summer Temperatures in Tokyo since 1721." *Geographical Reports of Tokyo Metropolitan University* 31 (1996): 157–65.

Mikami, Takehiko. "Quantitative Climatic Reconstruction in Japan Based on Historical Documents." Bulletin of the National Museum of Japanese History 81 (1999): 41–50.

Mikami, T., M. Zaiki, G. P. Können, and P. D. Jones. "Winter Temperature Reconstruction at Dejima, Nagasaki, Based on Historical Meteorological Documents during the Last 300 Years." In *Proceedings of the International Conference on Climate Change and Variability—Past, Present and Future*, edited by Takehiko Mikami, 103–106. Hachiōji: International Geographical Union Commission on Climatology and Tokyo Metropolitan University, 2000.

Miki Haruo. *Kyōto daijishin.* Shibunkaku Shuppan, 1979.

Mikuriya Takashi. *Seisaku no sōgō to kenryoku—Nihon seiji no senzen to sengo.* Tōkyō Daigaku Shuppansha, 1996.

Mimura, Janis. *Planning for Empire: Reform Bureaucrats and the Japanese Wartime State.* Ithaca, N.Y.: Cornell University Press, 2011.

Minami Manshū Tetsudō Kabushiki Kaisha. *Minami Manshū tetsudō ensen shashin chō.* Dairen (Dalian): Nanman Tetsudō, 1940.

Minami Manshū Tetsudō Kabushiki Kaisha Chihōbu Shōkōka. *Manshū daizu kasu to sono shiryōka ni tsuite.* Dairen (Dalian): Minami Manshū Tetsudō, 1932.

Minami Manshū Tetsudō Nōji Shikenjo. *Manshū daizu narabini mamekasu.* Gongzhuling: Minami Manshū Tetsudō Nōji Shikenjo, 1921.

Minami Manshū Tetsudō Shomubu Chōsaka. *Manshū ni okeru yubōgyō.* Dairen (Dalian): Minami Manshū Tetsudō Shomubu Chōsaka, 1924.

Mitchell, Timothy. "Carbon Democracy." *Economy and Society* 38, no. 3 (2009): 399–432.

Mitchell, Timothy. *Rule of Experts: Egypt, Techno-Politics, Modernity.* Berkeley: University of California Press, 2002.

Miwa Ryōichi. "Reorganization of the Japanese Economy." In *A History of Japanese Trade and Industry Policy*, edited by Sumiya Mikio, 151–249. Oxford: Oxford University Press, 2000.

Miyamoto, Matao. "Quantitative Aspects of Tokugawa Economy." In *Emergence of Economic Society in Japan, 1600–1859*, edited by Akira Hayami, Osamu Saito, and Ronald P. Toby, 36–84. Oxford: Oxford University Press, 2004.

Miyata Ichirō. "'Risō tsuikyū e no hi'—TVA shisō, minshuka, soshite jiritsu." In *Kaihatsu no jikan, kaihatsu no kūkan—Sakuma damu to chi'iki shakai no hanseiki*, edited by Machimura Takashi, 51–71. Tōkyō Daigaku Shuppankai, 2006.

Miyata Noboru and Takada Mamoru, eds. *Namazue: Shinsai to Nihon bunka.* Ribun Shuppan, 1995.

Miyataki Kōji. "Ima naze kankyōshi, saigaishi no shiten ka: Nihon kodaishi no tachiba kara." *Atarashii rekishigaku no tame ni* 259 (2005): 32–35.

Miyataki Kōji. "'Kankyōshi,' saigaishi ni fumidashita Nihon kodaishi kenkyū." *Rekishi hyōron* 626 (2002): 60–65.

Miyase Hiroshi. "Eirinjozai to sono shijō. *Taiwan no sanrin* 85 (1933): 129–34.

Moberg, Anders, Dmitry M. Sonechkin, Karin Holmgren, Nina M. Datsenko, and Wibjörn Karlén. "Highly Variable Northern Hemisphere Temperatures Reconstructed from Low- and High-Resolution Proxy Data." *Nature* 433 (2005): 613–17.

Molony, Barbara. *Technology and Investment: The Prewar Japanese Chemical Industry.* Cambridge, Mass.: Harvard University Press, 1990.

Moore, Aaron Stephen. *Constructing East Asia: Technology, Ideology, and Imperialism in Japan's Wartime Era, 1931–1945.* Stanford, Calif.: Stanford University Press, 2013.

Moore, Aaron Stephen. "'The Yalu River Era of Developing Asia': Japanese Expertise, Colonial Power, and the Construction of Sup'ung Dam." *Journal of Asian Studies* 72:1 (February 2013): 115–39.

Moore, Jason W. *"The Modern World-System* as Environmental History? Ecology and the Rise of Capitalism." *Theory and Society* 32 (2003): 307–77.

Morris-Suzuki, Tessa. *Re-Inventing Japan: Time, Space, Nation.* New York: M. E. Sharpe, 1998.

Murakushi Nisaburō. *Kokuritsu kōen seiritsushi no kenkyū: Kaihatsu to shizen hogo no kakushitsu o chūshin ni.* Hōsei Daigaku Shuppankyoku, 2005.

Murao Motonaga. *Nishin hiryō gaiyō.* Murao Motonaga, 1895.

Musha Kinkichi. *Jishin namazu.* 1957. Meiseki Shoten, 1995.

Nagai Risa. "Taiga no sōshitsu." In *"Manshū" no seiritsu: Shinrin no shōjin to kindai kūkan no keisei,* edited by Yasutomi Ayumu and Fukao Yōko, 19–60. Nagoya: Nagoya Daigaku Shuppankai, 2009.

Nagayama Kikuo. "Taiwansan katsuyōjyu no riyō kachi zōshin ni tsuite." *Taiwan sanrin kaihō* 62 (1931): 36–44.

Nakai Isao. "Dai kibō takuchi to sono ruikei." In *Kodai toshi no kōzō to tenkai,* edited by Kodai Tojōsei Kenkyū Shūkai Jikkō Iinkai, 185–216. Kodai Tojōsei Kenkyūkai dai 3 kai hōkokushū. Nara: Nara Kokuritsu Bunkazai Kenkyūsho, 1998.

Nakajima Chōtarō. "Kamogawa suigaishi (1)." *Kyōto daigaku bōsai kenkyūjo nenpō* 26 (B-2) (1983): 75–92.

Nakamura, Hachiro. "Urban Growth in Prewar Japan." In *Japanese Cities in the World Economy,* edited by Kuniko Fujita and Richard Child Hill, 26–52. Philadelphia: Temple University Press, 1993.

Nakamura Yoichi, Kazuyoshi Fukushima, Xinghai Jin, Motoo Ukawa, Teruko Sato, and Yayoi Hotta Yayoi. "Mitigation Systems by Hazard Maps, Mitigation Plans, and Risk Analyses Regarding Volcanic Disasters in Japan." *Journal of Disaster Research* 3, no. 4 (2008): 297–304.

Nakanishi Satoshi. "Bakumatsu-Meiji ki kinai hiryō shijō no tenkai." *Keizaigaku kenkyū* (Hokkaidō Daigaku) 47, no. 2 (1997): 281–98.

Nakanishi Satoshi. *Kinsei-kindai Nihon no shijō kōzō: "Matsumae nishin" hiryō torihiki no kenkyū.* Tōkyō Daigaku Shuppankai, 1998.

Nakata Setsuya. "Kazan bōsai sofuto gijutsu no saizensen." In *Kazan funka ni sonaete,* edited by Doboku Gakkai, 130–37. Doboku Gakkai, 2005.

Nara Kenritsu Kashihara Kōkogaku Kenkyujo. *Heijō-kyō sakyō gojō nibō jūgo, jūroku tsubo.* Nara: Nara-ken Kyōiku Iinkai, 2006.

Nara Kokuritsu Bunkazai Kenkyūsho and Asahi Shinbunsha Ōsaka Honsha Kikakubu. *Heijō-kyō ten.* Osaka: Asahi Shinbunsha Ōsaka Honsha Kikakubu, 1989.

National Police Agency of Japan. "Damage Situation and Police Countermeasures Associated with 2011 Tohoku District—Off the Pacific Ocean Earthquake." http ://www.npa.go.jp/archive/keibi/biki/higaijokyo_e.pdf (accessed January 14, 2014).

National Research Institute for Earth Science and Disaster Prevention (NIED). Database on Volcanic Hazard Maps and Reference Material (1983–): Contents of

Volcanic Hazard Maps of Japan, 2nd ed. http://vivaweb2.bosai.go.jp/v-hazard/articles-e.html (accessed October 28, 2013).

Nawa Tōichi. *Nihon bōseki gyō to genmen mondai kenkyū*. Osaka: Daidō Shoin, 1938.

Nihon Jinbun Kagakukai. *Sakuma damu: Kindai gijutsu no shakaiteki eikyō*. Tōkyō Daigaku Shuppankai, 1958.

Nihon Kenchiku Gakkai. *Hīto airando to kenchiku, toshi: Taisaku no bijon to kadai*. Nihon Kenchiku Gakkai, 2007.

Nihon Seitai Gakkai. *Gairaishu handobukku*. Tokyo: Chijin Shokan, 2002.

Niigata-ken, ed. *Niigata-ken no ayumi*. Niigata-shi: Niigata-ken, 1990.

Niigata Shishi Hensan Iinkai. *Niigata shishi, tsūshi hen*. Vol. 1. Niigata-shi: Niigata-shi, 1995.

Nishimura Makoto and Yoshikawa Ichirō, eds. *Nihon kyōkōshi kō*. Maruzen, 1936.

Nishino Machiko and Hamabata Etsuji. *Naiko kara no messēji—Biwako shūhen no shitchi saisei to seibutsu tayōsei hozen*. Sunrise Shuppan, 2005.

Noguchi Takehiko. *Ansei Edo jishin: Saigai to seiji kenryoku*. Chikuma Shobō, 1997.

Nōrinshō Daijin Kanbō Sōmuka. *Nōrin gyōsei shi*. Vol. 1. Nōrin Kyōkai, 1957.

Nunn, Patrick D. *Oceanic Islands*. Oxford: Blackwell, 1994.

Ó Gráda, Cormac. *Famine: A Short History*. Princeton, N.J.: Princeton University Press, 2009.

O'Bryan, Scott. *The Growth Idea: Purpose and Prosperity in Postwar Japan*. Honolulu: University of Hawai'i Press, 2009.

Oda Shizuo. "Kyūsekki jidai to Jōmon jidai no kazan higai." In *Kazanbai kōkogaku*, edited by Arai Fusao, 207–24. Kokon Shoin, 1993.

Ogasawara Kyōkai. *Ogasawara shotō gaishi: Nichi-Bei kōshō o chūshin to shite*. Ogasawara kyōkai, 1967.

Ogasawara Yoshihiko. "Kuni-kyō, Shigaraki-kyō, Naniwa-kyō." In *Heijō-kyō no jidai*, edited by Tanabe Ikuo and Satō Makoto, 216–37. Vol. 2, *Kodai no miyako*. Yoshikawa Kōbunkan, 2010.

Ogashima Minoru. *Nihon saiishi*. Nihon Kōgyōkai, 1894.

Ohkawa, Kazushi, Nobukiyo Takamaysu, and Yuzo Yamamoto, eds. *Estimates of Long-Term Economic Statistics of Japan since 1868*. Vol. 9. Toyo Keizai, 1966.

Okada Shin. "Taiwan ni okeru mokuzai jukyū zōrin." *Taiwan sanrin kaihō* 72 (1932): 9–17.

Okada Tomohiro. *Nihon shihonshugi to nōson kaihatsu*. Kyōto: Hōritsu Bunkasha, 1989.

Okamoto Sei. "Tokushu rinboku hinoki, benihi no jukyū oyobi torihiki ni tsuite: Hontō ni okeru mokuzai no jukyū to hinoki, benihi." *Taiwan no sanrin* 85 (1933): 124–29.

Okamoto Takuji. "The Reconstruction of the Electric Power Industry." In *Road to Self-Reliance 1952–1959*, edited by Nakayama Shigeru, Gotō Kunio, and Yoshioka Hitoshi, 414–53. Vol. 2, *A Social History of Science and Technology in Contemporary Japan*. Melbourne: Trans Pacific Press, 2005.

Ōkurashō Shuzeikyoku and Nonaka Jun, eds. *Dai Nihon sozeishi*. 4 vols. Chōyōkai, 1926–27.

Ono Yoshirō. *Mizu no kankyōshi: "Kyō no meisui" wa naze ushinawareta ka*. PHP Kenkyūjo, 2001.

Parker, D.E., T.P. Legg, and C.K. Folland. "A New Daily Central England Temperature Series, 1772-1991." *International Journal of Climatology* 12 (1992): 317-42.

Peluso, Nancy L. *Rich Forests, Poor People: Resource Control and Resistance in Java.* Berkeley: University of California Press, 1994.

Penna, Anthony N. *The Human Footprint: A Global Environmental History.* Malden, Mass.: Wiley-Blackwell, 2010.

Pflugfelder, Gregory M., and Brett L. Walker, eds. *JAPANimals: History and Culture in Japan's Animal Life.* Ann Arbor: Center for Japanese Studies University of Michigan, 2005.

Pomeranz, Kenneth. "Introduction." In *The Environment and World History*, edited by Edmund Burke III and Kenneth Pomeranz, 3–32. Berkeley: University of California Press, 2009.

Ponting, Clive. *A Green History of the World.* London: Penguin Books, 1991.

Ponting, Clive. *A New Green History of the World: The Environment and the Collapse of Great Civilizations.* New York: Penguin Books, 2007.

Pratt, Mary Louise. *Imperial Eyes: Travel Writing and Transculturation.* 2nd ed. London: Routledge, 2008.

Rajan, Ravi. *Modernizing Nature: Forestry and Imperial Eco-Development 1800–1950.* Oxford: Oxford University Press, 2006.

Redclift, Michael R. *Frontiers: Histories of Civil Society and Nature.* Cambridge, Mass.: MIT Press, 2006.

Richards, John F. *The Unending Frontier: An Environmental History of the Early Modern World.* The California World History Library. Berkeley: University of California Press, 2003.

Riebeek, Holli. "Urban Rain." NASA Earth Observatory, December 8, 2006. http://earthobservatory.nasa.gov/Features/UrbanRain/ (accessed February 17, 2011).

"Rinmu kaigi ni okeru shokusankyokuchō kunji." *Taiwan sanrin kaihō* 7 (1924): 2–4.

Roberts, Neil. *The Holocene: An Environmental History.* 2nd ed. Oxford: Blackwell Publishers, 1998.

Rossiter, A. "Lake Biwa as a Topical Ancient Lake." In *Ancient Lakes: Biodiversity, Ecology and Evolution*, edited by A. Rossiter and Hiroya Kawanabe, 571–98. Advances in Ecological Research 31. London: Academic Press, 2000.

Rubinger, Richard. *Popular Literacy in Early Modern Japan.* Honolulu: University of Hawai'i Press, 2007.

Saitō Satoshi. "Omowanu tokoro ni kodai shūraku." *Maibun Gunma* 45 (2006): 4–5.

Saito, Osamu. "Climate and Famine in Historic Japan: A Very Long-Term Perspective." In *Demographic Responses to Economic and Environmental Crises*, edited by Satomi Kurosu, Tommy Bengtsson, and Cameron Campbell, 272–81. Reitaku University, 2010. http://www.archive-iussp.org/members/restricted/publications/Kashiwa09/Chapter15.pdf (accessed October 22, 2013).

Saito, Osamu. "The Frequency of Famines as Demographic Correctives in the Japanese Past." In *Famine Demography: Perspectives from the Past and Present*, edited by T. Dyson and C. Ó Gráda, 218–39. Oxford: Oxford University Press, 2002.

Saitoh, Hidetoshi. "A Study of the Cattle Ploughing and the Size of the Rice Fields of the Ancient Period." *Bulletin of Gunma Archaeological Research Foundation* 17 (1999): 25–44. In Japanese with English title.

Saji Takanori. "Taiwan no eirin jigyō ni tsuite." *Taiwan sanrin kaihō* 79 (1932): 54–58.

Sakaguchi Makoto. "Kindai Nihon no daizu kasu shijō: Yunyū hiryō no jidai." *Rikkyō keizaigaku kenkyū* 57, no. 2 (2003): 53–70.

Sakaguchi Makoto. "Senkanki Nihon no ryūan shijō to ryūtsū rūto: Mitsui Bussan, Mitsubishi Shōji, Zenkōren o chūshin ni." *Rikkyō keizaigaku kenkyū* 59, no. 2 (2005): 153–77.

Sakaguchi, Yutaka. "Warm and Cold Stages in the Past 7600 Years in Japan and Their Global Correlation: Especially on Climatic Impacts to the Global Sea Level Changes and the Ancient Japanese History." *Bulletin of the Department of Geography, University of Tokyo* 15 (1983): 1–31.

Sakakibara Masazumi. "Shōwa 36-nen 6-gatsu Inadani no shūchū gōu ni yoru kōzui ni ki'in shita Sakuma damu-ko ni okeru ka no daihassei." *Nagano daigaku fūdobyō kiyō* 7:2 (1965): 130–41.

Sakō Tsuneaki. *Nihon hiryō zensho.* 2nd ed. Yūrindō, 1894.

"Sakuma damu ni kankōsen." *Asahi shinbun,* August 7, 1959.

"Sakuma damu no kibo." *Asahi shinbun,* August 31, 1975.

Samuels, Richard. *3.11: Disaster and Change in Japan.* Ithaca, N.Y.: Cornell University Press, 2013.

Sanka Isao. *Daizu no saibai.* Dairen (Dalian): Minami Manshū Tetsudō Kōgyōbu Nōmuka, 1924.

Sano Shizuyo. *Chū-kinsei no sonraku to mizube no kankyōshi: Keikan, seigyō, shigen-kanri.* Yoshikawa Kōbunkan, 2008.

Satō Atsushi. *Nihon no chi'iki kaihatsu.* Miraisha, 1965.

Satō Jin. *"Motazaru kuni" no shigenron—jizoku kanō na kokudo o meguru mō hitotsu no chi.* Tōkyō Daigaku Shuppankai, 2011.

Scheffer, Marten. *Critical Transitions in Nature and Society.* Princeton Studies in Complexity. Princeton, N.J.: Princeton University Press, 2009.

Scheiner, Irwin. "Benevolent Lords and Honorable Peasants: Rebellion and Peasant Consciousness in Tokugawa Japan." In *Japanese Thought in the Tokugawa Period, 1600–1868,* edited by Tetsuo Najita and Irwin Scheiner, 39–62. Chicago: University of Chicago Press, 1978.

Schwartz, Glen M., and John J. Nichols, eds. *After Collapse: The Regeneration of Complex Societies.* Tucson: University of Arizona Press, 2006.

Scott, James C. *Seeing Like a State: How Certain Schemes to Improve the Human Condition Have Failed.* New Haven, Conn.: Yale University Press, 1998.

Shapiro, Judith. *China's Environmental Challenges.* Malden, Mass.: Polity Press, 2012.

Shepherd, John. *Statecraft and Political Economy on the Taiwan Frontier, 1600–1800.* Stanford, Calif.: Stanford University Press, 1993.

Shiga Chōshi Henshū Iinkai. *Shiga chōshi.* Vol. 2. Shiga (Shiga Prefecture): Shiga Chōshi Henshū Iinkai, 1999.

Shiga Kenritsu Nōji Shikenjō. *Biwako engan mizumo no riyō to sono hikō.* Ōtsu: Shiga Kenritsu Nōji Shikenjō, 1939.

Shiga-ken. *Chūsei-Kinsei.* Vol. 3, *Shiga kenshi.* Ōtsu: Shiga-ken, 1928.

Shiga-ken Kyōiku Iinkai. *Biwako sōgō kaihatsu chiiki minzoku bunkazai tokubetsu chōsa hōkokusho.* Vols. 1–5. Ōtsu: Shiga-ken Kyōiku Iinkai, 1979–1983.

Shiga-ken Suisan Shikenjō. *Biwako suisan zōshoku jigyō seiseki hōkoku.* Vol. 1. Ōtsu: Shiga-ken Suisan Shikenjō, 1923.

Shimada Toshio. "Kenchiku shizai no risaikuru." In *Koto hakkutsu,* edited by Tanaka Migaku, 156–58. Iwanami shinsho 468. Iwanami Shoten, 1996.

Shimazaki Minoru. "Dengen kaihatsu sokushin hō: Sakuma damu no baai." *Jurisuto* 533 (1973): 61–66.

Shimokōbe Atsushi. *Sengo kokudo keikaku e no shōgen.* Nihon Keizai Hyōronsha, 1994.

Shimoyama Satoru. "Volcanic Disasters and Archaeological Sites in Southern Kyushu, Japan." In *Natural Disasters and Cultural Change,* edited by Robin Torrence and John Grattan, 326–41. London: Routledge, 2002.

Shinano Mainichi Shinbunsha Kaihatsukyoku Shuppanbu. *Kōka yonen Zenkōji daijishin.* Nagano-shi: Shinano Mainichi Shinbunsha, 1977.

Shinkōsha. *Nihon chiri fuzoku taikei.* Vol. 1. Shinkōsha, 1931.

Shinsai Yobō Chōsakai. *Dai-Nihon jishin shiryō.* 2 vols. Shibunkaku, 1973.

Shiraishi Taiichiro. "Mitsudera." In *Ancient Japan,* edited by Richard Pearson, 220–22. Washington, D.C.: George Braziller and Arthur Sackler Gallery, Smithsonian Institution, 1992.

Shiratori Kōji. "Inbanuma ni okeru 'Mokutori' no jittai." *Shizen to bunka* 3 (1996): 35–40.

Shizuoka-ken. *Shizuoka kenshi tsūshihen.* Vol. 6. Shizuoka: Shizuoka-ken, 1997.

Shutō Nobuo. "Meiji Sanriku jishin tsunami." In *Nihon rekishi kasai jiten,* edited by Kitahara Itoko, Matsuura Ritsuko, and Kimura Reō, 373–83. Yoshikawa Kōbunkan, 2012.

Shutō Nobuo. "Shōwa Sanriku jishin tsunami." In *Nihon rekishi kasai jiten,* edited by Kitahara Itoko, Matsuura Ritsuko, and Kimura Reō, 459–67. Yoshikawa Kōbunkan, 2012.

Simkin, Tom, Lee Siebert, and Russell Blong. "Volcano Fatalities—Lessons from the Historical Record." *Science* 291, no. 5502 (2001): 255.

Simmons, I. G. *Global Environmental History: 10,000 BC to AD 2000.* Edinburgh: Edinburgh University Press, 2008.

Sivaramakrishnan, K. *Modern Forests: Statemaking and Environmental Change in Colonial Eastern India.* Stanford, Calif.: Stanford University Press, 1999.

Smithsonian National Museum of Natural History Global Volcanism Program. "Sakura-jima." http://www.volcano.si.edu/world/volcano.cfm?vnum=0802-08=&volpage=photos (accessed June 24, 2011).

Smits, Gregory. "Conduits of Power: What the Origins of Japan's Earthquake Catfish Reveal about Religious Geography." *Japan Review* 24 (2012): 41–65. http://shinku.nichibun.ac.jp/jpub/pdf/jr/JN2402.pdf.

Smits, Gregory. *Seismic Japan: The Long History and Continuing Legacy of the Ansei Edo Earthquake.* Honolulu: University of Hawai'i Press, 2013.

Smits, Gregory. *When the Earth Roars: Lessons from the History of Earthquakes in Japan.* Lanham, Md.: Rowman and Littlefield, 2014.

"So Much for the Myth of Safety." *Nikkei Weekly,* special issue (Summer 2011): 5, 12.

Soda Tsutomu. "Evidence of Earthquakes and Tsunamis Recovered at Archaeological Sites in Japan." *Kōkogaku jānaru* 557 (2008): 21–26. In Japanese with English title.

Soda Tsutomu. "Gunma-ken no shizen to fūdo." In *Gunma kenshi 1: Genshi-kodai 1,* edited by Gunma Kenshi Hensan Iinkai, 39–129. Maebashi: Gunma-ken, 1990.

Soda Tsutomu. "Kofun jidai ni okotta Haruna-san Futatsudake no funka." In *Kazanbai kōkogaku,* edited by Arai Fusao, 128–50. Kokon Shoin, 1993.

Soda Tsutomu, Masami Izuho, and Hiroyuki Sato. "Human Adaptation to the Environmental Change Caused by the Gigantic AT Eruption (28–30ka) of the Ito Caldera in Southern Kyushu, Japan." Paper presented at the International Field Conference and Workshop on Tephrochronology, Volcanism and Human Activity, Kirishima City, Kagoshima, Japan, May 9–17, 2012.

Solomon, Susan, Dahe Qin, Martin Manning, Melinda Marquis, Kristen Averyt, Melinda M.B. Tignor, Henry LeRoy Miller, Jr., and Zhenlin Chen, eds. *Climate Change 2007: The Physical Science Basis (Contribution of Working Group I to the Fourth Assessment Report of the Intergovernmental Panel on Climate Change).* Cambridge: Cambridge University Press, 2007.

Sōmushō Tōkeikyoku. *Nihon no chōki tōkei keiretsu.* http://www.stat.go.jp/data/chouki / (accessed October 31, 2013).

Sōmushō Tōkeikyoku. *Statistical Handbook of Japan 2010.* http://www.stat.go.jp/data /handbook/index.htm (accessed February 20, 2010).

Suginami-ku Kyōiku Iinkai. Untitled Japanese-language environmental model school pamphlet for Ogikubo Elementary School. Suginami-ku Kyōiku Iinkai, 2009.

Suginami-ku Toshi Seibibu Midori Kōenka. *Suginami-ku midori no kihon keikaku.* Suginami-ku Toshi Seibibu Midori Kōenka, 2010.

Suginami-ku Toshigata Suigai Taisaku Kentō Senmonka Iinkai. *Arata na toshi-gata suigai no gensai ni idomu.* Suginami-ku Seisaku Keieibu Kiki Kanrishitsu, 2006.

"Suiryoku hatsuden no kinjitō no 'ima' o tazuneru—Sakuma damu unten kaishi gojū shūnen." *Kōken* 519 (2006): 118–21.

Suzuki, Bokushi. *Snow Country Tales: Life in the Other Japan.* Translated by Jeffrey Hunter and Rose Lesser. New York: Weatherhill, 1986.

Suzuki Tōzō and Koike Shōtarō, eds. *Fujiokaya nikki.* Vol. 15, Kinsei shomin seikatsu shiryō. San'ichi Shobō, 1995.

Sweda, Tatsuo, and Shinichi Takeda. "Construction of an 800-year-long *Chamaecyparis* Dendrochronology for Central Japan." *Dendrochronologia* 11 (1994): 79–86.

Szonyi, Michael. *Cold War Island: Quemoy on the Front Line.* Cambridge: Cambridge University Press, 2008.

Tainter, Joseph A. *The Collapse of Complex Societies.* New Studies in Archaeology. Cambridge: Cambridge University Press, 1988.

Taiwan Sanrinkai. *Taiwan no ringyō.* Taipei: Taiwan Sanrinkai, 1933.

Taiwan Sōtokufu Eirinkyoku. *Ringyō ippan.* Taipei: Eirinkyoku, 1919.

Taiwan Sōtokufu Eirinkyoku. *Taiwan rin'ya hōki.* Taipei: Eirinkyoku, 1919.

Taiwan Sōtokufu Keimukyoku. *Ribanshi kō.* Taipei: Keimukyoku, 1918–38.

Taiwan Sōtokufu Minseibu Shokusanka. *Taiwan Sōtokufu minseikyoku shokusanbu hōbun*. Shokusanka, 1898–99.

Taiwan Sōtokufu Naimukyoku. *Taiwan kanyū rin'ya seiri jigyō hōkokusho*. Taipei: Naimukyoku, 1926.

Taiwan Sōtokufu Ribanka. *Riban gaiyō*. Taipei: Banmu Honsho, 1913.

Taiwan Sōtokufu Shokusankyoku. *Shinrin keikaku jigyō hōkokusho II*. Taipei: Shokusankyoku, 1937.

Taiwan Sōtokufu Shokusankyoku. *Taiwan no rin'ya*. Taipei: Shokusankyoku, 1911.

Taiwan Sōtokufu Shokusankyoku. *Taiwan ringyō no kihon chōsasho*. Taipei: Shokusankyoku, 1931.

Taiwan Sōtokufu Shokusankyoku. *Taiwan rin'ya chōsa jigyō hōkoku*. Taipei: Shokusankyoku, 1915.

Taiwan Sōtokufu Shokusankyoku. *Taiwan shuyō rinboku seichōryō chōsasho*. Taipei: Shokusankyoku, 1922.

"Taiwan zōrin jigyō no enkaku oyobi genjō I." *Taiwan nōjihō* 123 (1917): 17–25.

Takada Mamoru. "Namazue no chosakutachi: chosha, gakō o meguru bakumastsu bunka jōkyō." In *Namazue: Shinsai to Nihon bunka*, edited by Miyata Noboru and Takada Mamoru, 34–51. Ribun Shuppan, 1995.

Takahashi Manabu. "Kodai matsu ikō ni okeru chikei kankyō no henbō to tochi kai-hatsu." *Nihonshi kenkyū* 380 (1994), 38–48.

Takahashi Manabu. "Kodai matsu ikō ni okeru rinkai heiya no chikei kankyō to tochi kaihatsu: Kawachi heiya no shimabatake kaihatsu o chūshin ni." *Rekishi chirigaku* 36:1 (1994) 1–15.

Takekoshi Yosaburō. *Taiwan tōchishi*. Hakubunkan, 1905.

Takeuchi, K., R. D. Brown, I. Washitani, A. Tsunekawa, and M. Yokohari, eds. *Satoyama: The Traditional Rural Landscape of Japan*. Tokyo: Springer, 2003.

Takeuchi, Kuniyoshi. "Disaster Management and Sustainability: Challenges of IRDR." International Conference on Science and Technology for Sustainability 2011— Building Up Regional to Global Sustainability: Asian Vision, September 14–16, 2011, Kyoto, Japan. http://www.scj.go.jp/ja/int/kaisai/jizoku2011/pdf/presentation/prese20.pdf (accessed December 3, 2011).

Talmadge, Eric. "Overpopulated Tokyo Becoming a Heat Island." *Seattle Times*, August 26, 2002. http://community.seattletimes.nwsource.com/archive/?date=20020826&slug=heatisland26 (accessed September 22, 2010).

Tanabe Ikuo and Satō Makoto, eds. *Heijō-kyō no jidai*. Vol. 2, *Kodai no miyako*. Yoshikawa Kōbunkan, 2010.

Tanaka Hiroyuki. *Bakumatsu no Ogasawara: Ōbei no hogeisen de sakaeta midori no shima*. Chūō Kōronsha, 1997.

Tanaka Iori. "Hokkaidō seigan ni okeru 20 seiki no engan suion oyobi nishin gyokak-uryō no hensen." *Hokkaidō suisan shikenjō hōkoku* 62 (2002): 41–55.

Tanaka Migaku. *Heijō-kyō*. Iwanami Shoten, 1984.

Tanaka, Nobuyuki, Keita Fukasawa, Kayo Otsu, Emi Noguchi, and Kumito Koike. "Eradication of the Invasive Tree Species *Bischofia javanica* and Restoration of Native

Forests on the Ogasawara Islands." In *Restoring the Oceanic Island Ecosystem: Impact and Management of Invasive Alien Species in the Bonin Islands*, edited by Isamu Okochi and Kazuto Kawakami, 161–70. London: Springer, 2010.

Tashiro Kiichi, Nakagawa Takeshi, and Hoshino Hitoshi. "Sakuma damu ni okeru ryūboku shori." *Denryoku doboku* 263 (1996): 33–39.

Tavares, Antonio. "The Japanese Colonial State and the Dissolution of the Late Imperial Frontier Economy in Taiwan, 1886–1909." *Journal of Asian Studies* 64, no. 2 (2005): 361–85.

Teng, Emma. *Taiwan's Imagined Geography: Chinese Colonial Travel Writing and Pictures, 1683–1895*. Cambridge, Mass.: Harvard University Asia Center, 2004.

Thomas, Julia Adeney. *Reconfiguring Modernity: Concepts of Nature in Japanese Political Ideology*. Berkeley: University of California Press, 2001.

Thompson, Edgar T. "Population Expansion and the Plantation System." *American Journal of Sociology* 41, no. 3 (1935): 314–26.

Tōkamachi-shi Hakubutsukan Tomo no Kai Komonjo Gurūpu. *Tamura Tani ke shiryō (1)*. Tōkamachi kyōdo shiryō sōsho 14. Tōkamachi, Niigata Prefecture: Tōkamachi Hakubutsukan, 2006.

Tokyo Metropolitan Government. "City Profile." http://www.metro.tokyo.jp/ENGLISH /PROFILE/overview03.htm (accessed February 20, 2010).

Tokyo Metropolitan Government. *Environmental Whitepaper 2006*. http://www2 .kankyo.metro.tokyo.jp/kouhou/env/eng_2006/ (accessed February 11, 2011).

Tokyo Metropolitan Government. *Environmental Whitepaper 2010*. http://www.kankyo .metro.tokyo.jp/en/documents/white_paper_2010.html (accessed July 15, 2012).

Tokyo Metropolitan Government. *Tokyo Fights Pollution: An Urgent Appeal for Reform*. TMG Municipal Library, No. 4. Tokyo Metropolitan Government, 1971.

Tokyo Metropolitan Government Bureau of the Environment. *Basic Policies for the Ten-Year Project to Green Tokyo*. Tokyo Metropolitan Government, 2006.

Tokyo Metropolitan Government Bureau of the Environment. "Guidelines for Heat Island Control Measures [Summary Edition]. Tokyo Metropolitan Government, 2005. http://www.kankyo.metro.tokyo.jp/en/attachement/heat_island.pdf (accessed October 28, 2013).

Tokyo Metropolitan Government Bureau of the Environment. "Thermal Environment Map of Tokyo." Tokyo Metropolitan Government, 2005. http://www.metro.tokyo.jp /ENGLISH/TOPICS/2005/ftf56100.htm (accessed October 28, 2013).

Tōkyō Shinbun. *Kaette kita Ogasawara—Chichijima, Hahajima—Ā Iōtō*. Tōkyō Shinbunsha, 1968.

Tōkyō-fu Dobokubu. *Ogasawara no shokubutsu*. Tōkyō-fu, 1935.

Tōkyō-fu Eirinkyoku. *Ogasawara-jima kokuyūrin shokubutsu gaikan*. Tōkyō-fu, 1929.

Tōkyō-fu Ogasawaratō-chō. *Ogasawara-jima no gaikyō oyobi shinrin*. Ogasawara-mura: Tōkyō-fu Ogasawaratō-chō, 1914.

Tōkyō-fu. *Ogasawarajima sōran*. Tōkyō-fu, 1929.

Tōkyō-to. *2016 nen Tokyo Orinpikku Pararinpikku kankyō gaidorainu*. Pamphlet. Tōkyō-to, 2006.

Tōkyō-to Kankyōkyoku. "Hīto airando genshō o saguru: Atsukunaru Tōkyō." http ://www2.kankyo.metro.tokyo.jp/heat2/index.htm (accessed September 21, 2010).

Tōkyō-to Kankyōkyoku. *Shiodome kankyō eikyō hyōkasho, Shiryō hen.* Tōkyō-to, 1999.

Tōkyō-to Kankyōkyoku. Tōkyō-to kenchikubutsu shōenerugī seinō hyōkasho no aramashi. Tōkyō-to, 2010.

Tōkyō-to Midori no Tōkyō Bokin Jittai Iinkai. *Midori no Tōkyō bokin.* Pamphlet. Tōkyō-to, 2006.

Totman, Conrad. *Early Modern Japan.* Berkeley: University of California Press, 1993.

Totman, Conrad. *The Green Archipelago: Forestry in Preindustrial Japan.* Berkeley: University of California Press, 1989.

Totman, Conrad. *A History of Japan.* 2nd ed. The Blackwell History of the World. Oxford: Blackwell Publishers Ltd., 2005.

Totman, Conrad. *Pre-Industrial Korea and Japan in Environmental Perspective.* Handbook of Oriental Studies/Handbuch Der Orientalistik, Section 5: Japan, Vol. 11. Leiden: Brill, 2004.

Toyama Kiyonori. "Ogasawara no Afurika maimai." *Ogasawara kenkyū nenpō* 11 (1985): 8–16.

Toyo Tokinari. *"Honchō jishinki," "Raikō jishin juraiki,"* hoka. Vol. 49, Edo josei bunko, unpaginated. Ōzorasha, 1994.

Tsing, Anna. *Friction: An Ethnography of Global Connection.* Princeton, N.J.: Princeton University Press, 2005.

Tsude Hiroshi. "Kuroimine." In *Ancient Japan,* edited by Richard Pearson, 223–25. Washington, D.C.: George Braziller and Arthur Sackler Gallery, Smithsonian Institution, 1992.

Tsuji Tomoe. *Ogasawara shotō rekishi nikki.* Vol. 2. Kindai bungeisha, 1995.

Tsukahara Tōgo, Zaiki Masumi, Matsumoto Keiko, and Mikami Takehiko. "Nihon no kiki kansoku no hajimari: Dare ga, dono yōna jōkyō de hajimeta no ka." *Gekkan chikyū* 27, no. 9 (2005): 713–20.

Tsuru Shigeto. *Ikutsu mo no kiro o kaiko shite: Tsuru Shigeto jiden.* Iwanami Shoten, 2001.

Tucker, Richard P., and Edmund Russell, eds. *Natural Enemy, Natural Ally: Toward an Environmental History of War.* Corvallis: Oregon State University Press, 2004.

"T.V.A. of Japan: Gigantic Sakuma Dam Nearing Completion." *New Japan* 8 (1955): 54–55.

Ueyama Kazuo. "Dai-ichiji taisen mae ni okeru Nihon kiito no tai-Bei shinshutsu." *Jōsai keizai gakkai shi* 19, no. 1 (1983): 39–103.

Ukawa Motoo, Eisuke Fujita, Eiji Yamamoto, Yoshimitsu Okada, and Masae Kikuchi. "The 2000 Miyakejima Eruption: Crustal Deformation and Earthquakes Observed." *Earth Planets Space* 52, no. 8 (2000): 19–26.

Umehara Takeshi, Itō Shuntarō, and Yasuda Yoshinori, eds. *Kōza: Bunmei to kankyō.* 15 vols. Asakura Shoten, 1995–96.

Usami Tatsuo, ed. *"Nihon no rekishi jishin shiryō" shūi: Tenmu Tennō 13-nen yori Shōwa 58-nen ni itaru.* Nihon Denki Kyōkai, 1998.

Usami Tatsuo, ed. *Saikō 2-nen yori Shōwa 21-nen ni itaru*. Vol. 3, "*Nihon no rekishi jishin shiryō" shūi*. Watanabe Tansa Gijutsu Kenkyūjo, 2005.

Usami Tatsuo, ed. *Seimu Tennō 3-nen yori Shōwa 39-nen ni itaru*. Vol. 2, "*Nihon no rekishi jishin shiryō" shūi*. Yamato Tansa Gijutsu Kabushikigaisha, 2002.

USGS Cascades Volcano Observatory. "Iceland Volcanoes and Volcanics: Laki—1783 Eruption." http://vulcan.wr.usgs.gov/Volcanoes/Iceland/description_iceland _volcanics.html (accessed June 24, 2011).

USGS Earthquake Hazards Laboratory. "Magnitude 9.0—Near the East Coast of Honshu, Japan." http://earthquake.usgs.gov/earthquakes/recenteqsww/Quakes /usc0001xgp.php#tsunami (accessed June 25, 2011).

USGS Volcano Hazard Program. "Volcanic Ash: Effects and Mitigation Strategies." http://volcanoes.usgs.gov/ash/agric/index.html#pasture (accessed June 24, 2011).

Vandergeest, Peter, and Nancy Lee Peluso. "Empires of Forestry: Professional Forestry and State Power in Southeast Asia, Part 1." *Environment and History* 12 (2006): 31–64.

Wada Tsunanori. *Hokkai no onpa*. Sapporo: Saisankaku, 1903.

Wakamizu Suguru. *Edokko kishitsu to namazue*. Kadokawa Gakugei Shuppan, 2007.

Wakana Hiroshi. "Gendai uotsuki-rin to 'Nishin yama ni noboru': Miura Masayuki, Otaki Shigenao ra no 'Mori to umi' ni kansuru fukusōryū." *Muroran kōgyō daigaku kiyō* 51 (2001): 147–58.

Walker, Brett L. *The Conquest of Ainu Lands: Ecology and Culture in Japanese Expansion, 1590–1800*. Berkeley: University of California Press, 2001.

Walker, Brett L. *The Lost Wolves of Japan*. Weyerhaeuser Environmental Books. Seattle: University of Washington Press, 2005.

Walker, Brett L. "Meiji Modernization, Scientific Agriculture, and the Destruction of Japan's Hokkaido Wolf." *Environmental History* 9, no. 2 (2004): 248–74.

Walker, Brett L. *Toxic Archipelago: A History of Industrial Disease in Japan*. Weyerhaeuser Environmental Books. Seattle: University of Washington Press, 2010.

Walker, Brian, and David Salt. *Resilience Thinking: Sustaining Ecosystems and People in a Changing World*. Washington, D.C.: Island Press, 2006.

Walthall, Anne. "Village Networks: *Sōdai* and the Sale of Edo Nightsoil." *Monumenta Nipponica* 43, no. 3 (1988): 279–303.

Warrick, Richard A. "Volcanoes as Hazard: An Overview." In *Volcanic Activity and Human Ecology*, edited by Payson D. Sheets and Donald K. Grayson, 161–94. New York: Academic Press, 1979.

Washitani Izumi and Yahara Tetsukazu. *Hozen seitaigaku nyūmon*. Bun'ichi Sōgō Shuppan, 1996.

Watanabe Isao. "Gaikokuzai no yunyū." *Dainippon sanrin kaihō* 477 (1922): 32–44.

Watanabe Isao. *Gaizai yunyū no kyokusei to sono taisaku*. Teikoku Sanrinkai, 1925.

Watkins, Susan Cotts, and Jane Menken. "Famines in Historical Perspective." *Population and Development Review* 11, no. 4 (1985): 647–75.

Watsuji Tetsurō. *Fūdo: Ningengaku teki kōsatsu*. Iwanami Shoten, 1935.

White, Richard. *The Organic Machine: The Remaking of the Columbia River*. New York: Hill & Wang, 1991.

Williams, Michael. *Americans and Their Forests: A Historical Geography*. Cambridge: Cambridge University Press, 1989.

Wolman, David. "Mount Fuji Overdue for Eruption, Experts Warn." *National Geographic News*, July 17, 2006. http://news.nationalgeographic.com/news/2006/07/060717-mount-fuji.html (accessed June 24, 2011).

World Commission on Dams. *Dams and Development: A New Framework for Decision-Making*. London and Sterling, Virginia: Earthscan Publications Ltd., 2000.

World Heritage Committee. "Convention Concerning the Protection of the World Cultural and Natural Heritage, 35th Session." Paris: UNESCO, 2011. http://whc.unesco.org/archive/2011/whc11-35com-20e.pdf (accessed July 20, 2011).

Worster, Donald. *Dust Bowl: The Southern Plains in the 1930s*. New York: Oxford University Press, 1979.

Worster, Donald. *Rivers of Empire: Water, Aridity, and the Growth of the American West*. New York: Pantheon Books, 1985.

Yamagata Ishinosuke. *Ogasawarajima shi*. Tōkyōdō, 1906.

Yamaji Katsuhiko. *Taiwan no shokuminchi tōchi: "Mushu no yabanjin" to iu gensetsu no tenkai*. Nihon Tosho Sentā, 2004.

Yamamoto, Yoshika. "Measures to Mitigate Urban Heat Islands." *Science and Technology Trends Quarterly Review* 8 (2006): 65–83. http://www.nistep.go.jp/achiev/ftx/eng/stfc/stt01 (accessed September 10, 2010).

Yamamura, Kozo. "Returns on Unification Economic Growth in Japan, 1550–1650." In *Japan before Tokugawa: Political Consolidation and Economic Growth, 1550–1650*, edited by Nagahara Keiji, John W. Hall, and Kozo Yamamura, 327–72. Princeton, N.J.: Princeton University Press, 1981.

Yamasato Hitoshi. "Katsukazan no bunrui to kazan katsudōdo reberu no dōnyū." In *Kazan funka ni sonaete*, edited by Doboku Gakkaishi, 138–45. Doboku Gakkai, 2005.

Yamazaki Satoshi. "Henkan 15 nengo no Ogasawara no shokubutsu." *Ogasawara kenkyū nenpō* 8 (1984): 5–13.

Yamazaki Yoshio. "Taiwan ni okeru mokuzai jukyū ni tsuite," *Taiwan sanrin kaihō* 3 (1923): 2–6.

Yasutomi Ayumu. "Kokusai shōhin to shiteno Manshū daizu." In *"Manshū" no seiritsu: Shinrin no shōjin to kindai kūkan no keisei*, edited by Yasutomi Ayumu and Fukao Yōko, 291–325. Nagoya: Nagoya Daigaku Shuppankai, 2009.

Yatagai Masayoshi. "Taiwan no shinrin to sono kaihatsu." *Taiwan no sanrin* 79 (1932): 28–33.

Yoffee, Norman, and George L. Cowgill, eds. *The Collapse of Ancient States and Civilizations*. Tucson: University of Arizona Press, 1988.

Yokohama Shōkin Ginkō Chōsabu. *Hokkaidō nishin gyogyō ni tsuite*. Yokohama Shōkin Ginkō Tōkyō Shiten Fuzoku Insatsubu, 1943.

Yonemoto, Marcia. *Mapping Early Modern Japan: Space, Place, and Culture in the Tokugawa Period, 1603–1868*. Asia-Local Studies/Global Themes. Berkeley: University of California Press, 2003.

Yoshikawa Kenji. "Ogasawara no Afurika maimai—Shinrakusha no seitaigaku." *Ogasawara kenkyū nenpō* 1 (1977): 49–56.

Yoshikoshi Akihisa. "Toshi no rekishiteki suimon kankyō." In *Toshi no suimon kankyō,* edited by Arai Tadashi, Shindō Shizuo, Ichikawa Arata, and Yoshikosi Akihisa, 201–52. Kyōritsu Shuppan, 1987.

Yoshino Masatoshi. *Kodai Nihon no kikō to hitobito.* Gakuseisha, 2011.

Yumoto Takakazu, ed. *Shirīzu Nihon rettō no san-man go-sen nen: Hito to shizen no kankyōshi.* 6 vols. Bun'ichi Sōgō Shuppan, 2011.

Zaiki, M., G. P. Können, T. Tsukahara, P. D. Jones, T. Mikami, and K. Matsumoto. "Recovery of Nineteenth-Century Tokyo/Osaka Meteorological Data in Japan." *International Journal of Climatology* 26 (2006): 399–423.

Contributors

Gina L. Barnes (Ph.D., University of Michigan, 1983) is Professor Emeritus of Durham University and Professorial Research Associate at SOAS, University of London. As an archaeologist, she has written extensively on state formation in Japan and Korea. She taught East Asian archaeology at Cambridge University (1981–1995) and at Durham University (1996–2006) with the addition of Japanese language and anthropology at Durham. She was head of the Department of East Asian Studies from 1997 to 2000. Her book, *The Rise of Civilization in East Asia* (Thames & Hudson 1993, 1999), is widely used as a textbook, and a new edition, *Archaeology of East Asia*, will be published by Oxbow Books in 2015. Recently she has extended her interests, obtaining a B.Sc. in geosciences at the Open University to support her current writings on Japanese geology.

Bruce L. Batten (Ph.D., Stanford University, 1989) is Professor of Japanese History at J. F. Oberlin University in Tokyo and the former director of the Inter-University Center for Japanese Language Studies in Yokohama. He is a specialist on ancient and medieval Japan and is the author of *To the Ends of Japan: Premodern Frontiers, Boundaries, and Interactions* (University of Hawai'i Press, 2003) and *Gateway to Japan: Hakata in War and Peace, 500–1300* (University of Hawai'i Press, 2006). He is currently writing a book-length survey of Japanese environmental history and prehistory.

Philip C. Brown (Ph.D., University of Pennsylvania, 1981) is Professor of History at The Ohio State University, Columbus, Ohio. He is a specialist in early modern and modern Japanese history and focuses on developments affecting rural Japan. He is author of *Central Authority and Local Autonomy in the Formation of Early Modern Japan: The Case of Kaga Domain* (Stanford University Press, 1993) and *Cultivating Commons: Joint Ownership of Arable Land in Early Modern Japan* (University of Hawai'i Press, 2011). His current research examines Japan's changing response to flood and landslide risk in the nineteenth through twentieth centuries.

Eric G. Dinmore (Ph.D., Princeton University, 2006) is Associate Professor of History at Hampden-Sydney College. His research probes into the history of modern Japanese developmentalism and focuses on linkages among economic thought, social policy, and the natural environment after 1945. His most recent publication in this vein is "Concrete Results? The TVA and the Appeal of Large Dams in Occupation-Era Japan," which appeared in the winter 2013 issue of *The Journal of Japanese Studies*. He is writing a book-length study on natural and human resources anxieties in twentieth-century Japan, entitled *A Small Island Nation Poor in Resources*.

Toshihiro Higuchi (Ph.D., Georgetown University, 2011) is Assistant Professor at the Hakubi Center for Advanced Research, University of Kyoto. He studies environmental problems in Japan's modern international relations. His published works in English include "The Biological Blowback of Empire? The Collapse of the Japanese Empire and the Influx of the 'Deadly Environment,' 1945–1952," in *Comparative Imperiology*, edited by Kimitaka Matsuzato (Hokkaido University Press, 2010) and "An Environmental Origin of Antinuclear Activism in Japan, 1954–1963: The Politics of Risk, the Government, and the Grassroots Movement," *Peace & Change* 33, no. 3 (June 2008).

Junpei Hirano (Ph.D., Tokyo Metropolitan University, 2009) is Postdoctoral Fellow at the National Research Institute for Earth Science and Disaster Prevention in Japan. He is a climatologist. His major research interest is climatic reconstruction in Japan based on historical weather documents.

Kuang-chi Hung (Ph.D., Harvard University, 2013) is Postdoctoral Researcher at Needham Research Institute, sponsored by the D. Kim Foundation for the History of Science and Technology in East Asia. Since beginning his graduate studies at the National Taiwan University's Department of Forestry, Hung has published widely, both in Chinese and in English, on subjects ranging from the management of natural resources to environmental history. His current project is about nineteenth-century biogeography. He is particularly interested in a biogeographical pattern that played an important role in nineteenth-century evolutionary thinking and debates: namely, the remarkable similarity between the flora of East Asia and that of eastern North America. Hung is now preparing a book manuscript about how naturalists of the nineteenth century discovered this pattern, and how their discovery gave shape to later thinking about world environmental history.

Tatsunori Kawasumi (Ph.D., Ritsumeikan University, 2003) is Associate Professor of Geography in the College of Letters, Ritsumeikan University, Kyoto. He is affiliated with the Art Research Center and the Institute of Disaster Mitigation for Urban Cultural Heritage at Ritsumeikan. A specialist in geographic information systems applications, he has participated widely in interdisciplinary research projects that explore Japan's cultural heritage. Among his recent publications are "The Environmental Archaeology of Flood Disasters: The History of Riverine Floods as Read from Surface Geological Information" (in Japanese), *Nihonshi kenkyū* 579 (2012): 3–19; and, with coauthors Keiji Yano and Tomoki Nakaya, *Analysis of Landscapes of Ancient Cities with 3D Urban*

Models: Relationship between Shapes of Mountains and the City's Central Axes Observed in Virtually Reconstructed Nagaoka-kyō and Heian-kyō (Nakanishiya Shuppan, 2011).

Takehiko Mikami (Ph.D., University of Tokyo, 1977) is Professor of Geography at Teikyo University and Professor Emeritus at Tokyo Metropolitan University. He is a climatologist with a background in physical geography. He has studied climatic changes in historical times of Japan based on various kinds of documentary sources, such as long-term lake freezing date records and daily weather records in old diaries. He is also interested in urban climate studies with meteorological observational data in the Tokyo metropolis. He has organized several international symposiums and conferences focusing on the theme of climatic changes, including the "International Symposium on the Little Ice Age Climate" in 1991 and the "International Conference on the Climatic Change and Variability" in 1999.

Scott O'Bryan (Ph.D., Columbia University, 2000) is Associate Professor with appointments in the Department of History and the Department of East Asian Languages and Cultures at Indiana University. He is a historian of the modern era and the author of *The Growth Idea: Purpose and Prosperity in Postwar Japan* (University of Hawai'i Press, 2009). His current research lies at the intersection of environmental and urban history.

Osamu Saito (M.A., Keio University, 1970, and D. Econ., Keio University, 1987) is Professor Emeritus, Hitotsubashi University, Tokyo. Although his major research interest is in economic history and historical demography, his publications include one environmental history article: "Forest History and the Great Divergence: China, Japan, and the West Compared," *Journal of Global History* 4, no. 3 (2009), and a Japanese-language book on the same topic, *An Economic History of the Environment: Forests, Markets and the State* (Tokyo: Iwanami Shoten, 2014).

Shizuyo Sano (D. Lit., Kyoto University, 2007) is Professor of Cultural History at Doshisha University, Kyoto. She is the author of *Medieval–Early Modern Villages and the Environmental History of Waterfronts* (in Japanese; Yoshikawa Kōbunkan, 2008). Her major research interest is the historical geography of Japanese wetlands. She is interested in the use and traditional management of marine resources in particular. The most recent of her publications is "Chū-kinsei no 'mizube' no komonzu: Biwako, Yodogawa no Yoshitai o megutte (Medieval and Early Modern Waterfront Commons: Concerning the Reed Zones of the Yodo River and Lake Biwa)," in Akimichi Tomoya, ed., *Nohon no komonzu shisō (Japanese Ideas of the Commons)*, Iwanami Shoten, 2014.

Gregory Smits (Ph.D., University of Southern California, 1992) is Associate Professor of History and Asian Studies at Pennsylvania State University. As a specialist in Japanese and East Asian history, he initially focused on the intellectual history of the Ryukyu Kingdom (present-day Okinawa Prefecture, Japan). More recently he has been researching the history of earthquakes in Japan, both with respect to the effect of earthquakes on history and from the standpoint of the history of science. He is the author of *Seismic Japan: The Long History and Continuing Legacy of the Ansei Edo Earthquake* (University of Hawai'i Press, 2013), *When the Earth Roars: Lessons from the History of Earthquakes in Japan* (Rowman and Littlefield, 2014), and numerous articles on the history and culture of earthquakes.

Colin Tyner (M.A., J. F. Oberlin University, 2004; M.A., University of California, Santa Cruz, 2008) is a Ph.D. candidate in modern East Asian history at the University of California, Santa Cruz, and teaches as an adjuct professor at both Hitotsubashi University and Temple University in Japan. He is currently writing a history on how the Ogasawara Islands have been routed into the global histories of commercial whaling, industrial agriculture, and conservation.

Masumi Zaiki (Ph.D., Tokyo Metropolitan University, 2004) is Associate Professor at Seikei University, Tokyo. Her academic specialty is historical climatology based on old meteorological records taken before the foundation of official meteorological networks in Southeast and East Asia. Her major publications are "Reconstruction of Historical Pressure Patterns over Japan Using Two-Point Pressure-Temperature Datasets since the 19th Century," *Climatic Change* 95 (2009): 231–48; "Meteorological Networking and Academic Research in Meteorology at the Southern Frontier of the Empire," *East Asian Science, Technology and Society: An International Journal (EASTS)* 1 (2007): 183–203; and "Recovery of Nineteenth-Century Tokyo/Osaka Meteorological Data in Japan," *International Journal of Climatology* 26 (2006): 399–423.

Index

museums
eco-museums, 38
Ibusuki Archaeological Museum, 27
Sakuma Electric Power, 127, 131

Nagano Prefecture, 24, 39, 65, 123, 129, 144, 148
namazu (catfish) prints, 66–67
natural disasters/events, 9, 21, 32, 38, 39, 214, 247
mitigation, 21, 32, 34–38
See also earthquakes; floods and flooding;
landslides; tsunami; volcanoes; typhoons
natural levees, 50, 52, 107
natural resources, 100, 119
colonial/overseas, 114, 117
energy, 114, 115
of lakes, 76, 78, 90
marine, 75, 160
See also fertilizer; fish and shellfish; forests
nature, views of, 11, 12, 18n22
nutrients/nutrient cycles, 83, 84, 85, 86, 87, 88,
89, 92, 94n18, 94n21, 139–45, 147–48, 150,
152–53. See also fertilizer

oil
vegetable, 147, 152, 153, 168
whale, 159

panarchy, 13–14
plains, 5, 97
Echigo Plain and vicinity, 97, 102–5, 107, 110,
plate 7
Kanto Plain, 97, 103
near Lake Biwa, 79
near Nara, 50–51, 54, 55, plate 2, plate 3
Sendai and vicinity, 21, 247–48
in Taiwan, 176, 182
Zenkōji, 65
plate tectonics. See earthquakes
plateaus
Yamato, 48, 50
Ponting, Clive, 2
pollution, 1, 251
air, 231, 237

export of, 3
industrial, 3, 9, 10
mitigation of, 238–43
noise, 127
radiation, 3, 248, 250
thermal, 235, 237–38, 242
urban, 49, 55, 230
water, 49, 84
See also waste
population
controls and fertility, 8, 9, 228n19
Echigo, 102
growth, 5, 7–8, 9, 52, 78–79, 100–1, 163–64, 188,
221–23, 226, 230, 235, 236
Heijō-kyō, 49Hikone, 78–79
Japan, 3, 100, 112n10, 112n11, 121t, 213–14,
221–23, 227n3
Lake Biwa area, 76, 89
of non-human species, 80, 90, 144–45, 152–53,
160–61
Ogasawara/Bonin Islands, 163–64, 167
Ōtsu, 87, 89
Sakuma district, 125, 128
Tokyo/Edo, 100, 235
urban, 43, 44, 49, 52, 54, 56, 87, 89, 235, 236
See also famine

railroads, 28, 124, 141, 149, 152, 181–82, 248.
See also transport/shipping
rainfall, 5, 8, 60–61, 64, 70n14, 97, 108, 134n69,
200, 201, 202, 203, 204, 213, 217, 218, 221, 233–34
reclamation/restoration
of arable, 6, 7, 75, 100, 112n16, 123, 152–53, 162,
221
of lakes, 91
regime shift, 13–14, 15
resilience, 12, 13, 18n19, 75, 92, 96, 97, 98, 111, 140,
154, 189, 231, 251
of built capital, 243
social/human, 58, 69, 214, 220, 221, 223, 225
resource exploitation/extraction/utilization, 3, 5,
8, 13, 32, 91, 140–41, 144, 149–50, 153, 160, 170,
175, 180

Walker, Brett L., 10
war and warfare, 2
   Ōnin War, 99
   Russo-Japanese War, 146, 148, 149, 184
   Sino-Japanese War (First), 147
   Warring States/Sengoku, 7, 98, 99, 101, 102,
      111, 214, 221, 222, 226
   World War I, 149, 150–52, 156n63, 177, 183
   World War II, 8–9, 52, 91, 110, 114, 117–18,
      125, 132n10, 132n15, 139, 168, 169,
      188, 236

waste
   human, 54–55, 89–90, 92, 139
   industrial, 3, 129
   solid, 231
   wastewater/sewage, 49, 79, 84, 89, 92
   See also pollution
water supply, 44, 45, 48, 97
weather. *See* climate
whaling, 159, 160
wind, 37, 62, 68, 79–80, 81, 84, 87, 166, 200,
   233, 241–42